Guest Editors

Andreja Đuka
Department of Forestry
Institute of Forest Engineering
University of Zagreb
Zagreb
Croatia

Ivica Papa
Department of Forestry
Institute of Forest Engineering
University of Zagreb
Zagreb
Croatia

Editorial Office
MDPI AG
Grosspeteranlage 5
4052 Basel, Switzerland

This is a reprint of the Special Issue, published open access by the journal *Sustainability* (ISSN 2071-1050), freely accessible at: www.mdpi.com/journal/sustainability/special_issues/T0QW4H30CP.

For citation purposes, cite each article independently as indicated on the article page online and using the guide below:

Lastname, A.A.; Lastname, B.B. Article Title. *Journal Name* **Year**, *Volume Number*, Page Range.

ISBN 978-3-7258-3698-7 (Hbk)
ISBN 978-3-7258-3697-0 (PDF)
https://doi.org/10.3390/books978-3-7258-3697-0

Contents

About the Editors

Andreja Đuka

Associate professor Andreja Đuka, Ph.D. is currently employed at the Institute of Forest Engineering at the Faculty of Forestry and Wood Technology, University of Zagreb. She defended her doctoral thesis in 2014, and in 2019 she was promoted to assistant professor. Her interests are harvesting operations, environmentally acceptable technologies, and forest accessibility. She has been involved in several scientific (harvesting operations) and educational projects (Erasmus+), as well as expert studies (in forest accessibility). Currently she leads a project granted by the Croatian Science Foundation, "Quantity and structure of biomass of silver fir and European spruce in the context of climate change" (BiomassInChange). The project deals with occurrence of sanitary felling due to bark beetle outbreaks in the forests of silver fir and European spruce in Gorski Kotar region. She is involved in lectures, practical lessons, and field courses at bachelor, master, and doctoral levels. Moreover, for the past two years, she works as a vice dean of the Forestry Department at the Faculty of Forestry and Wood Technology, University of Zagreb.

Ivica Papa

Associate professor Ivica Papa, Ph.D., is currently employed at the Institute of Forest Engineering at the Faculty of Forestry and Wood Technology, University of Zagreb. He defended his doctoral thesis in 2014, and in 2019 was promoted to assistant professor. His key interests are planning and designing of forest roads, forest accessibility, and forest road optimization with special reference to all the advantages and disadvantages of using modern technology (workstation, dual-frequency GPS, and LiDAR) in comparison to the classic field data-collecting method (theodolite, leveller, and sub-meter). He also participates in scientific and professional projects ("Energy Efficiency of Timber Truck Transportation in the Hilly and Mountainous Forest and Public Roads", where he participated in research related to the consumption of truck assemblies with regard to the horizontal and vertical alignment (characteristics) of public and forest roads; "Optimization of timber harvesting systems and forest traffic infrastructure on strategic-tactical planning level", where he participated in the study of main characteristics of primary and secondary forest road networks and studied the importance of planning forest roads at a strategic and tactical level). Currently he is a team member on a project granted by the Croatian Science Foundation "Quantity and structure of biomass of silver fir and European spruce in the context of climate change: (BiomassInChange)". The project deals with occurrence of sanitary felling due to bark beetle outbreaks in the forests of silver fir and European spruce in Gorski Kotar region. He is involved in lectures, practical lessons, and field courses at bachelor, master, and doctoral levels. For the past two years he works at as the head of the graduate study program "Techniques, Technologies and Management in Forestry" at the faculty.

Preface

The core of planning in forestry is sustainability and the promotion of five important 5E-criteria: environmental, economic, energy-efficient, ergonomic, and esthetic. This is due to the various roles forests have in everyday life. Forestry directly and indirectly affects society, and lately, the importance of forests is more obvious than ever, as we are experiencing climatic extremes and climate change. Efficient management in forestry includes cross-disciplinary actions as it combines various experts from forest operations, silviculture, ecology, forest protection, wildlife management, forest inventory, modelling, and remote sensing to forest genetics and dendrology, all with the same goal—achieving and maintaining healthy and stable forest eco-systems.

Andreja Đuka and Ivica Papa
Guest Editors

sustainability MDPI

Communication

Technodiversity—An E-Learning Tool as an Additional Offer for the Master's Degree and In-Company Training

Jörn Erler [1,*], Raffaele Spinelli [2] and Andreja Đuka [3]

1 Institute of Forest Utilization and Forest Technology, Dresden University of Technology, 01069 Dresden, Germany
2 Forest Harvesting Laboratory, Italian National Research Council, 00185 Rome, Italy
3 Department of Forest Engineering, Faculty of Forestry and Wood Technology, University of Zagreb, 10000 Zagreb, Croatia
* Correspondence: joern.erler1@tu-dresden.de

Abstract: In November 2021, a project proposal submitted by the TU Dresden and in which seven other partner institutions are involved was approved in the ERASMUS+ program Action Type KA220-HED. The aim of the project is to develop an e-learning tool that can be used to teach forest technology at the Master's level. Project work develops along four main tasks: (1) Facts and methods (theoretical contents), (2) Scientific audiovisuals (descriptive contents), (3) E-learning platform (structure) and (4) Didactics (implementation). In this article, the advantages and disadvantages of e learning are discussed and the development of the course contents (facts and methods) is presented in detail.

Keywords: e-learning course; forest operations; ECTS; ERASMUS+

check for updates

Citation: Erler, J.; Spinelli, R.; Đuka, A. Technodiversity—An E-Learning Tool as an Additional Offer for the Master's Degree and In-Company Training. *Sustainability* **2022**, *14*, 15851. https://doi.org/10.3390/su142315851

Academic Editor: Cláudio Moisés Santos e Silva

Received: 7 November 2022
Accepted: 24 November 2022
Published: 28 November 2022

1. Introduction

Forest technology is part of the curriculum at most forestry training centers. Compared to the basic disciplines (botany, soil science, chemistry, etc.), forest technology is an applied subject that deals with the technical implementation of forest goals under operational settings. A typical course offers a good description of forestry machines and equipment and of their use in forestry work, and it also covers ergonomics, work safety and accident prevention. It often includes the basics of forest road construction, logistics and other special topics, according to local conditions and perceived needs. In the previously usual sequence of diploma studies, the preliminary diploma was for the scientific basics, while applied subjects such as silviculture, business administration, politics, etc. were read in the last semesters before the diploma. The subjects of forest technology and work science are among the applied subjects and therefore usually only appeared relatively late in the curriculum.

As a result of the switch to bachelor's and master's degrees in the course of the Bologna process [1], a significant shift had to be made. The new didactic strategy required that the bachelor's degree should be designed in such a way that it would fully qualify students for the profession. Therefore, all essential instruction on forest technology had to be integrated into the bachelor's degree at an early stage. For the master's degree, however, the scientifically oriented, methodical approach remained, which could vary greatly depending on the individual Institutions. Unfortunately, in many places, the number of students enrolled in the master's courses was too small for yielding a critical mass of people who would opt for a teaching module with high technical content and the forest technology course would not be included in the teaching offer. Since there are no signs that this situation will improve in the future, a general lack of technical content can be predicted in the Master's programs. Therefore, statutory teaching in forest technology tends to be reduced to the simple basics and may become impoverished in its scientific content.

At the same time, the proportion of forest entrepreneurs offering technical services such as timber harvesting and forest regeneration is increasing across Europe. They too are looking for qualified training programs for their employees. Technology development is so rapid that knowledge becomes quickly obsolete and must be regularly updated. That problem is less severe with the basics of forestry technology, as taught in the bachelor's degree, but becomes urgent when it comes to applied subjects and methodological in-depth studies, as offered in the master's programs. However, that specific target group is often too busy for attending regular lectures at a university during the week, and is best reached through other teaching formats for which the Internet offers ideal conditions. This was the reason why eight research institutions across Europe have joined forces to create an e-learning offer for forest technology training at the Master's level. Institutions are as follows:

⇒ Dresden University of Technology, Germany; Jörn Erler is the project leader and guides the working group that is responsible for the course's theoretical contents (facts and methods); he gets administrational support from Christina Spirow at the European Project Center, Dresden.
⇒ Transylvania University, Romania; Stelian Borz coordinates the filming of audiovisual units and assembles them into scientific videos.
⇒ National Research Council, Italy; Raffaele Spinelli and Marco Simonetti build the e-learning platform to overcome technical barriers.
⇒ Poznań University of Life Sciences, Poland; Piotr Mederski transforms the tool into an e-learning system.
⇒ University of Natural Resources and Life Sciences, Austria; Karl Stampfer is responsible for the testing under mountain conditions.
⇒ Faculty of Forestry and Wood Technology University of Zagreb, Croatia; Andreja Đuka deals with the Mediterranean view and is responsible for communication.
⇒ FCBA Technological Institute, France; Nathalie Mionetto bridges to the practitioners as target group.
⇒ Swedish University of Agricultural Sciences, Sweden; Ola Lindroos and Mikael Lundbäck bring the Nordic focus on the fully mechanized methods.

2. Experiences with E-Learning in the Academic Environment

Initially, the term e-learning only meant that electronic aids were integrated into the teaching and learning process. Various concepts have emerged over the past 20 years or so, which for example are discussed and didactically evaluated in [2,3]. Findeisen et al. [4] emphasize that here not only a new way of imparting knowledge is used, but that the type of communication and the didactics are also shifting and diversifying. At this point it is useful to describe the difference between the classic face-to-face teaching approach, blended learning and e-learning in the narrower sense. In the classic teaching approach, the teacher first conveys content in face-to-face lessons, which the learner then has to repeat and learn. The opposite has proven to be advantageous with blended learning, where the learner first acquires the teaching content with the help of electronic media, and then can ask questions, conduct exercises etc. in the following face-to-face sessions, flipped classroom [5].

With e-learning in the narrower sense, there are no face-to-face phases; the entire teaching module is carried out remotely. Of course, there can be phases in which teachers and learners are connected, but that occurs exclusively via electronic media (e.g., in a video conference). Teaching modules are first created by the teacher and uploaded to the Internet, for the learner to download and use at a later time. In the strict implementation of the concept, regular two-way communication is categorically excluded; those who are looking for it must offer specific question times.

Köhler et al. [6] show that the European Council already recognized the potential of e-learning in its Lisbon Declaration of 2000. In 2011, the European Commission declared that individual learning through electronic means was a key policy issue. In [7], a spirit of optimism can be felt, which evidently prevailed among schools and universities at the beginning of the century, building the expectation of a speedy introduction of those new teaching and learning opportunities. In a comparative experiment [8] showed that success in a final exam depended significantly on how often learners took part in a face-to-face event. She therefore recommended using e-learning components only as a supplement to conventional teaching (i.e., in blended learning). Riegg Cellini et al. [9] confirmed that students who carried out their studies in presence "on campus" performed significantly better than students who studied the same subjects online at a distance. Only in the case of "shorter technical certificates" was that difference not so sharp. Längin [10], on the other hand, was already able to prove in the early days of e-learning that the willingness of teachers and learners to try out e-learning methods increased significantly when external conditions made it difficult to practice classic teaching methods, and then the advantages of a spatially separate teaching with online formats obviously stood out. For forestry and even specifically for the area of forest ergonomics, he reported that blended learning proved itself in South Africa and was suitable for closing gaps in university education. Awan et al. [11] pointed out that e-learning offers have improved significantly in recent years and that use of the new didactic approach has increased as a result. They emphasized that students who could make their own experiences with e-learning modules overcame their prejudice and reached a better appreciation of the teaching formats. Giron-García et al. [12] addressed the question of whether the learning style of students who are confronted with e-learning offers changed, and whether that change had an impact on their motivation. Eventually, they could not find any clear trend and concluded that motivation to learn depends rather on the quality of the teaching modules, not their format.

The global COVID-19 pandemic has led to a significant acceleration in the adoption of digital exchange formats. Erdmann et al. [13] showed that not only has the diversity and professionalism of the e-learning offerings increased under that new pressure, but that teachers and students also reached a more differentiated view of electronic teaching forms. Egger and Witzel [14] assumed that the recent dramatic experience would lead to a permanent improvement in online formats, increasing their importance, especially in further education. This is quite revealing, especially if one considers that just before the pandemics the first of the two authors was committed to maintaining conventional teaching formats [15]. Cacault et al. [16] found limited confirmation to the common concern that students would no longer participate to face-to-face events if given the e-learning opportunity. In fact, they found that e-learning led to a further differentiation among the students, reducing the performance of low-ability students on average, while triggering positive effects in high-ability students. In particular, they reported that live-streaming options had less impact on live course attendance than expected, since they were only seized if that choice resulted in a clear reduction of expenses. Finally, Flake et al. [17] pointed out the potential that e-learning has for further education and training. With ISO 29993 [18] there are now standards that serve to ensure quality of e-learning.

3. Technodiversity as an E-Learning Tool for Forest Technology

The research partners, who all deal with questions of forest technology and six of whom work at universities and two at research institutes, developed a concept for a forest technology e-learning package with the following characteristics:

- ⇒ The e-learning package is primarily aimed at students in the forestry master's courses, but could also be usable by forest practitioners as a further training offer.
- ⇒ The possibility of digital knowledge transfer are fully developed in order to encompass all the variety and diversity of forestry challenges and solutions in Europe.
- ⇒ Teaching is based on a mix of basic verbal information, graphic visualization, short films and interactive sessions.
- ⇒ English language is chosen as the common language because it seems neither possible nor necessary to translate the entire course offer into all the official languages of the EU.

⇒ Access should be free of charge for everyone with standard software. Costs will only be incurred if a service is provided for which an administrative fee has to be levied (e.g., for issuing an official certificate or similar).

The e-learning course received the suggestive name Technodiversity. This name was deliberately chosen for creating a parallel to the concept of biodiversity and stress the importance of choosing among a diverse offer of techniques and technologies to match Europe's diverse operating conditions in terms of physical environment, economic conditions and social expectations. The project is planned to last 29 months and is funded by the European Union as part of ERASMUS+ Action Type KA220-HED. Work started in November 2021 and will continue until the end of March 2024.

3.1. Objectives

The aim of the project is to show and explain the technological diversity in harvesting operations and to promote it through targeted training. For this purpose, four main components are being developed:

⇒ Facts and methods: tutorials about work techniques and an explicit method for their selection based on work conditions and operational goals. This component also includes a glossary on forest operations.

⇒ Scientific audiovisuals: Video clips illustrating all typical harvesting methods under regional forest situations.

⇒ Knowledge platform: An electronic platform to host, organize and allow easy access to the wide array of teaching materials offered by the course.

⇒ E-learning course: Implement the e-learning course in existing master programs at the participating universities and used for professional life-long learning.

3.2. Target Groups

Four different target groups will be addressed:

1. Students on master's degree courses are the primary future learners. In three summer-schools the e-learning course will be tested and discussed among young specialists (PhD-students) and interested students on master's level under different environmental conditions (Poland, Croatia and Austria).
2. Forestry professionals are the second target group. They will be informed about the availability of the new course through articles in technical journals and posters/information points at sector fairs such as EUROFOREST in France in June 2023 and KWF fair in 2024.
3. Scientists will be specifically addressed in one open session organized at the FORMEC international forest engineering conference held in Italy in 2023. Here the subjects of education and harmonization of scientific methods and terminology will be discussed. The ultimate goal of that initiative is to get support from the scientific community and get the new e-learning course adopted by as many universities as possible, even if not project partners.
4. Deans and study coordinators, who are responsible for curricula planning at the partner universities will be informed with the projects' content. After familiarizing with the content of the e-learning course and the final test rules, they will get the opportunity to decide about the implementation of the e-learning course in their local curricula.

3.3. First Results: Facts and Methods

The core content of the learning package is a coherent array of Power Point presentations. All project participants contributed their knowledge to the course according to the following development process:

⇒ When applying for the project, the rough framework was already defined and agreed by all applicants.

⇒ The first author of this article was responsible for the first draft of the presentations. In that first draft, the topics were placed in a logical order, a selection was made as to which scientific content was to be conveyed, and a first didactic suggestion was made.

⇒ The second Author then worked intensively on that first draft to improve content consistency, linguistic style and general readability, as well as relevance for the target groups.

⇒ After that, all other participants contributed their views, asked critical questions and suggested alternative texts. This step was particularly relevant in order to meet the requirement to actively involve all project participants in the result and to actually reflect the diversity in Europe.

⇒ During that step a number of questions were raised that could not be clarified in the written circular procedure. A multi-day workshop was then held for this purpose. For most questions, a common solution could be found. However, such topics on which no agreement could be reached among the project participants were shown in the presentations with all positions, so that learners will be able to form their own opinions.

⇒ Finally, the presentations were adapted to the status of the discussion and edited graphically and linguistically.

The presentations currently cover the following topics:

A. Basics of technodiversity
   1. Do we need diversity in forest techniques?
   2. Role of forest techniques in forestry
   3. Criteria for assessment
   4. Technique selection process (three steps model)
   5. Responsibility for the choice

B. Terminology and models
   1. Functions of timber harvesting
   2. Sub-functions of timber harvesting
   3. Harvesting equipment
   4. Steps and degrees of mechanization
   5. Functiograms
   6. Harvesting processes
   7. Process chaining

C. Economic assessment
   1. Economic criteria
   2. Engineering formula
   3. Performance of a subsystem
   4. Additional costs
   5. Total system costs

D. Ecological assessment
   1. Risks, side-effects and damage
   2. Damage on forest soils
   3. Repair and prevention
   4. Avoidance of soil damage
   5. Processes on trafficable areas

E. Social assessment
   1. Social criteria
   2. Societal compatibility
   3. Standard methods
   4. Safety and ergonomics
   5. Social suitability

F. Standard solutions for
   1. Northern Europe
   2. Atlantic broadleaf trees
   3. South-Western pine stands
   4. Mediterranean stands
   5. Alpine regions
   6. Eastern mixed stands
G. Path to the optimal solution
   1. Singular problem
   2. Selection of options
   3. Assessing options
   4. Extracting the best option
   5. Target-driven optimality.

Parallel to the Power Point teaching modules, a glossary is being created, in which terms are defined and explained using various examples. That is necessary due to different ways in which the same term is used and the different terms used to define the same thing by different professionals and in different countries. When creating the project glossary, existing glossaries already published and used in English-speaking countries are being taken into account. Here, too, the same principle applies: standardization and consensus if possible, but presentation of differences and diversity if necessary. The project participants do not want to add to the confusion with one more glossary, but rather make an attempt at consolidation in the hope that it will help clarify the terminology.

4. Instead of Conclusions

In summary, the following statements can currently be made about e-learning for-mats in the strict sense:

⇒ In training, they are not so much suitable for imparting basic knowledge, but for deepening existing knowledge and for further training.
⇒ Above all, they are offered by teachers and accepted by learners when the ad-vantages of the digital format become clearly recognizable for both sides.
⇒ The didactics of digital media must differ from the didactics in the lecture hall and must make full use of the possibilities offered by dedicated software.
⇒ The offers must be easily accessible; their costs must be significantly lower than those that a face-to-face event would incur.

The success of technology training is often dependent on psychological impact. Field trips are best suited for that effect, but they are usually limited to the region in which the training takes place and are therefore are not an option for a European project such as Technodiversity. Therefore, our project tries to obtain the same effect through the use of short instructional videos. That expectation is supported by the experience that e-learning is primarily perceived and accepted by the target group as advantageous if the special possibilities of digital communication are also used [4]. One might think that there are enough videos on the Internet that deal with forest technology topics. Some of them are also extremely valuable, especially if they deal specifically with working processes and provide tips for the practitioner. However, they are usually too specific for training at master's level and can hardly be used in teaching, simply because of the time it takes to look at them. At the other extreme are the many videos of forestry machines in use, in which the focus is apparently on 'showing particularly impressive situations', often reinforced by emotional music, so that psychological impact becomes the main goal and technical content is only secondary. Between these two categories are commercial videos, offered by equipment manufacturers and primarily aimed at attracting customers: although rich in technical information, those videos are weakened by a strong commercial overtone that is all, but impartial. All those categories can hardly be defined as ideal for use in scientific training at master's level.

For that reason, the project work plan included the production of short videos of around 5 min, in which machines and sub-processes are shown, explained and evaluated in a targeted manner. Those videos, which are offered in different categories (machining principles, technical features, machine in action, effects, risks and side effects...), are based on a uniform script style developed by Stelian Alexandru Borz (Brasov, Romania) in collaboration with Mikael Lundbäck (Umeå, Sweden). According to that plan, each project partner will shoot a few clips, which will be then edited and equipped with English texts in Romania according to a uniform pattern. In this way, a complete range of explanatory videos will be made available for describing all the sub-processes in the lectures, for a total of more than 100 short films, all featuring the same consistent style and format.

With 38 teaching modules (Power Point), a glossary and probably more than 100 videos and other audiovisual offers it is no longer possible to set up a linear educational path that every learner must complete. Such a wealth of training materials will be integrated in a platform on which learners will develop their own path, according to capacity and needs. Such a platform is being created by Raffaele Spinelli and Marco Simonetti from National Research Council (CNR), Italy. To that purpose, they are using the proven learning software MOODLE in connection with 5hp and LUMI, so that all materials mentioned above can be integrated in a way that is logical, user-friendly and didactically appealing.

An e-learning tool must not only provide information, but also organize the learning process towards a final exam. Such process will match the formal requirements for a master level course and conform with the European standards and examination regulations. That way a student in Padua, like her fellow student in Vilnius, can choose this course and have it recognized as an equivalent exam achievement.

On the one hand, the project requires quality control in which the study ability and the level of performance are ensured. For this purpose, summer schools are being organized within the scope of the project, for all target groups. Furthermore, the success of the new e-learning tool requires recognition by colleagues from other forest faculties not involved in the project. Piotr Mederski from Poznan, Poland, is responsible for quality assurance and market launch. The further steps mentioned above will occupy the project team in the coming months. The end of the project is set for March 2024. In order to reflect the diversity and also get to know other opinions, the project team decided to involve other partners as well. Although their later arrival excludes them from accessing to the project financing, they are included in most exchanges so that they can actively contribute to the development process. If you are interested in this, please contact us at the following e-mail address: joern.erler1@tu-dresden.de.

Author Contributions: Conceptualization, J.E.; methodology, R.S.; validation, J.E. and R.S.; investigation, J.E.; resources, R.S.; writing—original draft preparation, J.E. and R.S.; writing—review and editing, A.Đ.; visualization, A.Đ.; project administration, J.E.; funding acquisition, J.E. All authors have read and agreed to the published version of the manuscript.

Funding: This research was funded by the Erasmus+ program of the European Union, Action Type KA220-HED—Cooperation partnerships in higher education under the Grant Agreement N°: 2021-1-DE01-KA220-HED-000032038 Technodiversity.

Institutional Review Board Statement: The study did not require ethical approval.

Informed Consent Statement: Informed consent was obtained from all subjects involved in the study.

Conflicts of Interest: The authors declare no conflict of interest.

References

1. Klein, M. *Die Studienabschlüsse Bachelor und Master: Entwicklung und Auswirkung der neuen Studienstruktur auf das Deutsche Hochschulsystem*; VDM Verlag Dr. Müller: Saarbrücken, Germany, 2007; p. 123.
2. Kergel, D.; Heidkamp-Kergel, B. *E-Learning, E-Didaktik und Digitales Lernen*; Springer Fachmedien Wiesbaden: Wiesbaden, Germany, 2020; p. 116.
3. Niegemann, H.M.; Weinberger, A. *Handbuch Bildungstechnologie: Konzeption und Einsatz digitaler Lernumgebungen*; Springer: Berlin/Heidelberg, Germany, 2020; p. 722.

4. Findeisen, S.; Horn, S.; Seifried, J. *Lernen durch Videos: Empirische Befunde zur Gestaltung von Erklärvideos*; MedienPädagogik, Zeitschrift für Theorie und Praxis der Medienbildung; Pestalozzianum: Zürich, Switzerland, 2019; pp. 16–36.
5. Wipper, A.; Schulz, A. *Digitale Lehre an der Hochschule: Vom Einsatz digitaler Tools bis zum Blended-Learning-Konzept*; Verlag Barbara Budrich: Toronto, ON, Canada, 2021; p. 122.
6. Schoop, E.; Köhler, T.; Börner, C.; Schulz, J. *Consolidating eLearning in a Higher Education Institution: An Organisational Issue Integrating Didactics, Technology, and People by the Means of an eLearning Strategy*; TUD Press: Dresden, Germany, 2016; p. 12.
7. Bremer, C.; Kohl, K.E. *E-Learning-Strategien und E-Learning-Kompetenzen an Hochschulen*; W. Bertelsmann: Bielefeld, Germany, 2004; p. 114.
8. Schumny, U. *Effektivität von Vorlesungsbegleitendem E-Learning in der Dermatologie in Bezug auf den Lernerfolg*; Dissertation Friedrich-Schiller-University Jena: Söflingen, Germany, 2022.
9. Riegg Cellini, S.; Grueso, H. Student Learning in Online College Programs. *AERA Open* **2021**, *7*, 23328584211008105.
10. Längin, D. Blended Learning in der Forstlichen Hochschullehre–eine Untersuchung im Südlichen Afrika. Ph.D. Thesis, Albert-Ludwig-Universität Freiburg, Breisgau, Germany, 2006.
11. Awan, R.K.; Afshan, G.; Memon, A.B. Adoption of E-Learning at Higher Education Institutions: A Systematic Literature Review. *Multidiscip. J. Educ. Soc. Technol. Sci.* **2021**, *8*, 74–91.
12. Girón-García, C.; Gargallo-Camarillas, N. Multimodal and Perceptual Learning Styles: Their Effect on Students' Motivation in a Digital Environment. *EuroCALL Rev.* **2020**, *28*, 23. [CrossRef]
13. Erdmann, A.; Estrada Presedo, A.; De Miguel Valdés, M. Digital Transformation of Universities: The Influence of COVID-19 and Students' Perception. *Multidiscip. J. Educ. Soc. Technol. Sci.* **2021**, *8*, 19. [CrossRef]
14. Egger, R.; Witzel, S. *Hybrid, Flexibel und Vernetzt?—Möglichkeiten, Bedingungen und Grenzen von Digitalen Lernumgebungen in der Wissenschaftlichen Weiterbildung*; Springer VS: Wiesbaden, Germany, 2022; p. 276.
15. Egger, R.; Eugster, B. *Lob der Vorlesung. Vorschläge zur Verständigung über Form, Funktion und Ziele Universitärer Lehre*; Springer VS: Wiesbaden, Germany, 2020; p. 268.
16. Cacault, M.P.; Hildebrand, C.; Laurent-Lucchetti, J.; Pellizzari, M. Distance learning in higher education: Evidence from a randomized experiment. *J. Eur. Econ. Assoc.* **2021**, *19*, 2322–2372. [CrossRef]
17. Flake, R.; Malin, L.; Meinhard, D.B.; Müller, V. Digitale Bildung in Unternehmen: Wie KMU E-Learning Nutzen und Welche Unterstützung sie Brauchen. Institut der deutschen Wirtschaft Köln. Available online: https://www.iwkoeln.de/studien/regina-flake-lydia-malin-david-b-meinhard-valerie-mueller-wie-kmu-e-learning-nutzen-und-welche-unterstuetzung-sie-brauchen.html (accessed on 16 October 2022).
18. ISO 29993:2017; Branchenübergreifender Seervicestandard für Aus- und Weiterbildung, Kurzdarstellung TÜV Süd. Available online: https://www.iso.org/standard/70357.html (accessed on 1 September 2022).

sustainability

MDPI

Communication

Evaluating an Innovative ICT System for Monitoring Small-Scale Forest Operations: Preliminary Tests in Mediterranean Oak Coppices

Rodolfo Picchio [1], Rachele Venanzi [1,*], Aurora Bonaudo [1], Lorenzo Travisani [1], Vincenzo Civitarese [2] and Francesco Latterini [3]

1 Department of Agriculture and Forest Science, University of Tuscia, Via San Camillo de Lellis, 01100 Viterbo, Italy; r.picchio@unitus.it (R.P.); aurora.bonaudo@unitus.it (A.B.); travisanilorenzo@gmail.com (L.T.)

2 Consiglio per la Ricerca in Agricoltura e l'Analisi dell'Economia Agraria, Centro di Ricerca Ingegneria e Trasformazioni Alimentari, Via della Pascolare 16, 00015 Monterotondo, Italy; vincenzo.civitarese@crea.gov.it

3 Institute of Dendrology, Polish Academy of Sciences, Parkowa 5, 62-035 Kórnik, Poland; latterini@man.poznan.pl

* Correspondence: venanzi@unitus.it

Abstract: The application of modern technologies to increase the overall sustainability of forest operations is known as precision forest harvesting. Precision forest harvesting can be a very powerful tool; however, it requires modern forest machinery, which is expensive. Given that most of the forest operators in the Mediterranean area are small-scale businesses, they do not have the resources to purchase costly equipment; thus, the application of precision forest harvesting is affected. Bearing this in mind, in this study, we aimed to test the accuracy of the GNSS receiver on which an innovative Information and Communication Technology (ICT) system developed to monitor small-scale forest operations is based. We tested the GNSS's accuracy by comparing the extraction routes recorded during coppicing interventions in two forest sites located in Central Italy with those obtained with a more high-performing GNSS receiver. We also used linear mixed-effects models (LMMs) to investigate the effects on the GNSS positioning error of topographic features, such as the slope, elevation, aspect and Topographic Position Index (TPI). We found that the average positioning error was about 2 m, with a maximum error of about 5 m. The LMMs showed that the investigated topographic features did not significantly affect the positioning error and that the GNSS accuracy was strongly related to the specific study area that we used as a random effect in the model (marginal coefficient of determination was about 0.13 and conditional coefficient of determination grew to about 0.59). As a consequence of the negligible canopy cover after coppicing, the tested GNSS receiver achieved satisfactory results. It could therefore be used as a visualising tool for a pre-planned extraction route network, allowing the operator to follow it on the GNSS receiver screen. However, these results are preliminary and should be further tested in more experimental sites and various operational conditions.

Keywords: sustainable forest operations; forwarder; strip roads; terrain features

Citation: Picchio, R.; Venanzi, R.; Bonaudo, A.; Travisani, L.; Civitarese, V.; Latterini, F. Evaluating an Innovative ICT System for Monitoring Small-Scale Forest Operations: Preliminary Tests in Mediterranean Oak Coppices. *Sustainability* **2024**, *16*, 4629. https://doi.org/10.3390/su16114629

Academic Editors: Lihu Dong and Xander Wang

Received: 22 April 2024
Revised: 27 May 2024
Accepted: 28 May 2024
Published: 29 May 2024

1. Introduction

One of the most important things to take into account in the effort to achieve Sustainable Forest Management (SFM) [1–5], a primary goal of the European Forest Strategy [6], is the implementation of Sustainable Forest Operations (SFOs) [7–9]. The term "SFOs" refers to the use of logging techniques that have reasonable costs, have minimal environmental effects, and guarantee worker safety [10,11]. Precision forestry is defined as the use of contemporary technology, including Information and Communication Technology (ICT), Geographic Information Systems (GISs), Global Navigation Satellite Systems (GNSSs), Unmanned Aerial Vehicles (UAVs), and a variety of sensors, to enhance the overall sustainability of forest management [12–14]. It is one of the most powerful tools for addressing the target of Sustainable Forest Management [15–17].

The term "precision forestry" refers to an interdisciplinary and transdisciplinary idea that allows for the creative application of cutting-edge technology in specific forest sector situations [18]. There are many uses for various smart technologies within the context of precision forestry, given the high degree of multidisciplinarity that characterises the forest sector. When performing forest inventories, satellites or unmanned aerial vehicles (UAVs) can be used to measure the aboveground biomass [19–21] and track pests and drought [22–24]. Geographic Information Systems (GISs) can be used to plan harvesting operations in a variety of ways, such as choosing the best extraction technology [25,26], designing skid trail networks that are optimised in advance [27–29], or creating maps of soil trafficability [30–33].

Monitoring harvesting operations is another function that cutting-edge and intelligent technologies may carry out within the context of precision forestry. When a fully mechanised harvesting system with state-of-the-art machinery is used, some data on the machine functioning and the features of the harvested wooden material can be obtained, such as the dendrometric features, species, or location [34]. Additionally, data, including the fuel consumption, productivity, and position, can be seen remotely through the manufacturer's portal. The Controller Area Network (CAN-bus) system, which is implemented in nearly all modern forest harvesters and processors, makes all these applications realisable by transmitting the machine status and the parameters to the on-board computer (OBC) via the Standard for Forest Machine Data and Communication (StanForD) [35,36].

However, these systems are only available on modern forest machines. Nonetheless, it is very important to develop systems that are just as efficient and effective at a cost that is feasible for small-scale forest business. Small-scale forest businesses make up the majority of those operating on the territory in several modern countries, including Italy [37]. To the best of our knowledge, no scientific paper has focused on the development of a system to monitor forest operations within the context of small-scale forestry. Small-scale forestry is defined as forest operations carried out by small or medium enterprises. These businesses typically use machinery that is not specifically developed for forestry. Most of the machinery used by these businesses is machinery originally purposed for agricultural use, which is adapted for forest work. It is worth noting that the type of agricultural machinery is usually of an old model. There is a previous study that made an effort to find a solution to this issue. A monitoring system was developed for forwarding operations performed by old models of forwarders with no CAN-bus system and for forestry-fitted farm tractors equipped with a trailer [38]. It included a GNSS portable receiver with an Android platform-specific app and a weighing system mounted on the loading deck of the machine to assess the amount of extracted timber in each working cycle. In addition to tracking the working hours and productivity, the system can record the strip road pattern [38]. In this preliminary research, we wanted to address the following question: "Is the GNSS precision of the developed ICT system enough to guide the driver of the machine along a pre-planned strip road network?".

It is known that the correct functioning of a GNSS system is often hampered in a forest environment [39–41]. The use of GNSS in forestry is frequently impacted by a multi-pathing error, which happens when satellite signals are reflected or diffracted by surrounding objects or surfaces or occluded by the canopy or other solid objects. This problem exists even with the application of real-time kinematics correction techniques, which may normally increase the position accuracy, even at a sub-centimetric level, but only in the presence of robust satellite coverage [42–44].

Taking all the above into account, we established this research with the goal of testing the efficiency of the previously developed monitoring system for forwarding operations [38], with a particular focus on the level of precision of the GNSS receiver. We tested the system in two operational contexts in Central Italy's Turkey oak (*Quercus cerris* L.) coppice forests. We further tried to investigate the effects on the GNSS's precision of topographical features such as the slope, aspect, elevation, and Topographic Position Index.

We also aimed to understand if this system is reliable in the harsh topographic conditions in which forest operations are performed in the context of Mediterranean forestry.

2. Materials and Methods

2.1. Study Areas

We selected two forest sites located in Central Italy (Figure 1), one in the municipality of Avigliano Umbro (Umbria, Italy, 777769 E, 4726803 N WGS84UTM32N) and the other in the municipality of Bracciano (Latium, Italy, 761256 E, 4671354 N WGS84UTM32N). The two sites are both Turkey oak coppices harvested in the harvesting season 2022–2023 and are highly representative of oak coppices in Central Italy. The sites were specifically selected because they presented a wide range of slopes (from 0% to almost 30%, representing the upper limit where forwarders are generally applied) and complex topography, thus resulting in optimal testing conditions for our scope. In both the study areas, the forest operations consisted in motor-manual felling and processing by chainsaw and extraction by an old model of forwarder (Avigliano) and forestry-fitted farm tractor equipped with a trailer (Bracciano).

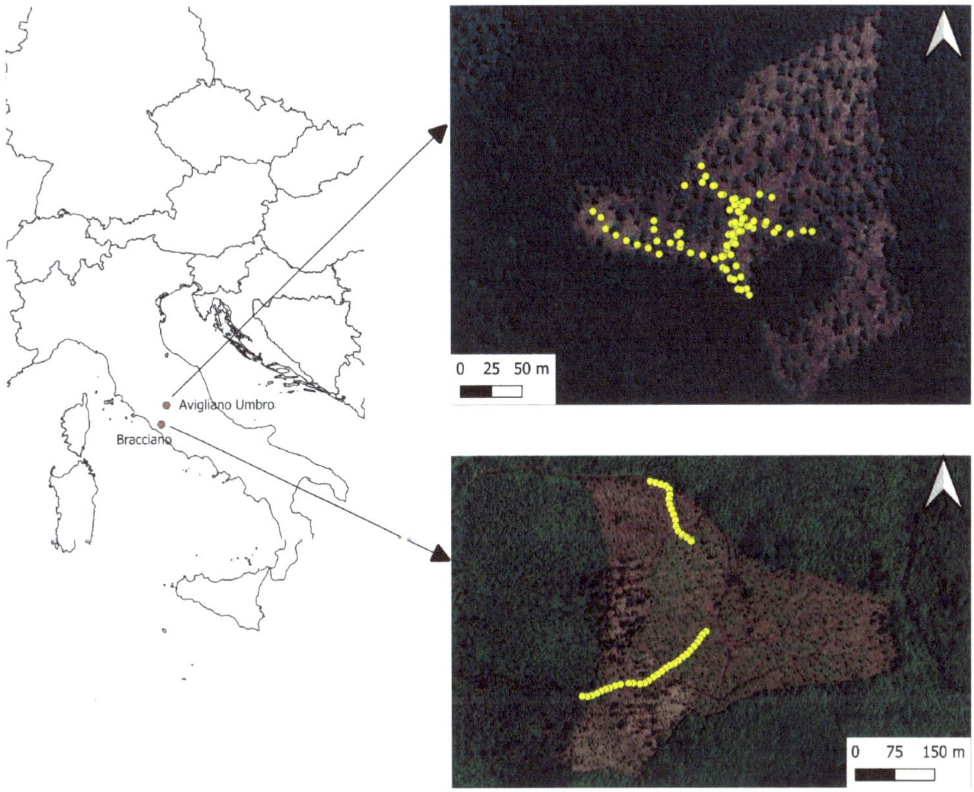

Figure 1. Locations of the two study areas. The sampling points are in yellow (see Section 2.2).

2.2. Field Surveys

A full description of the tested ICT system for monitoring forest operations on forest machines not equipped with a CAN-bus system is reported in [38]. Briefly, it is made up of a weighting system located on the forwarding deck, a GNSS receiver and a dedicated app to record and describe the various working elements. Concerning the GNSS, the system is based on a GNSS Android 4G receiver Mobile Mapper 50 (Spectra Geospatial, Westminster,

CO, USA) with a declared GNSS precision up to 1 m. The overall implementation cost of the ICT system ranges from EUR 4000 to 8000, depending on the precision of the installed weighing system and of the performance of the GNSS receiver.

The ICT system has two purposes. Firstly, it records the duration of the work session, then it measures the weight of the timber extracted. The work productivity is calculated automatically. Secondly, it acts as a navigator where a GIS-planned strip road network can be uploaded to the Mobile Mapper 50 software. The driver can visualise and follow the work progress on the screen of the GNSS receiver. At the same time, the ICT system records in .gpx format all the strip roads established by the machine. In our experiment, we did not develop a GIS-planned strip road network, but we allowed the drivers to establish the extraction routes according to their experience. A technical flow chart of the ICT system is reported in Figure 2.

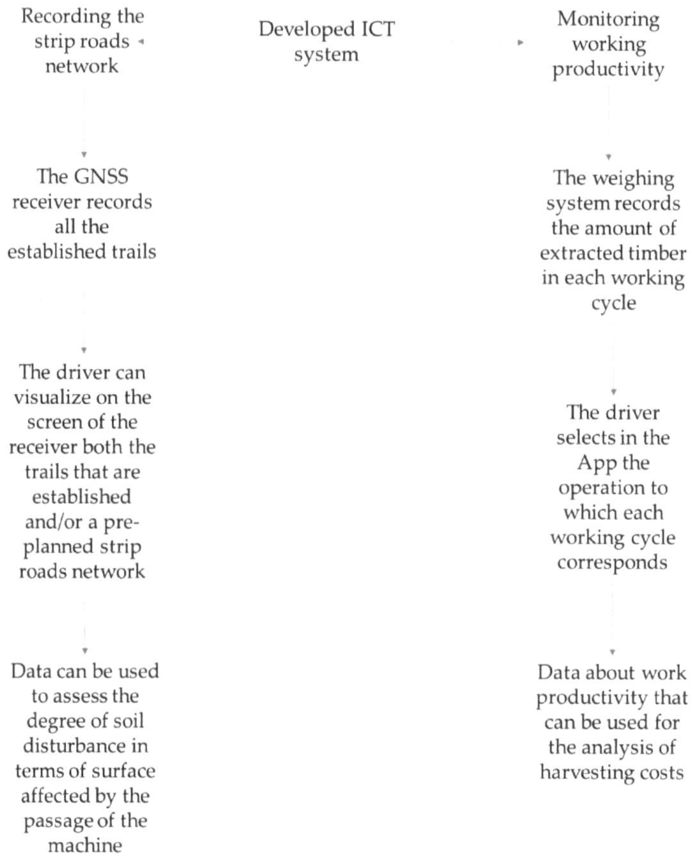

Figure 2. Technical flow chart of the developed ICT system.

During the harvesting operations in both study areas, the forwarder operator was provided with the ICT system, which recorded all the machine's movements, thus creating a vector file of the strip road network established for carrying out the extraction operation in each study area.

After the end of the harvesting operations, the created vector files of the strip roads were converted into a point shapefile and uploaded to a more complex and high-performing GNSS receiver, Spectra SP20 (Spectra Geospatial, Westminster, CO, USA), which, thanks to the real-time kinematics correction techniques, can reach a centimetric precision. The Spectra SP20 receiver was equipped with Mobile Mapper 50 software as well.

At this stage, the field surveys started. With the Spectra SP20, one operator identified 85 points located along the strip roads in Avigliano and 45 points located along the strip roads in Bracciano. The operator reached each point and waited 3 min so as to be sure of being in the exact location of the recorded point. Then, the operator used a tape measure to check the distance between the location of the point and the centre of the actual strip road.

In this way, we obtained 130 measurements for the distance between the point along the strip roads recorded by the ICT system and the real location of the strip road. These distances were used as a measure of the error of the ICT GNSS and as dependent variables in the developed models.

2.3. Investigated Terrain Features and Statistical Analysis

To assess the reliability of the ICT system in different operating conditions, we wanted to understand how the topography of the logging site affects the precision of the GNSS receiver. We therefore used a 10 m resolution Digital Elevation Model (DEM) [45,46] to extract the elevation (m a.s.l.) and calculate the terrain slope (%), terrain aspect and Topographic Position Index (TPI) [47,48] in correspondence to the survey points. The Topographic Position Index is a terrain categorisation technique that compares the height of each data point to that of the surrounding points. A point index value will be positive if it is higher than the surrounding area, as on ridges and hilltops, and negative for sunken features such as valleys. All the GIS analyses were carried out using QGIS 3.28 software [49], while the statistical analysis was performed using R software [50].

We used linear mixed-effects models (LMMs) to investigate the effects of the topographical variables on the GNSS error. In the models, we used the positioning error as the dependent variable, the elevation, slope, aspect and TPI as fixed effects, and the study area (Avigliano or Bracciano) as a random effect. A random effect was included to account for the spatial-related data dependency and for the fact that the surveys were carried out on two different days, one per each study area.

We started by fitting a full model with all the dependent variables by using the package lme4 [51], and we checked the Variance Inflation Factor (VIF) to exclude excessive correlation among the fixed-effects with the car package [52]. No variable showed a VIF higher than 10, so we could exclude collinearity among the fixed effects. Subsequently, we used the dredge() function from the MuMin package [53] to select the best model; in particular, we selected the model with the lowest Akaike's Information Criterion, corrected for the small sample size (AICc). In this way, we developed a further LMM by using only the elevation and TPI as fixed effects. Finally, we used the MuMin package [53] to calculate the marginal coefficient of determination (R^2m) and conditional coefficient of determination (R^2c), respectively indicating the amount of variance explained by the fixed effects and that explained by both the fixed and random effects [54]. We used the package ggeffects [55] to calculate and visualise the marginal response and marginal means, representing the mean values of a given fixed effect assuming a constant level of all the other predictors and no random effect (global estimate).

3. Results

Descriptive statistics concerning the investigated variables and the GNSS positioning error are reported in Table 1. The average positioning error result was 1.71 m in Avigliano (maximum error 4.80 m) and 2.30 m in Bracciano (maximum error 5.20 m). In both the study areas, the minimum positioning error was practically negligible (0.01 m). Particularly for the Avigliano study area, the slope range covered practically all the slopes where ground-based logging is generally applied by forwarding, ranging from 0% to 25%.

Table 1. Descriptive statistics concerning the investigated terrain features and the GNSS positioning error for the two study areas.

		Avigliano	**Bracciano**
Elevation (m)	Average	390	407
	Max	397	425
	Min	382	399
Slope (%)	Average	14.55	10.99
	Max	25.19	16.6
	Min	0.76	5.68
Topographic position index	Average	0.15	0.04
	Max	1.41	0.50
	Min	−0.35	−0.47
Positioning error (m)	Average	1.71	2.30
	Max	4.80	5.20
	Min	0.01	0.01

The results of the linear mixed-effects model are shown in Table 2. The standard deviation related to the random effect in the model (study area) was 1.21 m. Both the elevation and TPI did not show a statistically significant influence ($p < 0.05$) on the GNSS positioning error, while the slope and aspect were even excluded by the final model after checking the AICc. The marginal coefficient of determination R^2m was 0.136, which increased to 0.592 when considering the effect of the study area-related random effect. Therefore, it is evident that there is a strong effect of the study area on the GNSS positioning error, which is not explainable by the topography.

Table 2. Summary of the linear mixed-effects model investigating the effects on the GNSS positioning error of the elevation and Topographic Position Index.

	Linear Mixed Model		
	Random effects		
Groups	Name	Variance	Std.Dev
Area	(Intercept)	1.464	1.21
Residual		1.311	1.145
Number of observations	130, groups	area, 2	
	Fixed effects		
	Estimate	Std.Error	t value
(Intercept)	26.693	7.017	3.804
Elevation	−0.062	0.017	−3.551
TPI	0.369	0.272	1.357
R^2m		R^2c	
0.1363706		0.5920209	

The relationship between the GNSS positioning error and the topographic variables is reported graphically in Figure 3A (error vs. elevation) and Figure 3B (error vs. TPI). The regression line for the elevation is negative, while for the TPI it is positive; however, for both variables, the data distribution clearly confirms the lack of influence on the GNSS positioning error, as already highlighted by the results of the model.

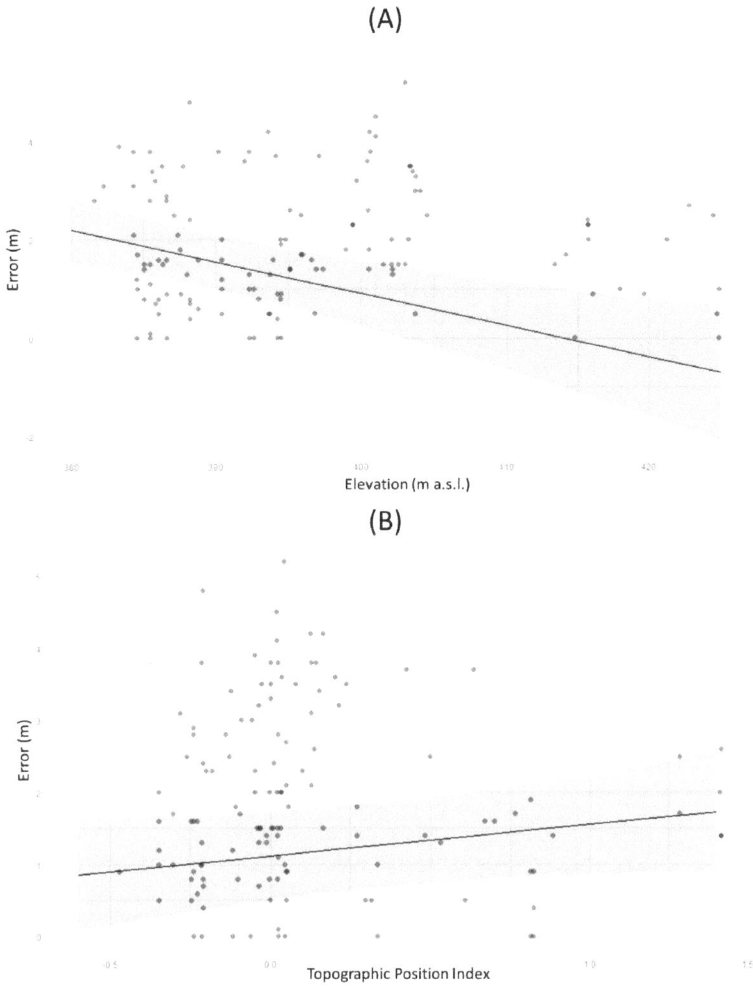

Figure 3. (**A**) Effect of the elevation on the GNSS positioning error. (**B**) Effect of the Topographic Position Index on the GNSS positioning error. Both variables did not show significant effects on the GNSS error. Dots indicate data distribution. Grey area indicates the confidence intervals.

4. Discussion

The GNSS is globally recognised as the key tool for positioning an object on the Earth's surface [56–58]. However, it is well known that forests represent a challenging environment for GNSS position accuracy, with positioning errors that generally prove to be two to four times higher than in open areas using the same GNSS receiver [39,59]. The main reason for the decreased GNSS precision in forests is related to the canopy cover, with increasing canopy density generally leading to bigger errors [39]. In our case study, both the operator driving the forwarder and recording the positions with the ICT system and the operator checking its accuracy with the SP20 receiver worked in conditions of negligible canopy cover. This took place after a coppicing intervention releasing about 100 standards per hectare. Therefore, the influence of the canopy on the GNSS receiver accuracy can be excluded. As a result, the positioning errors were on average much lower than what was shown in previous similar studies in forest environments, where the average error was about 2 m and the max error about 5 m (Table 1) [59]. This level of the positioning error is obviously not enough to consider developing an automatic driving system [60], but it is enough to perform a very important task in the framework of sustainable forest operations. Indeed, with this position accuracy, it is possible to use the GIS to pre-plan an optimised strip road network, upload it to the GNSS receiver and allow the operator to visualise it on the screen and follow it, similarly to what happens in modern forest machines [28]. In this way, the operator can avoid zones with excessive slope or areas of soil particularly sensitive to machinery-induced compaction [30].

Previous studies have shown that that topography can have an influence on the GNSS positioning accuracy as well. For instance, Zimbelman and Keefe (2018) found that topographic features can affect the number of signal losses from the GNSS technology paired with radio frequency (RF) transmission (GNSS-RF) [42]. However, the study [42] also found that the positioning error was significantly related to the stand features (proxy of canopy cover) exclusively and not to the terrain features such as the slope, aspect and presence of concave. Our results confirmed these findings. Indeed, we did not reveal any significant effect for any of the investigated terrain features, with the slope and aspect even being excluded from the final model, and the TPI and elevation showing non-significant results (Table 2, Figure 3). This means that the system can be used in the target operational context (Mediterranean forestry) without being affected by the topographic features of the forest site.

Furthermore, it is worth noting that the developed ICT system is applicable in any small-scale forestry operations in which timber extraction is carried out by forwarding; that is to say, by using a loading deck. Being based on a weighing system, a GNSS receiver and a specific app, it can be installed on any machine equipped with a loading deck used for forest operations.

Obviously, our study presents some limitations. First of all, it should be observed that this paper can be considered a preliminary study. In fact, the test that was carried out was a preliminary one, taking into consideration only two study areas and both referring to one forest typology (oak coppice), although it was highly representative of the typical working conditions of forwarding extractions after coppicing in oak forests. It is interesting to note how the model's coefficient of determination grew substantially when including the effect of the study area (R^2c 0.59 vs. R^2m 0.13—Table 2). This highlights the importance of the specific study area in relation to the GNSS accuracy, most probably related to the localisation of the satellites in those areas on those specific days.

Further studies to test the overall reliability of the developed ICT system should therefore involve more study areas and also different forest typologies with higher canopy cover. However, the preliminary results obtained in this study are quite encouraging. After deeper investigation and further tests of the developed ICT system in alternative contexts, we do believe that the proposed solution can represent a valuable contribution to the implementation of precision forest harvesting in the framework of small-scale forestry.

5. Conclusions

In this preliminary research, we tested the reliability of one of the components that compose an innovative ICT system previously developed to monitor forest operations in the framework of small-scale forestry, i.e., the GNSS receiver. We tested the GNSS positioning error in two study areas, consisting of coppiced oak forests in Central Italy, by checking the precision of the recorded extraction routes with a better-performing GNSS receiver with RTK correction. We further used linear mixed-effects models to investigate the influence of the elevation, slope, terrain aspect and Topographic Position Index on the GNSS positioning error. We found that the positioning error was lower than in previous trials in a forest environment as a consequence of the negligible canopy cover that characterises a coppice forest after logging. We found that none of the investigated terrain features had a significant influence on the GNSS positioning error. Although this study can be considered a preliminary study, we confirmed that the previously developed ICT system is suitable for integration with a GIS pre-planned network of extraction routes. The operator of the forest machine can visualise it on the screen of the GNSS receiver so that it can be followed, thus avoiding zones of the parcels that may be unsafe or lead to impactful disturbance to the soil.

Author Contributions: Conceptualization, R.P.; methodology, R.P., L.T. and F.L.; formal analysis, F.L.; investigation, A.B. and L.T.; writing—original draft preparation, R.V., V.C., A.B., L.T. and F.L.; writing—review and editing, R.P., R.V. and F.L.; supervision, R.P.; funding acquisition, R.P. All authors have read and agreed to the published version of the manuscript.

Funding: This work was supported by the Italian Ministry of Agriculture (MASAF), AGRIDIGIT project, sub-project Selvicoltura (DM n. 36509/2018). It received funding from the European Union Next-GenerationEU (PIANO NAZIONALE DI RIPRESA E RESILIENZA (PNRR)—MISSIONE 4 COMPONENTE 2, INVESTIMENTO 1.4—D.D. 1032 17 June 2022, CN00000022).

Institutional Review Board Statement: Not applicable.

Informed Consent Statement: Not applicable.

Data Availability Statement: Data are available from the corresponding author upon reasonable request.

Acknowledgments: This study was carried out in cooperation with the Agritech National Research Center—WP 4.1—Task 4.1.4. This manuscript reflects only the authors' views and opinions; neither the European Union nor the European Commission can be considered responsible for them.

Conflicts of Interest: The authors declare no conflicts of interest.

References

1. Latterini, F.; Dyderski, M.K.; Horodecki, P.; Picchio, R.; Venanzi, R.; Lapin, K.; Jagodziński, A.M. The Effects of Forest Operations and Silvicultural Treatments on Litter Decomposition Rate: A Meta-Analysis. *Curr. For. Rep.* **2023**, *9*, 276–290. [CrossRef]
2. Marchi, E.; Chung, W.; Visser, R.; Abbas, D.; Nordfjell, T.; Mederski, P.S.; McEwan, A.; Brink, M.; Laschi, A. Sustainable Forest Operations (SFO): A New Paradigm in a Changing World and Climate. *Sci. Total Environ.* **2018**, *634*, 1385–1397. [CrossRef] [PubMed]
3. Latterini, F.; Mederski, P.S.; Jaeger, D.; Venanzi, R.; Tavankar, F.; Picchio, R. The Influence of Various Silvicultural Treatments and Forest Operations on Tree Species Biodiversity. *Curr. For. Rep.* **2023**, *9*, 59–71. [CrossRef]
4. Labelle, E.R.; Hansson, L.; Högbom, L.; Jourgholami, M.; Laschi, A. Strategies to Mitigate the Effects of Soil Physical Disturbances Caused by Forest Machinery: A Comprehensive Review. *Curr. For. Rep.* **2022**, *8*, 20–37. [CrossRef]
5. Vančura, K.; Šimková, M.; Vacek, Z.; Vacek, S.; Gallo, J.; Šimůnek, V.; Podrázský, V.; Štefančík, I.; Hájek, V.; Prokůpková, A.; et al. Effects of Environmental Factors and Management on Dynamics of Mixed Calcareous Forests under Climate Change in Central European Lowlands. *Dendrobiology* **2022**, *87*, 79–100. [CrossRef]
6. European Union. New European Forest Strategy for 2030. Available online: https://eur-lex.europa.eu/legal-content/EN/TXT/?uri=CELEX:52021DC0572 (accessed on 15 August 2021).
7. Britto, P.C.; Jaeger, D.; Hoffmann, S.; Robert, R.C.G.; Vibrans, A.C.; Fantini, A.C. Impact Assessment of Timber Harvesting Operations for Enhancing Sustainable Management in a Secondary Atlantic Forest. *Sustainability* **2019**, *11*, 6272. [CrossRef]
8. Santopuoli, G.; Temperli, C.; Alberdi, I.; Barbeito, I.; Bosela, M.; Bottero, A.; Klopčič, M.; Lesinski, J.; Panzacchi, P.; Tognetti, R. Pan-European Sustainable Forest Management Indicators for Assessing Climate-Smart Forestry in Europe. *Can. J. For. Res.* **2021**, *51*, 1741–1750. [CrossRef]
9. Oliver, C.D.; Nassar, N.T.; Lippke, B.R.; McCarter, J.B. Carbon, Fossil Fuel, and Biodiversity Mitigation with Wood and Forests. *J. Sustain. For.* **2014**, *33*, 248–275. [CrossRef]
10. Schaaf, A.A.; García, C.G.; Ruggera, R.A.; Tallei, E.; Vivanco, C.G.; Rivera, L.; Politi, N. Influence of Logging on Nest Density and Nesting Microsites of Cavity-Nesting Birds in the Subtropical Forests of the Andes. *For. Int. J. For. Res.* **2022**, *95*, 73–82. [CrossRef]
11. Hoffmann, S.; Schönauer, M.; Heppelmann, J.; Asikainen, A.; Cacot, E.; Eberhard, B.; Hasenauer, H.; Ivanovs, J.; Jaeger, D.; Lazdins, A.; et al. Trafficability Prediction Using Depth-to-Water Maps: The Status of Application in Northern and Central European Forestry. *Curr. For. Rep.* **2022**, *8*, 55–71. [CrossRef]
12. Keefe, R.F.; Zimbelman, E.G.; Picchi, G. Use of Individual Tree and Product Level Data to Improve Operational Forestry. *Curr. For. Rep.* **2022**, *8*, 148–165. [CrossRef]
13. Kovácsová, P.; Antalová, M. Precision Forestry-Definition and Technologies. *Sumar List* **2010**, *134*, 603–611.
14. Holopainen, M.; Vastaranta, M.; Hyyppä, J. Outlook for the Next Generation's Precision Forestry in Finland. *Forests* **2014**, *5*, 1682–1694. [CrossRef]
15. Venanzi, R.; Latterini, F.; Civitarese, V.; Picchio, R. Recent Applications of Smart Technologies for Monitoring the Sustainability of Forest Operations. *Forests* **2023**, *14*, 1503. [CrossRef]
16. Tsioras, P.A.; Żak, J.; Karaszewski, Z. RFID Implementations in the Wood Supply Chains: State of the Art and the Way to the Future. *Drewno Pr. Nauk. Doniesienia Komun. Wood Res. Pap. Rep. Announc.* **2022**, *65*, 209. [CrossRef]

17. Putzenlechner, B.; Koal, P.; Kappas, M.; Löw, M.; Mundhenk, P.; Tischer, A.; Wernicke, J.; Koukal, T. Towards Precision Forestry: Drought Response from Remote Sensing-Based Disturbance Monitoring and Fine-Scale Soil Information in Central Europe. *Sci. Total Environ.* **2023**, *880*, 163114. [CrossRef] [PubMed]
18. Ziesak, M. Precision Forestry-An Overview on the Current Status of Precision Forestry. A Literature Review. In Proceedings of the Precision Forestry in Plantations, Semi-Natural and Natural Forests, IUFRO Precision Forestry Conference Technical University, Munich, Germany, 5–10 March 2006; pp. 5–10.
19. David, R.M.; Rosser, N.J.; Donoghue, D.N.M. Improving above Ground Biomass Estimates of Southern Africa Dryland Forests by Combining Sentinel-1 SAR and Sentinel-2 Multispectral Imagery. *Remote Sens. Environ.* **2022**, *282*, 113232. [CrossRef]
20. Tian, Y.; Huang, H.; Zhou, G.; Zhang, Q.; Tao, J.; Zhang, Y.; Lin, J. Aboveground Mangrove Biomass Estimation in Beibu Gulf Using Machine Learning and UAV Remote Sensing. *Sci. Total Environ.* **2021**, *781*, 146816. [CrossRef]
21. Xu, D.; Wang, H.; Xu, W.; Luan, Z.; Xu, X. LiDAR Applications to Estimate Forest Biomass at Individual Tree Scale: Opportunities, Challenges and Future Perspectives. *Forests* **2021**, *12*, 550. [CrossRef]
22. Ecke, S.; Dempewolf, J.; Frey, J.; Schwaller, A.; Endres, E.; Klemmt, H.-J.; Tiede, D.; Seifert, T. UAV-Based Forest Health Monitoring: A Systematic Review. *Remote Sens.* **2022**, *14*, 3205. [CrossRef]
23. Haghighian, F.; Yousefi, S.; Keesstra, S. Identifying Tree Health Using Sentinel-2 Images: A Case Study on Tortrix Viridana L. Infected Oak Trees in Western Iran. *Geocarto Int.* **2022**, *37*, 304–314. [CrossRef]
24. Zhao, J.; Liu, D.; Cao, Y.; Zhang, L.; Peng, H.; Wang, K.; Xie, H.; Wang, C. An Integrated Remote Sensing and Model Approach for Assessing Forest Carbon Fluxes in China. *Sci. Total Environ.* **2022**, *811*, 152480. [CrossRef] [PubMed]
25. Latterini, F.; Stefanoni, W.; Venanzi, R.; Tocci, D.; Picchio, R. GIS-AHP Approach in Forest Logging Planning to Apply Sustainable Forest Operations. *Forests* **2022**, *13*, 484. [CrossRef]
26. Marčeta, D.; Petković, V.; Ljubojević, D.; Potočnik, I. Harvesting System Suitability as Decision Support in Selection Cutting Forest Management in Northwest Bosnia and Herzegovina. *Croat. J. For. Eng.* **2020**, *41*, 251–265. [CrossRef]
27. Sterenczak, K.; Moskalik, T. Use of LIDAR-Based Digital Terrain Model and Single Tree Segmentation Data for Optimal Forest Skid Trail Network. *IForest* **2015**, *8*, 661–667. [CrossRef]
28. Picchio, R.; Latterini, F.; Mederski, P.S.; Tocci, D.; Venanzi, R.; Stefanoni, W.; Pari, L. Applications of GIS-Based Software to Improve the Sustainability of a Forwarding Operation in Central Italy. *Sustainability* **2020**, *12*, 5716. [CrossRef]
29. Görgens, E.B.; Mund, J.-P.; Cremer, T.; de Conto, T.; Krause, S.; Valbuena, R.; Rodriguez, L.C.E. Automated Operational Logging Plan Considering Multi-Criteria Optimization. *Comput. Electron. Agric.* **2020**, *170*, 105253. [CrossRef]
30. Latterini, F.; Venanzi, R.; Tocci, D.; Picchio, R. Depth-to-Water Maps to Identify Soil Areas That Are Potentially Sensitive to Logging Disturbance: Initial Evaluations in the Mediterranean Forest Context. *Land* **2022**, *11*, 709. [CrossRef]
31. Schönauer, M.; Hoffmann, S.; Maack, J.; Jansen, M.; Jaeger, D. Comparison of Selected Terramechanical Test Procedures and Cartographic Indices to Predict Rutting Caused by Machine Traffic during a Cut-to-Length Thinning Operation. *Forests* **2021**, *12*, 113. [CrossRef]
32. Schönauer, M.; Prinz, R.; Väätäinen, K.; Astrup, R.; Pszenny, D.; Lindeman, H.; Jaeger, D. Spatio-Temporal Prediction of Soil Moisture Using Soil Maps, Topographic Indices and SMAP Retrievals. *Int. J. Appl. Earth Obs. Geoinf.* **2022**, *108*, 102730. [CrossRef]
33. Schönauer, M.; Väätäinen, K.; Prinz, R.; Lindeman, H.; Pszenny, D.; Jansen, M.; Maack, J.; Talbot, B.; Astrup, R.; Jaeger, D. Spatio-Temporal Prediction of Soil Moisture and Soil Strength by Depth-to-Water Maps. *Int. J. Appl. Earth Obs. Geoinf.* **2021**, *105*, 102614. [CrossRef]
34. Suvinen, A.; Saarilahti, M. Measuring the Mobility Parameters of Forwarders Using GPS and CAN Bus Techniques. *J. Terramech.* **2006**, *43*, 237–252. [CrossRef]
35. Bacescu, N.M.; Cadei, A.; Moskalik, T.; Wiśniewski, M.; Talbot, B.; Grigolato, S. Efficiency Assessment of Fully Mechanized Harvesting System through the Use of Fleet Management System. *Sustainability* **2022**, *14*, 16751. [CrossRef]
36. Olivera, A.; Visser, R. Development of Forest-Yield Maps Generated from Global Navigation Satellite System (GNSS)-Enabled Harvester StanForD Files: Preliminary Concepts. *N. Z. J. For. Sci.* **2016**, *46*, 1–10. [CrossRef]
37. Abbas, D.; Di Fulvio, F.; Marchi, E.; Spinelli, R.; Schmidt, M.; Bilek, T.; Han, H.S. A Proposal for an Integrated Methodological and Scientific Approach to Cost Used Forestry Machines. *Croat. J. For. Eng.* **2021**, *42*, 63–75. [CrossRef]
38. Civitarese, V.; Figorilli, S.; Acampora, A.; Sperandio, G.; Assirelli, A.; Scarfone, A.; Bascietto, M. Innovative System for Monitoring and Mapping Loads in Logs Forwarding. In Proceedings of the European Biomass Conference and Exhibition Proceedings, Online, 26–29 April 2021; pp. 265–267.
39. Feng, T.; Chen, S.; Feng, Z.; Shen, C.; Tian, Y. Effects of Canopy and Multi-Epoch Observations on Single-Point Positioning Errors of a GNSS in Coniferous and Broadleaved Forests. *Remote Sens.* **2021**, *13*, 2325. [CrossRef]
40. Balestra, M.; Tonelli, E.; Vitali, A.; Urbinati, C.; Frontoni, E.; Pierdicca, R. Geomatic Data Fusion for 3D Tree Modeling: The Case Study of Monumental Chestnut Trees. *Remote Sens.* **2023**, *15*, 2197. [CrossRef]
41. Wempe, A.M.; Keefe, R.F. Characterizing Rigging Crew Proximity to Hazards on Cable Logging Operations Using GNSS-RF: Effect of GNSS Positioning Error on Worker Safety Status. *Forests* **2017**, *8*, 357. [CrossRef]
42. Zimbelman, E.G.; Keefe, R.F. Real-Time Positioning in Logging: Effects of Forest Stand Characteristics, Topography, and Line-of-Sight Obstructions on GNSS-RF Transponder Accuracy and Radio Signal Propagation. *PLoS ONE* **2018**, *13*, e0191017. [CrossRef]

43. D'EON, R.G.; DELPARTE, D. Effects of Radio-Collar Position and Orientation on GPS Radio-Collar Performance, and the Implications of PDOP in Data Screening. *J. Appl. Ecol.* **2005**, *42*, 383–388. [CrossRef]
44. Bastos, A.S.; Hasegawa, H. Behavior of GPS Signal Interruption Probability under Tree Canopies in Different Forest Conditions. *Eur. J. Remote Sens.* **2013**, *46*, 613–622. [CrossRef]
45. Tarquini, S.; Vinci, S.; Favalli, M.; Doumaz, F.; Fornaciai, A.; Nannipieri, L. Release of a 10-m-Resolution DEM for the Italian Territory: Comparison with Global-Coverage DEMs and Anaglyph-Mode Exploration via the Web. *Comput. Geosci.* **2012**, *38*, 168–170. [CrossRef]
46. Tarquini, S.; Isola, I.; Favalli, M.; Mazzarini, F.; Bisson, M.; Pareschi, M.T.; Boschi, E. TINITALY/01: A New Triangular Irregular Network of Italy. *Ann. Geophys.* **2007**, *50*, 407–425. [CrossRef]
47. Wilson, J.P.; Gallant, J.C. Secondary Topographic Attributes. In *Terrain Analysis: Principles and Applications*; John Wiley & Sons, Inc.: Hoboken, NJ, USA, 2000; pp. 87–131.
48. De Reu, J.; Bourgeois, J.; Bats, M.; Zwertvaegher, A.; Gelorini, V.; De Smedt, P.; Chu, W.; Antrop, M.; De Maeyer, P.; Finke, P.; et al. Application of the Topographic Position Index to Heterogeneous Landscapes. *Geomorphology* **2013**, *186*, 39–49. [CrossRef]
49. QGIS Development Team. Quantum GIS Version 3.28.9 Firenze. Available online: https://qgis.org/pl/site/forusers/download.html (accessed on 20 February 2024).
50. R Development Core Team. R: A Language and Environment for Statistical Computing; R Foundation for Statistical Computing: Vienna Austria. Available online: https://www.r-project.org/ (accessed on 6 October 2023).
51. Bates, D.; Mächler, M.; Bolker, B.; Walker, S. Fitting Linear Mixed-Effects Models Using Lme4. *J. Stat. Softw.* **2015**, *67*, 48. [CrossRef]
52. Fox, J.; Weisberg, S. *An R Companion to Applied Regression*; Sage Publications: Thousand Oaks, CA, USA, 2018; ISBN 1544336489.
53. Bartoń, K. *MuMIn: Multi-Model Inference*; R Package Version 1.47.5; Springer-Verlag: New York, NY, USA, 2023.
54. Nakagawa, S.; Schielzeth, H. A General and Simple Method for Obtaining R 2 from Generalized Linear Mixed-effects Models. *Methods Ecol. Evol.* **2013**, *4*, 133–142. [CrossRef]
55. Lüdecke, D. Ggeffects: Tidy Data Frames of Marginal Effects from Regression Models. *J. Open Source Softw.* **2018**, *3*, 772. [CrossRef]
56. Becker, R.M.; Keefe, R.F. Development of Activity Recognition Models for Mechanical Fuel Treatments Using Consumer-Grade GNSS-RF Devices and Lidar. *Forestry* **2022**, *95*, 437–449. [CrossRef]
57. Becker, R.; Keefe, R.; Anderson, N. Use of Real-Time GNSS-RF Data to Characterize the Swing Movements of Forestry Equipment. *Forests* **2017**, *8*, 44. [CrossRef]
58. Lopatin, E.; Väätäinen, K.; Kukko, A.; Kaartinen, H.; Hyyppä, J.; Holmström, E.; Sikanen, L.; Nuutinen, Y.; Routa, J. Unlocking Digitalization in Forest Operations with Viewshed Analysis to Improve GNSS Positioning Accuracy. *Forests* **2023**, *14*, 689. [CrossRef]
59. Tomaštík, J.; Saloň, Š.; Piroh, R. Horizontal Accuracy and Applicability of Smartphone GNSS Positioning in Forests. *Forestry* **2017**, *90*, 187–198. [CrossRef]
60. Lindroos, O.; La Hera, P.; Häggström, C. Drivers of Advances in Mechanized Timber Harvesting—A Selective Review of Technological Innovation. *Croat. J. For. Eng.* **2017**, *38*, 243–258.

sustainability

MDPI

Article

Impact of Chainsaw Power on Fuel and Oil Consumption

Slavica Antonić [1,*], Milorad Danilović [1], Dušan Stojnić [1] and Srđan Dražić [2]

1 Faculty of Forestry, University of Belgrade, Kneza Višeslava 1, 11000 Belgrade, Serbia
2 The Public Enterprise "Šume Republike Srpske" a.d. Sokolac, Research Develop and Project Center, Grčka 21, 78000 Banja Luka, Bosnia and Herzegovina
* Correspondence: slavica.antonic@sfb.bg.ac.rs; Tel.: +381-11-30-53-833

Abstract: This research was carried out in a selection stand of beech and fir, where logging was performed with chainsaws of different powers. Two chainsaws of 3.9 kW were used for work on one area, while on the other one the applied tools were a combination of a smaller and a larger saw (2.5 kW and 3.9 kW). After logging, the consumption of energy products (fuel (gasoline and motor oil mix ratio 50:1) and chainsaw oil) was measured for each individual chainsaw by the method of refueling. It was found that the power of the chainsaw, i.e., the combination of chainsaws of different powers, significantly influences the consumption of energy products in a beech–fir mixed stand. By using a combination of one larger and one smaller chainsaw instead of two larger chainsaws, savings of about 26% for fuel and 24% for oil were achieved. The prediction analysis found that by using two chainsaws (of lower and higher power) instead of two higher power chainsaws (currently common mode of work) in beech stands for an estimated volume of about 1 million m^3 year^{-1}, the savings could be about 54,000 L of fuel and about 19,000 L of oil. In monetary value, this is a saving of about EUR 120,000 year^{-1}, i.e., of about 2,500,000 MJ year^{-1} of energy.

Keywords: chainsaws; fuel consumption; oil consumption; felling; logging

Citation: Antonić, S.; Danilović, M.; Stojnić, D.; Dražić, S. Impact of Chainsaw Power on Fuel and Oil Consumption. *Sustainability* **2023**, *15*, 2795. https://doi.org/10.3390/su15032795

Academic Editors: Ivica Papa and Andreja Đuka

Received: 14 December 2022
Revised: 15 January 2023
Accepted: 18 January 2023
Published: 3 February 2023

1. Introduction

It is estimated that a volume of about 3.7 billion m^3 of timber is felled around the globe annually, of which about 824 million m^3 is felled in Europe [1]. In developed countries, a large part of the wood volume is cut by machinery. For example, in Scandinavian countries, machine felling accounts for about 90% of the wood volume (Latvia, Finland and Sweden), in Germany about 50%, Belarus 41%, the Czech Republic 29% and Poland 20% [2]. However, in some countries this percentage is far lower. For example, in Slovakia it is only 5% [2] and in Croatia 3–4% [3], while in less developed countries these values are even lower (e.g., Ukraine 1%, Romania 2%) [2].

According to the research of [4], about 98.5% of the total amount of logged wood in Serbia is cut with chainsaws, while only about 1.5% is logged with a harvester. However, in the period since 2011, several harvesters have been procured in Serbia, so the assumption is that now the ratio is 90:10, in favor of chainsaws. That means that about 4.5 million m^3/year of wood volume from state and private forests are cut by chainsaws, which speaks volumes about their high importance.

Although the number of engaged harvesters is on the rise, chainsaws are still dominant. The reasons for their use are a number of advantages over machinery which are, in essence, reflected in lower unit costs. These costs are directly affected by daily operating costs, which include, but are not limited to, the purchase value of the asset (which directly affects depreciation), fuel and oil costs, replacement of spare parts, maintenance costs, etc. In addition, chainsaws are sometimes the only tools for logging and tree felling, as well as for the cutting of assortments in some specific conditions, such as on steep slopes, or in case of a large tree diameter [5–7]. The main reasons for the limited use of harvesters are the specifics of natural conditions, different ways of forest management as well as the financial capacity of companies to procure funding for that purpose.

While efficiency has increased through mechanization [8], forest operations are still the major producer of emissions in the wood value chain. In that regard, harvesting is the most critical phase, due to high consumption of fossil fuels [9–12]. Felling and processing alone require over 1.1 L diesel per m^3 of roundwood [13] and when this amount is multiplied by the volume harvested in industrialized countries, consumption soars to over 1 billion liters of diesel per year [11].

Fuel consumption of cut-to-length (CTL) harvesters accounts for 38% of the total fuel used in the technological cycle, which is higher than the amount consumed during forwarding (35%) and transportation (27%) [11]. As with most production systems, logging assumes energy input that is largely provided by fossil fuels, thus contributing to greenhouse gas emissions (GHG) [14–16].

Author [17] state that determining the consumption of energy and energy carriers (fuel and oils) for chainsaws is of great importance for work in practice. Several researchers were involved in the research of energy consumption of chainsaws in logging [17–22].

However, [23] states that fuel consumption in logging is still not sufficiently researched (unlike in transport) and that the existing measurement technologies are mostly untested, because although several studies have been conducted, more comprehensive research is needed to investigate how to manage forests in a more efficient way while reducing fuel consumption.

The aim of this paper is to determine the consumption of energy products and determine the economic benefit in logging and wood assortment production in a mixed beech and fir stand for different combinations of chainsaws, which primarily differ in power, i.e., to determine which combination of chainsaws is more efficient from the aspect of fuel consumption (two saws of the same power—Husqvarna XP 372—versus two saws of which one has lower and the other one higher power—Husqvarna XP 372 and Husqvarna H 545).

2. Materials and Methods

2.1. Object of Research

Data collection was performed in a hilly mountainous area, in a selection forest of beech (*Fagus silvatica* L.) and fir (*Abies alba* Mill.), while the conditions for sample plot 1 (hereinafter: SP1) and for sample plot 2 (SP2) were the same. The ratio of beech to fir was approximately 45:55. Recording of data for the purposes of this paper was performed in the period of winter logging (November–March), at the teaching base Goč of the University of Belgrade, Faculty of Forestry, Serbia.

Characteristics of the sample plots (tree species, average tree volume, average diameter of the felled tree at breast height and the percentage share of roundwood in the total volume) are shown in Table 1. The percentage share of firewood in the total volume ranged from 9–17%, but in addition to technical roundwood, long firewood was also produced (their sum value makes a supplement up to 100%).

Table 1. Basic characteristics of the sample plots.

Sample plot	Altitude (m)	Terrain Slope (%)	Species	Processed Number of Trees	Average Tree Volume (m^3 $Tree^{-1}$)	Average Diameter of Felled Tree at Breast Height ($D_{1,3}$) (cm)	The Percentage Share of Firewood in the Total Volume (%)
SP1	1150	10–20	Beech	9	6.41	67	9
			Fir	23	4.75	56	15
SP2			Beech	16	5.63	73	17
			Fir	24	3.52	62	10

2.2. Chainsaw Characteristics

In Serbia, in over 90% of cases, trees are logged by chainsaw. In most cases, two workers with chainsaws work together, most often both of them have chainsaws of the same power and size, and they are mostly chainsaws with a power of over 4 kW. A combination of a lower and higher power saw is very rare in practice.

The chainsaws used in this study are Husqvarna XP 372 (3.9 kW) and Husqvarna H 545 (2.5 kW) (Husqvarna, Stockholm, Sweden) with X-Torq® technology (higher fuel efficiency and lower exhaust emissions compared to conventional engines). As can be seen from Table 2, the Husqvarna 372 XP chainsaw is larger in size and has a higher power than the Husqvarna H 545 chainsaw.

The Husqvarna 372 XP chainsaws were used in SP1 (both saws were the same). They were 3 months old, while the Husqvarna H 545 chainsaw was 1 month old (applied in SP2), along with the Husqvarna 372 XP. New guides and chains were mounted on all saws. Chainsaws used a gasoline and motor oil mix in the ratio 50:1.

Table 2. Basic technical characteristics of the applied Husqvarna chainsaws.

Technical Characteristics	Husqvarna 372 XP	Husqvarna H 545
Cylinder capacity (cm^3)	70.7	50.1
Output power (kW)	3.9	2.5
Recommended guide bar length (cm)	70	50
Chainsaw weight without the cutting assembly (kg)	6.3	4.9
Special characteristic	-	X-Torq® technology

2.3. Work Methods

Two workers operated the saws in each sample plot. In SP1, both workers operated the same chainsaws (two chainsaws Husqvarna 372 XP), while in SP2, one worker operated a higher power saw (Husqvarna 372 XP), and the other one operated a lower power saw (Husqvarna H545). Workers were different for both sample plots, but they worked at the same time on both sample plots.

In SP1, both workers took turns in all work operations (felling, bucking, delimbing technical roundwood and firewood processing), while in SP2, the worker with the larger chainsaw performed work operations of tree felling, technical roundwood and firewood bucking, while the worker who operated the smaller saw mainly performed the delimbing of technical roundwood, firewood processing and firewood bucking (which was produced on this surface as long firewood and as a classic one-meter firewood).

2.4. Energy Consumption Measurement

Fuel consumption was measured for each tree individually by the method of energy products refueling the tank. Measuring cylinders of 250 and 500 mL with a 10 mL accuracy of reading, with engraved values of 10 to 250 (for oil) and 500 mL (for fuel), respectively, were used for the refueling of fuel and oil into tanks. The workers would start their workday with full tanks, and then after each felled and cut tree, they would refill the fuel and oils. The amount of refilled fuel or oil represents the amount of fuel or oil consumed for the previously cut tree. The data on fuel and oil consumption were entered in the corresponding manual. The manual also contains data on the tree (tree species and diameter at breast height), as well as data on the processed wood logs (diameter in the middle of length), from which the volume of logs was calculated according to the Huber formula.

$$V = \frac{{D_s}^2 \cdot \pi}{4} \cdot L$$

V—volume of log (m^3), D_s—diameter in the middle of the log (cm), L—length of log (m).

2.5. Data Analysis

The data were analyzed using descriptive statistics and Student's t-test to determine differences in mean values. Regression analysis was used to analyze fuel and lubricant consumption by diameter classes. All analyses were performed in Microsoft Excel 2013.

3. Results

3.1. Consumption of Energy

The average consumption of energy products per individual sample plot is shown in Figures 1 and 2. As can be seen, the average consumption of both fuel and oil was higher in SP1, where two saws of higher power were used, compared to SP2, where one saw had higher (3.9 kW) and the other one lower power (2.5 kW).

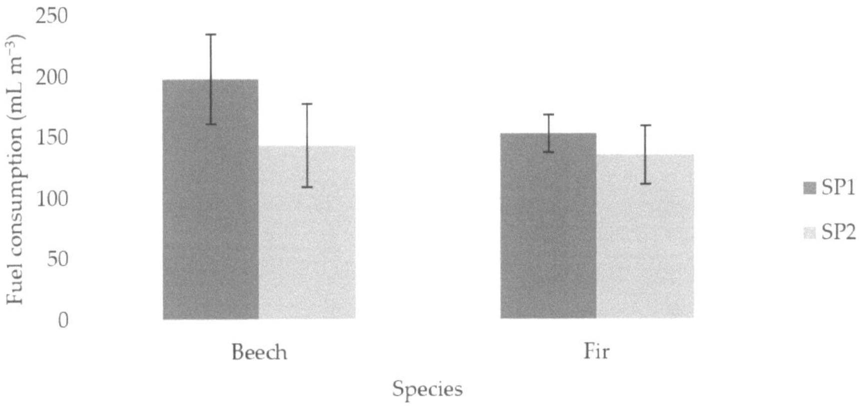

Figure 1. Fuel consumption (mL m^{-3}) by sample plots—mean and 95% confidence interval.

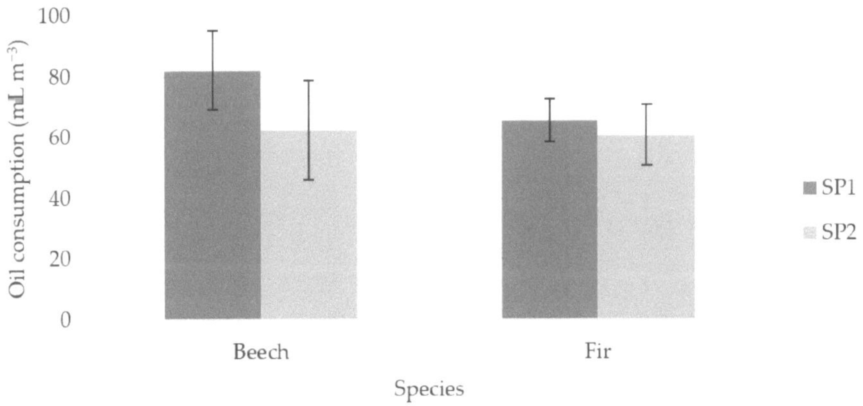

Figure 2. Oil consumption (mL m^{-3}) by sample plots—mean and 95% confidence interval.

The t-test established that at the 0.05 level of significance, there are statistically significant differences in fuel consumption between SP1 and SP2 for beech (t = 2.200, df = 11, p = 0.025), as well as for oil consumption (t = 2.178, df = 12, p = 0.020). However, there are no statistically significant differences between SP1 and SP2 for fir in terms of both fuel consumption (t = 2.021, df = 40, p = 0.178) and oil consumption (t = 2.026, df = 37, p = 0.261).

For further analyses, only the consumption of energy products for beech was taken into account, because there are obvious differences, but also due to the fact that beech is the most common species in Serbia.

With the increase in diameter at breast height, fuel and oil consumption logically grows. However, as can be seen from the figure, differences in energy product consumption between SP1 and SP2 make a drastic difference in higher diameter classes (from 67.5 cm upwards), while in lower diameter classes these differences are smaller (Figures 3 and 4). By applying two higher power saws (in SP1) and by applying one higher and one lower power saw (SP2), an obvious difference can be observed. This difference in energy product consumption reaches over 40% in the highest diameter classes (on average about 26% for fuel and 24% for oil).

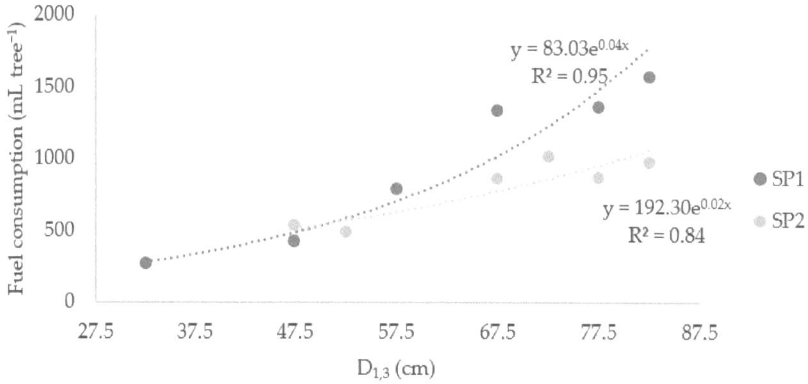

Figure 3. Fuel consumption of beech in mL tree^{-1}.

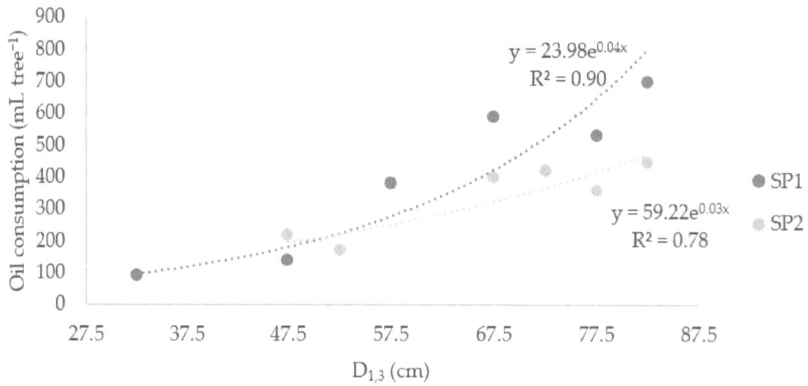

Figure 4. Oil consumption of beech in mL tree^{-1}.

The increase in the consumption of energy products expressed in mL tree^{-1} with an increasing diameter at breast height is shown by exponential functions, which are given below.

SP1—fuel consumption $y = 83.03e^{0.04x}$
SP1—oil consumption $y = 23.98e^{0.04x}$
SP2—fuel consumption $y = 192.30e^{0.02x}$
SP2—oil consumption $y = 59.22e^{0.03x}$

where

y—fuel consumption is expressed in mL tree^{-1}
x—diameter at breast height (cm).

This is also shown in the figures of the average consumption of energy products expressed in mL m^{-3} (Figures 5 and 6), whose balancing functions are linear, where the consumption of energy products is correlated with the diameter class, i.e., with the increase in diameter class, the average consumption expressed in ml m^{-3} of processed wood decreases.

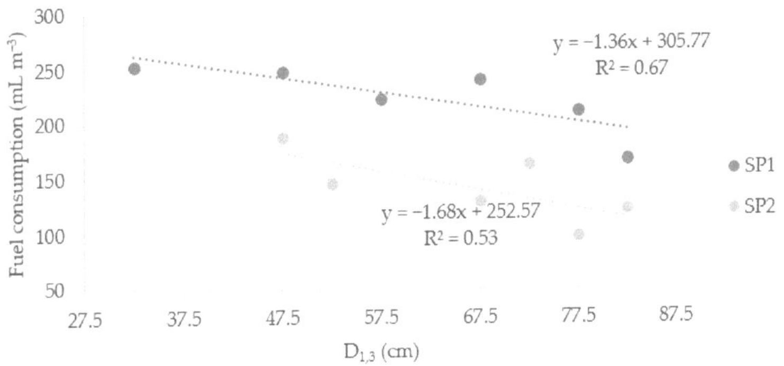

Figure 5. Fuel consumption of beech in mL m^{-3}.

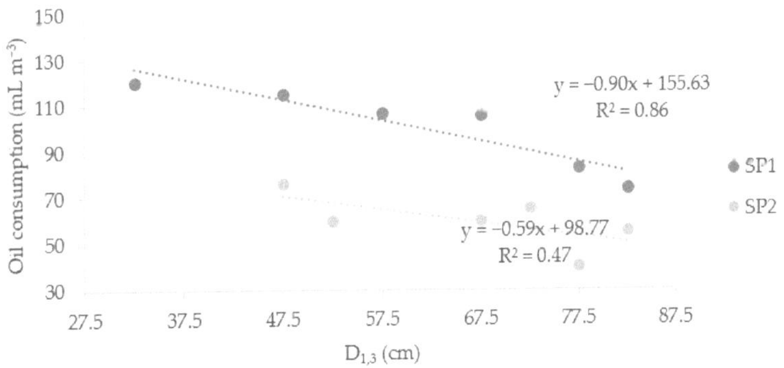

Figure 6. Oil consumption of beech in mL m^{-3}.

However, the correlation coefficient in SP2 for fuel and oil ($R^2 = 0.84$; $R^2 = 0.78$) shows that diameter is affected on fuel and oil consumption. Nevertheless, there is a clear difference in the consumption of energy products in SP1 and SP2, which is shown by dependency functions, where y, fuel consumption, is expressed in ml m^{-3} and x is diameter at breast height (cm).

The functions that show fuel consumption in the individual sample plots are given below:

SP1—fuel consumption y = −1.36x + 305.77
SP1—oil consumption y = −0.90x + 155.63
SP2—fuel consumption y = −1.68x + 252.57
SP2—oil consumption y = −0.59x + 98.77

where

y—fuel consumption is expressed in mL m^{-3}.
x—diameter at breast height (cm).

3.2. Economic Benefit

Beech accounts for about 50% of the total felled wood volume in state forests in Serbia [24]. Beech (*Fagus sylvatica* L.) is the most widespread broadleaf species in Europe and it plays a key role in the European forest sector, covering more than 11 million hectares throughout Europe [25]. Beech is predominantly widespread in mountain conditions. Fuel consumption in the selection stand of beech found in this study is on average 26% higher for fuel and 24% higher for oil in SP1 (two higher power saws) than in SP2 (one higher power saw and one lower power saw). If we make a prediction, according to which all the wood volume from state beech forests in Serbia (estimated at about 1 million m^3 $year^{-1}$) [24] would be cut down by two saws of higher and lower power compared to two saws of higher power (as is normally done), then the annual savings in absolute value would be about EUR 120,000 $year^{-1}$ (Table 3) (in 2022). Calculated as energy savings, that would be about 54,000 L of fuel and about 19,000 L of oil per year. Especially in times of energy crises, these seemingly small savings would be reflected in the reduction of energy product consumption, i.e., the reduction of dependence on the imports of energy products, but also in the emission of harmful gases. The wood volume from private forests, which is assumed to have approximately the same value, was not included in this analysis.

It should also be noted that the chainsaws that are currently in use mostly have a power of over 4 kW, so the difference in the consumption of energy products is probably even greater.

Table 3. Predictive analysis of energy carriers' consumption for different combinations of use of chainsaws in beech forests in Serbia.

Combination of Applied Chainsaws	Estimated Wood Volume of High Beech Forests (m^3 $Year^{-1}$) *	Average Fuel Consumption (L m^{-3})	Average Oil Consumption (L m^{-3})	Fuel Price by Volume (EUR m^{-3}) **	Price of Oil by Volume (EUR m^{-3}) **	Total Fuel Cost (EUR $Year^{-1}$)	Total Cost of Oils (EUR $Year^{-1}$)	Total Cost of Energy Carriers (EUR $Year^{-1}$)
Two saws Husqvarna 372 XP	1,000,000	0.19	0.08	0.31	0.17	310,000	170,000	480,000
Husqvarna 372 XP and Husqvarna H 545	1,000,000	0.14	0.06	0.23	0.13	230,000	130,000	360,000
Absolute difference		0.05	0.01	0.08	0.04	80,000	40,000	120,000
Percentage difference (%)						26	24	25

* Estimated value based on data from the monofigure "Beech in Serbia" (2005) [24]. ** The value of fuel was EUR 1.57 L^{-1}, while the value of oils was EUR 2.13 L^{-1}.

If the amount of potentially saved energy was recalculated according to [26] as the energy value of fuel, which is 33.75 MJ L^{-1}, this would also mean a lower consumption of energy products in beech stands in the total value of about 2,463,750 MJ $year^{-1}$. This issue is very important in times of the great worldwide energy crisis. Of course, this would consequently mean a significantly lower emission of harmful gases, to which special attention should be paid.

4. Discussion

Forest mechanization plays an important role in increasing labor productivity and reducing production costs. Chainsaws are the most commonly used tool for tree felling and can have both positive and negative environmental impacts on the forest ecosystem [27].

As already mentioned, in Serbia, trees are logged by chainsaw in over 90% of cases, where two workers with chainsaws work together (a power of saws over 4 kW). Very rarely, combinations of larger and smaller saws are used, assuming that this would lead to savings not only in the initial investment, but also in the consumption of energy products and also in lower emissions of harmful gases, etc., as already shown in the results.

The average fuel consumption for beech was 198 mL m^{-3} in SP1, and in SP2 it was almost 30% lower (0.144 L m^{-3}), while the average fuel consumption of fir in SP1 was 153 mL m^{-3} and in SP2 it was 0.136 L m^{-3}. The situation is similar for oil consumption—0.082 L m^{-3} and 0.063 L m^{-3} for beech (SP1 and SP2) and 0.066 L m^{-3} and 0.061 L m^{-3} for fir (SP1 and SP2).

When comparing the average fuel consumption expressed in L m^{-3} for beech with the results obtained by [18], it can be noted that it is higher than in the case of fuel consumption in our research for the same organizational form of work (2+0—two workers with chainsaw, without assistant) (0.295 L m^{-3} compared to 0.198 L m^{-3} in SP1 and 0.163 L m^{-3} in SP2). Converted into percentage values, this consumption in [18] is higher by about 32% compared to SP1, or about 44% compared to SP2.

Comparing the data of [21] for a fir and spruce stand, where the chainsaw Husqvarna 372 XP (3.9 kW) was used in the organizational form of work 1+0, and the data of this research, where the same chainsaw was used in the organizational form 2+0, we can conclude that a higher consumption of energy products was achieved in this research (SP1), where the fuel consumption was 0.153 L m^{-3} and the oil consumption reached 0.066 L m^{-3}, while in the research of [21] these values were 0.104 L m^{-3} for fuel and 0.023 L m^{-3} for oil. Calculated as a percentage, in terms of fuel consumption, this amount is slightly higher than 30%, while for oil this value is as high as over 60%. However, the factor with the highest impact on fuel consumption, i.e., diameter at breast height, was not taken into account, so these data are only partially comparable. The values obtained by other researchers [17,28,29] are almost completely incomparable with the data of this research study, both due to missing data and the incomparability of the existing data.

The problem with comparing the consumption of energy products in different studies is that the works often do not specify all the conditions (factors) that can have an impact on fuel consumption, as well as the fact that very rarely can comparable data regarding most impact factors be found. For this reason, the data obtained by other researchers can only serve for orientation, and in order to determine the norms, it is necessary to conduct research under specific conditions.

The consumption of energy products increased with an increase in diameter at breast height, i.e., with an increase in diameter class, which is logical. However, for larger diameter classes, the difference in consumption between SP1 and SP2 increases significantly, and in the highest diameter classes it reaches up to 40%. This can be explained by the fact that the worker who operates the lower power chainsaw (SP2) mainly performs the delimbing and firewood processing, thus saving energy. Unlike that, in SP1, these work operations are performed by a worker with a larger chainsaw, which can be considered irrational, since mostly diameters below 20 cm are dealt with in this case. This means that the application of one smaller and one larger saw is particularly significant for trees of large dimensions. The correlation coefficient in SP2 for fuel and oil shows that, in addition to diameter, other factors such as crown size probably influence fuel consumption, which were not taken into account in this study.

The manufacturer's website (Husqvarna) states that the Husqvarna H 545 chainsaw has a fuel consumption which is reduced by 20%, but it is not specified in relation to which type of chainsaw. Therefore, these studies have shown that this consumption is even lower, but in relation to the Husqvarna 372 XP chainsaw (almost 30% perceived as total average consumption). The Husqvarna H 545 chainsaw has X-Torq® technology (this technology uses pure air, instead of air mixed with fuel, to flush out exhaust fumes).

When the values from SP2 are compared with the data from SP1, where the data are recorded in the same conditions, it can be concluded that the organizational form of work in which one worker operates a higher power saw and the other operates a lower power saw is more efficient compared to the form of work in which both workers work with larger saws, if beech logging is performed.

The average oil consumption in relation to the fuel consumption in this study, observed by diameter classes, ranged from 0.42 to 0.45 of fuel consumption regardless of the tree species.

Currently, in Serbia, there are norms, which were researched by [20], which can serve as a basis for the orientation on energy product consumption for different tree species. However, in order to achieve any savings, and also to reduce the emission of harmful gases, a more detailed analysis is needed. This can only be achieved by determining the consumption of energy products in a specific situation, their analysis and possible comparisons, in order to improve the process while saving energy, but also by getting an overview of other aspects (efficiency and ergonomics).

Authors [21] state that fuel and oil savings could be achieved primarily by regular inspections of chainsaw working order, proper storage and distribution of energy products, the use of adequate accessories for fuel manipulation, as well as education and stimulation of workers. It is necessary to continue to pay a lot of attention to the rationalization of fuel and oil consumption, as well as to the reduction of their harmful effects on man and the environment (introduction and use of biodegradable fuels and oils).

In this research, it was found that by using the appropriate type of chainsaw, there can be a reduction in the consumption of energy products, and consequently a reduction in the emission of harmful gases. Fuel consumption in the selection beech stand is 27% higher for fuel and 23% for oil in SP1 (two higher power saws) compared to SP2 (one higher power saw and one lower power saw). If the wood volume from state beech forests in Serbia were cut down by two saws (about 1 million m^3) of higher and lower power compared to two saws of higher power, then the annual savings in absolute value would be about EUR 122,000 year^{-1} and about 2.5 million MJ year^{-1}, i.e., about 54,000 L of fuel and about 19,000 L of oil.

Despite the fact that these savings cannot be compared to those stated by [11], which refers to savings in fuel consumption in mechanized logging in the Nordic and Baltic regions alone (potential savings may amount to almost 50 million L diesel year^{-1}), they are still not negligible. This is primarily due to the fact that the global production of roundwood amounts to approximately 3.7 billion m^3 year^{-1} [1], of which over 1 billion originates from industrialized regions such as Europe, North America and Russia [30]. Thus, almost three-quarters of the total wood cut in the world comes from forests from less developed environments, which in a large number of cases are not able to afford cutting machines, so logging is performed with chainsaws and sometimes with hand saws. This fact should further motivate researchers to pay more attention to chainsaws, i.e., their consumption of energy products in other countries.

Although there is a tendency in Serbia to procure new machinery such as harvesters, it is certain that in the coming period logging with chainsaws will still take precedence over harvesters. As already stated, the investment in the supply of chainsaws is several hundred times lower than the investment in the supply of harvesters, and this is the main reason for the relatively small share of the use of harvesters. The consumption of energy products observed per unit of product (m^3 of processed wood) is sometimes even more than five times higher for harvesters than for chainsaws.

It is also a fact that there are currently no special regulations in Serbian forestry in which at least recommendations/obligations are made regarding the use of fuels and oils that are environmentally friendly, as well as regarding the type and quality of fuel and oils to be used (except those given in the commercial leaflets and on the website of chainsaw manufacturers). One way to reduce greenhouse gas emissions when using chainsaws is to use biofuels and bio-oil. This is due to the fact that there are no significant changes in the performance of the chainsaw when using mineral and bio-oil [31–33], while biofuels and oils have an advantage of rapid degradation in soil and lower ecotoxicity [31]. In addition, as an alternative to the use of petrol saws, electrically powered saws (battery powered) are available. Electric chainsaws are a good alternative to petrol chainsaws for use in small-scale forestry, since their use can be compared to petrol chainsaws in terms of efficiency; they provide a lower level of energy product consumption, as well as lower noise and vibration exposure in the hand–arm area [34,35]. Similarly to the situation in other fields, the results confirmed a positive impact of electrification on man and the environment. The development of electric tools for forestry is expected to intensify with the development of more powerful batteries, which are still considered the biggest bottleneck [35]. However, these saws are still under development and are almost unusable for commercial purposes, especially for larger diameter trees.

5. Conclusions

In this study, it was confirmed that the some of the factors with the highest impact on the consumption of energy products of a chainsaw are diameter at breast height and chainsaw power.

In the selection stand of beech and fir, it was found that lower energy product consumption is achieved with a combination of two chainsaws (one of higher and one of lower power—Husqvarna 372 XP 3.9 kW and Husqvarna H 545 2.5 kW), by on average 26% for fuel and 24% for oil.

If, theoretically, a wood volume of 1 million m^3 from Serbian state forests, dominated by beech, were cut down by two saws of higher and lower power compared to two saws of higher power, then the annual savings in absolute value would be about EUR 120,000 year^{-1}, i.e., about 2,500,000 MJ year^{-1} in terms of energy products.

Potential savings by using two saws (one of lower and one of higher power instead of two higher power saws) in beech stands would be 54,000 L year^{-1} for fuel and 19,000 L year^{-1} for oil.

Although there are seemingly small differences in the consumption of energy products (fuel consumption of 0.144 L m^{-3} versus 0.198 L m^{-3}), they make a big difference. These values are proportionally higher in countries where the use of chainsaws continues to dominate, which is estimated to be about three-quarters of the total wood cut in the world that comes from the forests of less developed environments. Apart from Northern Europe, Russia, North America and part of Western Europe, the rest of the world is dominated by logging with chainsaws, which speaks in favor of the importance of research studies such as this one and changes that should consequently be put into practice.

Author Contributions: Conceptualization, S.A. and M.D.; methodology, M.D.; software, S.D.; validation, S.A., D.S. and S.D.; formal analysis, S.A.; investigation, M.D.; resources, D.S.; data curation, S.A.; writing—original draft preparation, S.A.; writing—review and editing, D.S.; visualization, S.D.; supervision, M.D.; project administration, S.A.; funding acquisition, S.A. All authors have read and agreed to the published version of the manuscript.

Funding: This research received no external funding.

Institutional Review Board Statement: Not applicable.

Informed Consent Statement: Not applicable.

Data Availability Statement: The data presented in this study are available on request from the corresponding author.

Conflicts of Interest: The authors declare no conflict of interest.

References

1. Food and Agriculture Organization [FAO]. Forestry Production and Trade. *Food and Agriculture Data.* 2018. Available online: http://www.fao.org/faostat/en/#data/FO/visualize (accessed on 15 July 2021).
2. Moskalik, T.; Borz, S.; Dvořák, J.; Ferenčík, M.; Glushkov, S.; Muiste, P.; Lazdiņš, A.; Styranivsky, O. Timber Harvesting Methods in Eastern European Countries: A Review. *Croat. J. For. Eng.* **2017**, *38*, 231–241.
3. Landekić, M.; Martinić, I.; Mijoč, D.; Bakarić, M.; Šporčić, M. Injury Patterns among Forestry Workers in Croatia. *Forests* **2021**, *12*, 1356. [CrossRef]
4. Danilović, M.; Ćuprić, N. The state of forests in Serbia in terms of their utilization. In Proceedings of the First Serbian Forestry Congress—Future with Forests, Belgrade, Serbia, 11–13 November 2010; pp. 180–190.
5. Nordfjell, T.; Bacher, M.; Eriksson, L.; Kadlec, J.; Stampfer, K.; Suadicani, K.; Suwala, M.; Talbot, B. Operational factors influencing the efficiency in conversion. In *Norway Spruce Conversion-Options and Consequences*; Spiecker, H., Hansen, J., Klimo, E., Skovsgaard, J.P., Sterba, H., Von Teuffel, K., Eds.; European Forest Institute Research Report 18; Brill: Leiden, The Netherlands; Boston, MA, USA, 2004.
6. Silversides, U.; Sundberg, U. *Operational Efficiency in Forestry. Volume 2: Practice*; C. R. Kluwer Academic: Boston, MA, USA, 1989; p. 169. ISBN 0-7923-0063-7.
7. Behjou, F.K.; Majnounian, B.; Dvořák, J.; Namiranian, M.; Saeed, A.; Feghhi, J. Productivity and cost of manual felling with a chainsaw in Caspian forests. *J. For. Sci.* **2009**, *55*, 96–100. [CrossRef]
8. Berg, S.; Karjalainen, T. Comparison of greenhouse gas emissionfrom forest operations in Finland and Sweden. *Forestry* **2003**, *76*, 271–284. [CrossRef]
9. Dias, A.C.; Arroja, L. Environmental impacts of eucalypt and maritime pine wood production in Portugal. *J. Clean. Prod.* **2012**, *37*, 368–376. [CrossRef]
10. Morales, M.; Aroca, G.; Rubilar, R.; Acuna, E.; Mola-Yudego, B.; Gonzalez-Garcia, S. Cradle-to-gate life cycle assessment of Eucalyptus globulus short rotation plantations in Chile. *J. Clean. Prod.* **2015**, *99*, 239–249. [CrossRef]
11. Prinz, R.; Spinelli, R.; Magagnotti, N.; Routa, J.; Antti, A. Modifying the settings of CTL timber harvesting machines to reduce fuel consumption and CO 2 emissions. *J. Clean. Prod.* **2018**, *197*, 208–217. [CrossRef]
12. George, A.K.; Kizha, A.R.; Kenefic, L. Timber harvesting on fragile ground and impacts of uncertainties in the operational costs. *Int. J. For. Eng.* **2022**, *33*, 12–21. [CrossRef]
13. Athanassiadis, D. Energy consumption and exhaust emissions in mechanized timber harvesting operations in Sweden. *Sci. Total Environ.* **2000**, *255*, 135–143. [CrossRef]
14. Lijewski, P.; Merkisz, J.; Fuc, P.; Ziolkowski, A.; Rymaniak, L.; Kusiak, W. Fuel consumption and exhaust emissions in the process of mechanized timber extraction and transport. *Eur. J. For. Res.* **2017**, *136*, 153–160. [CrossRef]
15. Markewitz, D. Fossil fuel carbon emissions from silviculture: Impacts on net carbon sequestration in forests. *For. Ecol. Manag.* **2006**, *236*, 153–161. [CrossRef]
16. Vusić, D.; Šušnjar, M.; Marchi, E.; Spina, R.; Zečić, Z.; Picchio, R. Skidding operations in thinning and shel-terwood cut of mixed stands—Work productivity, energy inputs and emissions. *Ecol. Eng.* **2013**, *61*, 216–223. [CrossRef]
17. Popovici, R. Estimating chainsaw operating costs based on fuel, oils and spare parts. *Bull. Transilv. Univ. Bras.* **2013**, *6*, 63–68.
18. Danilović, M.; Đurić, S.; Antonić, S. Efekti rada na poslovima seče i izrade drvnih sortimenata u prorednim sečama bukve. [Effects of work on the operations of felling and wood assortment processing in a beech thinning]. *Glas. Šumar. Fak. Univ. Banjoj Luci* **2017**, *1*, 45–62. (In Serbian) [CrossRef]
19. Danilović, M. *Metodika Izrade Normi i Normativa Rada u Šumarstvu [Methodology for the Development of Norms and Work Norms in Forestry]. Monografija*; Istraživačko Razvojni i Projektni Centa: Banja Luka, Bosnia and Herzegovina, 2019. (In Serbian)
20. Danilović, M. *Norme i Normativi—Monografija [Norms—Monograph]*; Faculty of Forestry, University of Belgrade: Belgrade, Serbia, 2020. (In Serbian)
21. Halilović, V.; Gurda, S.; Sokolović, Dž.; Music, J.; Bajric, M.; Ganić, E. *Fuel and Oils Consumption in the Phase of Harvesting and Processing Wood in Public Enterprise Forest Offices Vareš Section 37*; Works of the Faculty of the Forestry University of Sarajevo; University of Sarajevo: Ilidža, Bosnia and Herzegovina, 2012; Volume 42, pp. 31–43.
22. Halilović, V.; Musić, J.; Bajrić, M.; Sokolović, Dž.; Knežević, J.; Kupusović, A. Analiza potrošnje goriva pri sječi i izradi stabala hrasta kitnjaka na području p.j. Šumarija "Zavidovići". [Fuel and lubricants consumption during timber felling and processing in the area of p.j. Forest office "Zavidovići". *Šumar. List* **2019**, *143*, 337–345. (In Croation) [CrossRef]
23. Ghaffariyan, R.; Kuehmaier, M. A Short Review of Fuel Consumption Rates of Whole Tree and Cut-To-Length Timber Harvesting Methods. *Curr. Investig. Agric. Curr. Res.* **2018**, *5*, 603–606. [CrossRef]
24. Udruzenje Šumarskih Inženjera i Tehničara Srbije. *Bukva u Srbiji. [Beech in Serbia: (Fagus moesiaca/Domin, Mally/Czeczott.)]*; Udruzenje Šumarskih Inženjera i Tehničara Srbije: Belgrade, Serbia, 2005.
25. Latterini, F.; Venanzi, R.; Stefanoni, W.; Sperandio, G.; Suardi, A.; Civitarese, V.; Picchio, R. Work Productivity, Costs and Environmental Impacts of Two Thinning Methods in Italian Beech High Forests. *Sustainability* **2022**, *14*, 11414. [CrossRef]
26. Galezia, A. Analysis of the duration of basic logging operations performed using a chainsaw. *For. Res. Pap.* **2014**, *75*, 25–30. [CrossRef]

27. Bernardi, B.; Macrì, G.; Falcone, G. Productivity and Life Cycle Assessment (LCA) of Tree Felling by Chainsaw in Thinning of Calabrian Pine Stands. *Environ. Sci. Proc.* **2022**, *22*, 10. [CrossRef]
28. Bajić, V.; Danilović, M. Potrošnja Goriva i Maziva pri Seči u Mladim Hrastovo-Grabovim Sastojinama [Fuel and Lubricant Consumption in Felling the Young Oak-Hornbeam Stands]. *Glas. Šumar. Fak.* **2002**, *86*, 59–66. (In Serbian) [CrossRef]
29. Sever, S.; Horvat, D. Prilog proučavanja potrošnje goriva pri prijevozu drva kamionskim kompozicijama [Fuel consumption study of forest truck assemblies]. *Meh. Šumar.* **1989**, *14*, 157–162. (In Croatian)
30. Eurostat. Roundwood Production. *Code: TAG 00072*. 2016. Available online: http://ec.europa.eu/eurostat/web/forestry/data/main-tables (accessed on 22 August 2021).
31. Skoupy, A.; Klvac, R.; Hosseini, S. Changes in the External Speed Characteristics of Chainsaw Engines with the Use of Mineral and Vegetable Oils. *Croat. J. For. Eng.* **2010**, *31*, 149–155.
32. Stanovský, M.; Schürger, J.; Jankovský, M.; Messingerová, V.; Hnilica, E.; Kučera, M. The Effect of Lubricating Oil on Temperature of Chainsaw Cutting System. *Croat. J. For. Eng.* **2010**, *34*, 83–90.
33. Orawiec, A.; Suryan, L.; Parmigiani, J. An Evaluation of the Performance of Chainsaw Lubricants. *Croat. J. For. Eng.* **2020**, *41*, 325–332. [CrossRef]
34. Colantoni, A.; Mazzocchi, F.; Cossio, F.; Cecchini, M.; Bedini, R.; Monarca, D. Comparisons between battery chainsaws and internal combustion engine chainsaws: Performance and safety. *Contemp. Eng. Sci.* **2016**, *9*, 1315–1337. [CrossRef]
35. Poje, A.; Potočnik, I.; Mihelič, M. Comparison of Electric and Petrol Chainsaws in Terms of Efficiency and Safety When Used in Young Spruce Stands in Small-Scale Private Forests. *Small-Scale For.* **2018**, *17*, 411–422. [CrossRef]

sustainability

MDPI

Article

Multi-Purpose Accessibility of Mountain Area Forests for the Purpose of Forest Management and Protection of the State Border

Doroteja Krivić-Tomić [1], Ivica Papa [1,*] and Mijo Kožić [2]

1 Department of Forest Engineering, Faculty of Forestry and Wood Technology, University of Zagreb, Svetošimunska Cesta 23, 10000 Zagreb, Croatia; krivic.dorotea@gmail.com
2 Croatian War College "Ban Josip Jelačić", Ilica 256b, 10000 Zagreb, Croatia; mijo.kozic@morh.hr
* Correspondence: ipapa@unizg.sumfak.hr

Abstract: The planning and implementation of surveillance of state territory in forested border areas, especially mountainous areas, is considered to be highly complex. This is illustrated by the example of the difficulties the European Union faced in controlling the 2015 European migration crisis. Thereby, Croatia has the difficult task of protecting the borders of the Union because a particular problem on the Western Balkan Route is the so-called bottleneck to Slovenia in the area of the Municipality of Donji Lapac, which consists of the green border with Bosnia and Herzegovina. Consequently, by using the example of planning multi-purpose forest roads, the aim of this paper is to propose the inclusion of the road network of border management units of mountain areas in the control system of the state's green border, which, in this paper, includes its surveillance and protection by land for the purpose of national security. The research was conducted on the example of the Visočica–Lisac border management unit in the Municipality of Donji Lapac. The results of the research indicate a possible solution to the control of the border management unit by establishing a two-level surveillance system. The higher level consists of strategically defined surveillance points and corresponding multi-purpose forest roads designed on a tactical level. At this level, the priority is protection or, more precisely, defense of the state border. The lower level consists of tactically determined surveillance points with corresponding multi-purpose forest roads designed on the operational level. In addition to protecting the state border, this level would also have the task of protecting the forest, that is, monitoring the area of the management unit.

Keywords: forest roads; Western Balkan Route; European migration crisis

Citation: Krivić-Tomić, D.; Papa, I.; Kožić, M. Multi-Purpose Accessibility of Mountain Area Forests for the Purpose of Forest Management and Protection of the State Border. *Sustainability* **2023**, *15*, 16935. https://doi.org/10.3390/su152416935

Academic Editor: Kalle Kärhä

Received: 3 October 2023
Revised: 18 November 2023
Accepted: 16 December 2023
Published: 18 December 2023

1. Introduction

Migration, especially illegal migration, is one of the biggest problems of the modern world [1] because the contemporary understanding of borders is bipolar. Namely, "[...] the semantics of borders reveals a dialectical relationship between a completely 'fixed borderline' [...] and a 'vague border area' [...]" [2] (p. 294). In other words, ref. [3] states that fixed, impassable barriers were built for national security purposes. Therefore, there is no doubt that they are considered a contemporary iron curtain [4] (p. 112). However, an issue is that borders are conventional, mentally imagined lines. They consist of intentionally placed border crossings and imagined bounded areas, the so-called *buffer* zones around them, which together form the porous border area [5] (p. 5). In the context of the 2015 European migration crisis, an example of such a border is the concept of the green-grey border [6], which exists between Bosnia and Herzegovina and Croatia.

Namely, the Balkans have been an area of migration between Europe, Asia, and Africa for centuries [2] (p. 294), whereby the final destination for most migrants is the countries of Western Europe [1] (p. 33). Therefore, disregarding the actual geographical and cultural borders of the Balkan Peninsula, the countries encompassed by this name are considered to be the border area of Western Europe [7], so it is not surprising that the Western Balkan

Route is one of the eight main migration routes to the European Union [8,9]. Consequently, the Union has been facing the phenomenon of so-called *transit migration* [10] in recent years.

The Western Balkan Route consists of a journey by sea to Turkey or Greece, by land to Macedonia or Bulgaria, and further to the European Union, through Serbia and Hungary, Serbia and Croatia, or Bosnia and Herzegovina and Croatia. The route was a secret at first. It was used to smuggle people through mountain terrain and forested areas of Southeast Europe to the desired country in the West [11]. The route is/was also preferred for drug and arms trafficking [1]. In 2015, after a large migrant wave towards the Austrian-Hungarian border, the route became a controlled transit corridor from Greece to Austria. After Hungary and Bulgaria completed the construction of wire fences on the focal parts of their borders in 2016, the corridor was closed. The migrant wave has developed a new strategy for entering the territory of the Union. The route was modified across Croatia's green border with Bosnia and Herzegovina [8] (p. 193). Since it is one of the longest-established land borders in Europe, Croatia, as a member of the Union, has been given the difficult task of safeguarding European sovereignty and cultural identity. Therefore, in addition to the national security system, this Union border is also under strong monitoring by the European Border and Coast Guard Agency (Frontex). As this is a predominantly green border between the two countries, this paper will provide an example of how the forestry profession can contribute to the improvement of Croatia's homeland security system. In other words, the aim of the paper is to show how the planning of multi-purpose forest roads in the road network of border management units in mountain areas can support the established system of monitoring the state's green border. Consequently, the paper starts with (a) the idea of determining the key points from which it is possible to monitor the management unit, (b) the idea of determining monitored cardinal points within the network of forest roads in the management unit, and (c) the idea of designing multi-purpose forest roads that, on the one hand, would serve for the needs of forest management and, on the other hand, for the protection (that is, surveillance and defense) of the state border.

2. Overview of Previous Research

In general, the security risk in forested areas can be broken down into traditional and extended understandings. According to the former, in addition to the risk of fire, it also refers to poaching, theft of wood assortments from landing sites, and illegal tree felling [12] (p. 158). A broader understanding of the security risk in forest areas is mainly related to the issue of national security. In this context, it can be said that it refers to the defense of the state border, illegal crossing of the green border, and various types of smuggling through the forest. Safety risks also include destructive activities of *off-road* motorcycling and illegal races with *quad* vehicles [13] (p. 1). Therefore, when it comes to the land control of forested (especially border) areas, the primary objective is to monitor forest transport infrastructure.

During the wartime period in the first half of the 20th century, forest areas were monitored by establishing a road maze and controlling road junctions and roads within forest complexes [14–16]. The movement of vehicles of any kind outside the existing transport communication could have been excluded with a high probability [14]. Likewise, in his opinion, the existing road network was also preferred when walking, and pedestrians, if they were away from the roads, at some point certainly strived to approach them. However, [14] (p. 31) does not agree that the forest road network should look like a maze, so he suggests organizing circular traffic. He believes that circular roads prevent congestion on roads in the forest and reduce the possibility of unnoticed movement for those who want to remain unnoticed.

Ref. [17], in their paper, emphasize that today's forest roads should strive for multi-functionality. Therefore, the order of importance of parameters, on the basis of which forest road routes are planned, has changed over time. According to them, forest road planning "[in] the past took into account only hydrology and slope factors. Today, socioeconomic factors are also included [. . .]" [17] (p. 18). Amani (2000) is of a similar opinion, arguing that multi-purpose forest roads condition multi-purpose forest management (as cited in [17]

(p. 13)). His intention is to state that, in addition to the prevailing practice of emphasizing the economic component, the political component should also be taken into account when managing forests in certain areas. Such an opinion is shared by Potočnik (1996) and Bjorklund (2006), as they state that national policy is also taken into account in forest management (as cited in [18] (p. 5389)). The significance rate of forest roads for security use in Slovenia is 3.09%, of which 1.79% refers to the police and 1.30% to the army [19] (p. 68).

By examining previous research on this topic, it can be established that the planning of the optimal network of primary forest transport infrastructure for the purpose of control (which includes surveillance and protection) of border areas is increasingly based on multi-criteria decision-making on the route of the future forest road. Most often used are the Delphi method, the Analytic Hierarchical Procedure, or the Fuzzy Set Theory [20] (p. 1768), which are based on weighting or scoring. In addition to multi-criteria assessments, it is important to mention that computer and linear programming [21] (p. 63) are also used in the planning of forest transport infrastructure because, in this manner, a complete overview of the most favorable position of conceptual routes in space is obtained. The collective name for such multi-criteria methods is *spatial multi-criteria evaluation* (SMCE) [20] (p. 1768). When planning the detailed layout of the road network in extremely large areas, ref. [22] (p. 1) explains that "[...] optimization of the layout of the forest road network can also be based on iterative algorithms [...]. This belongs to complex methods because, in addition to linear programming, it is also hierarchical and takes into account not only the issue of road construction but also the issue of timber harvesting.

When it comes to forestry surveillance of forest transport infrastructure, aerial photogrammetric or satellite images are less commonly used because they cannot always show in detail what is occurring on the ground due to the structure of the stand, that is, the degree to which the canopy covers the ground. Therefore, the so-called terrestrial methods are mostly applied, and they most often refer to wireless sensor networks [23] (p. 2). For example, in order to detect the location of illegal tree felling, sound sensors for the detection of chainsaw operation and vibration sensors for the detection of tree fall can be installed in forests [23] (p. 2). Furthermore, for the purpose of preventing theft of wood assortments, a surveillance system based on an accelerometer, geophone, and magnetometer is used, which provides information on the entry point of the vehicle into the forest area and the speed and direction of its movement [13] (p. 1). Similarly, an acoustic system used for the surveillance of the remote transport of wood and the movement of forestry officials' vehicles in the management unit is also applied. This system consists of a database of sounds of company and subcontractor vehicles in the process of timber harvesting [12] (pp. 158–162).

When monitoring the green border in order to prevent illegal crossings, the conventional surveillance system based on security checkpoints and patrols at specific time intervals is still widely used [24]. Therefore, one of the most radical examples of preventing illegal border crossings is still the application of the US Border Security Policy paradigm called *Prevention Through Deterrence* (PTD). In this case, border control uses forest transport infrastructure mapping technology. In other words, "[...] maps are made that distinguish between 'controlled' or enclosed parts and 'monitored' or unenclosed parts" [25] (p. 163). In more detail, in order to ensure the safety of urban zones in border areas, dedicated forest transport infrastructure planning directs migrants towards more remote (rural) areas, where environmental conditions act as a natural barrier to movement. On the one hand, the *spatial displacement effect* [26] is achieved in this manner, while, on the other hand, the implementation of the so-called *Tactical Advantage Act* [25] is enabled.

A step prior to mostly applying contemporary technology to green border surveillance is to combine conventional methods with sensor technology. In this case, surveillance consists of (a) *border patrol*, which controls a certain part of the border day and night; (b) *ambushes*, that is, patrols whose task is to prevent illegal crossings at night; (c) *keeping watch*, which refers to the daily surveillance of the border and terrain of the neighboring country through a system of observation posts; (d) stationary or mobile *thermal cameras*, which enable the detection of targets, recognition, and observation in all weather conditions; and (e) the application of a stationary or mobile *Askarad* radar system for monitoring, finding and classifying moving targets, and precisely locating them [27] (pp. 1395–1396).

An example of linear architecture in contemporary transport infrastructure surveillance technology for illegal green border crossings is the *Unattended Ground Sensor Network* (UGS). Namely, according to estimates, in a five-year period, 1000 migrants create an average of 722 m of forest trails, i.e., degrade 656 m^2 of forest land [28] (p. 403). Therefore, this network is suitable for preventing border crossings in areas remote from local border crossings. "[The] UGS network was developed to remotely detect, localize, identify, and classify targets for the purpose of [. . .] situational awareness, perimeter protection, border control, and surveillance of borders and targets" [29] (p. 2). To put it simply, the network consists of sensors hidden in the field that channel the data to the nearest central node (base) over long distances, and the information thus collected from all nodes is further sent to the center. The sensors can be magnetic, acoustic, infrared, radar, or capacitive. The UGS network is further connected to the *Intelligent Route Surveillance*—IRS system, based on a triple traffic classification: (a) non-motorized traffic (people and livestock), (b) small motorized traffic (motorcycles, cars, and vans), and (c) large motorized traffic (trucks and forest wheel and track machinery) [29] (p. 6).

However, for the purpose of monitoring forest transport infrastructure in mountainous areas, [30] (pp. 0849–0851) believes that the most suitable method is MWSN (*Multimedia Wireless Sensor Nodes*). According to this method, the area of the green border covered by the sensor node is called the *Field of View*—FOV. Each sensor node consists of (a) a *sensor unit* with a thermal camera for capturing images in space; (b) a *communication unit* that enables communication of the sensor node with other nodes in the network and sending information from the environment to the desired destination (e.g., the nearest police station); (c) a *power unit*, i.e., a battery; and (d) an *information processing and storage unit*. The principle of operation of the system is such that each node has the task of protecting the neighboring (adjacent) sensor node. Namely, in case of damage to the node, the next one, which is on the same route but in a different place, can replace the damaged node [30] (p. 0852).

Nevertheless, the advancement of sensor technology is aimed at creating a hierarchical architecture. One example of such a method of controlling forest road infrastructure is the platform for military surveillance of green borders, *FemtoNode*. The system consists of two sensor levels. The nodes of the so-called low sensors are scattered along the borderline, either on the ground or underground. At the moment when a vehicle crosses the borderline, the low sensors sound an alarm to activate the so-called high sensors, i.e., *Unmanned Aerial Vehicles*—UAVs equipped with, for example, vehicle recognition radars [30] (p. 0849).

An even more complex hierarchical architecture surveillance system is the application of the hybrid architecture of wireless sensor networks. An example of this is the *BorderSense* method with three layers of hierarchy, that is, three types of sensor nodes. The first are *scalar sensor* nodes equipped with a seismic sensor and arranged on the ground or underground. The second type consists of *multimedia sensor nodes* equipped with video cameras or night vision binoculars located on observation posts along the green part of the border. The last are *mobile sensor nodes*, i.e., mobile ground robots and unmanned aerial vehicles [24] (p. 468). This method is characterized by a high level of reliability, i.e., a minimum number of false alarms.

The technologies and methods used in the Croatian homeland security system are mostly classified and not publicly available. Therefore, on the basis of very limited documentation, it is only possible to assume that Croatia is in the stage of combining the application of conventional methods of green border control with the use of a linear architecture of various sensors. Consequently, in order for the primary forest transport infrastructure to be used for multiple purposes, it is desirable that its planning be based not only on forestry multi-criteria assessments of its optimization but also on the possibility of applying more complex methods of security systems for its surveillance.

3. Methodology

The area of the Municipality of Donji Lapac is one of the high-frequency crossing zones on the border between Bosnia and Herzegovina and Croatia. Namely, this part of the Croatian territory is the shortest route to Slovenia, which, at the time of the 2015 European migration crisis, was the Schengen border. In order to avoid border control, migrants stopped moving along the gray border, so demanding forested areas often became alternative routes. Therefore, passing through this part of the Western Balkan Route was called a "gamble" [31,32]. Nevertheless, within forest complexes, migrants will still resort to moving through forest transport infrastructure, at least part of the way. Consequently, the aim of this paper, based on the example of planning multi-purpose forest roads, is to provide a proposal for the inclusion of the road network of border management units in mountain areas in the surveillance system of the green border of the state.

The research starts with the following research questions:

1. Can the state border, as well as the territory of the management unit, be controlled if the strategic surveillance point is positioned on the highest peaks of the management unit?
2. Does the surveillance of forest roads within the management unit depend on defining the position of the monitored cardinal points and terrain relief?
3. Does the multi-purpose forest road designed on a tactical level affect the accessibility of the management unit, although this is not its primary purpose?
4. Does the length of the multi-purpose forest road designed on the operational level affect the size of the accessible area in the management unit?

The analysis was conducted on the example of the Visočica–Lisac management unit bordering Bosnia and Herzegovina. Within the framework of forest management, it is under the jurisdiction of the company Hrvatske šume d.o.o., i.e., the Forest Administration of Gospić and its component Forest Office of Donji Lapac. It is positioned between 44°28′40″ north latitude and 15°57′45″ and 16°08′25″ east longitude [33] (p. 11).

The total area of the Visočica–Lisac management unit is 4774.37 ha [33] (p. 11). The management unit is divided into 91 compartments with associated sub-compartments. The relief is characterized by mountain ridges, very steep slopes, karst fields, and deep ravines. In terms of vegetation, this management unit is characterized by uneven-aged beech forests and uneven-aged Austrian oak forests, i.e., the following forest communities: Austrian oak and Southern European flowering ash forest (*Fraxino orni-Quercetum cerridis* Stefanović 1968), beech forest with autumn moor grass (*Seslerio autumnalis-Fagetum sylvaticae* (Horvat) M. Wraber ex Borhidi 1963), mountain beech forest with large red dead nettle (*Lamio orvalae-Fagetum* (Horvat 1938) Borhidi 1963).

The road network has a total length of 58.91 km, of which 47.1 km are forest roads and 11.8 km are public roads. Given that 38.59 km is entered into the calculation of the accessibility of this management unit, its total accessibility is 8.07 km/1000 ha [33] (p. 84). Taking into account the minimum required classical accessibility of mountain areas of 15 m/ha from 1990 or 20 m/ha from 2012, as stated by [34], the aforementioned indicator of total accessibility leads to the conclusion that this management unit is poorly accessible.

The research is based on the M 1:25,000 topographic map and a digital relief model with a resolution of 15 × 15 m of the Visočica–Lisac management unit. First defined on the topographic map are strategic and tactical surveillance points. Strategic surveillance points are those positions from which it is possible to control the state border. They are based on the idea of a location to which the "Lisac" forest road leads, which was built in 2020 as a result of cooperation between the company Hrvatske šume d.o.o. and the Ministry of the Interior for the purpose of state border control. Considering that the peak to which the aforementioned road leads is the highest point of the southern part of the management unit, it is to be assumed that it has strategic significance because a stationary video surveillance system has been installed there. Furthermore, taking into account that, on Gologuz Peak (one of the highest in the northern part of the management unit), there is a transmitting facility, in this paper, it is defined as the second strategic surveillance point (Figure 1). Tactical surveillance points were defined based on the research of [35]. He uses the *photographic detection system*, hidden at road intersections, to provide the US Forest Service with information on the number and type of vehicles passing through forest complexes. In accordance with the above, the paper starts with the thesis that the tactical (planned and installed by foresters) points in the Visočica–Lisac management unit are defined with regard to the cardinal points in its forest road network. The proposal is that, from these positions, it is possible to control (a) the entry/exit points of the management unit and (b) the key intersections of the main and side forest roads in the transport network. The intersections were selected based on the position of the roads leading to/from the management unit. Consequently, the peaks of Pod Barićev at 869.75 m above sea level, in the area of Visočica (Tactical Point 1) and of Špija at 804.33 m above sea level, in the area of Lisac (Tactical Point 2) were selected as tactical surveillance points. The assumption is that, from them, it is possible to view the largest number of monitored cardinal points (Figure 1).

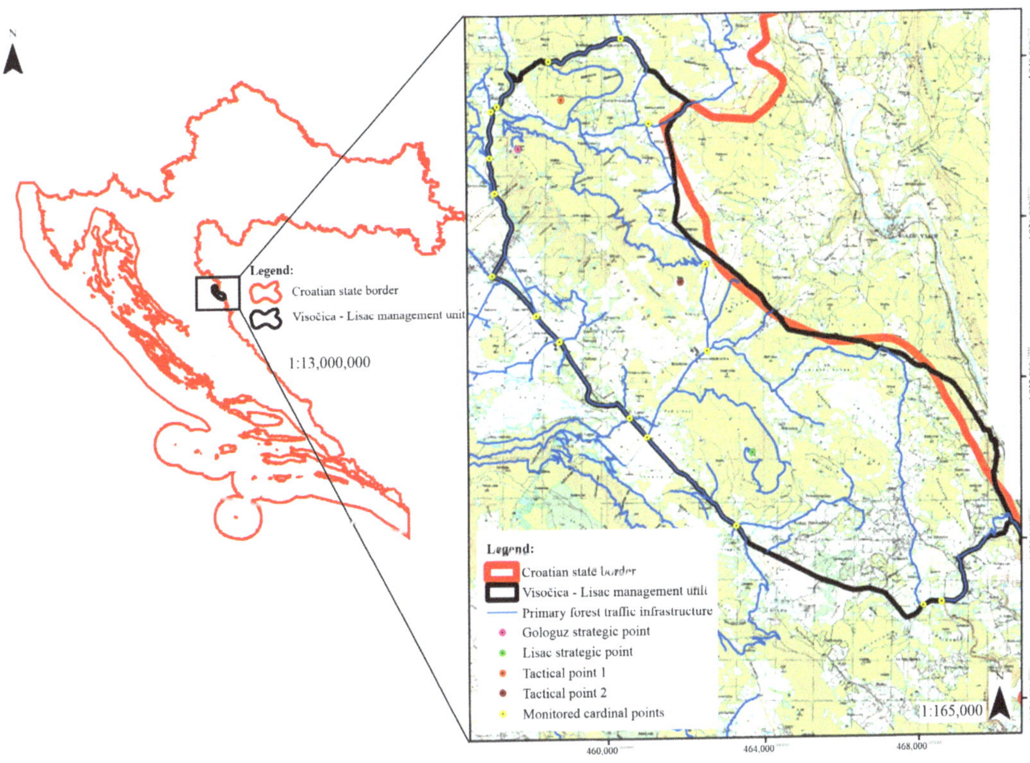

Figure 1. Surveillance points in the Visočica–Lisac management unit. Source: Prepared by the author.

The surveillance of the management unit, from the strategic and tactical points defined in this manner, was simulated on the basis of data on the stationary video surveillance system manufactured by FLIR Systems (*the stationary video surveillance system was acquired by Croatia in a public procurement procedure for the defense and security of the state. Therefore, the technical specifications are classified in accordance with the Croatian Data Confidentiality Act (Official Gazette NN 79/07, 86/12). For the purposes of this research, we were allowed to use them, but not to make them publicly available*), which Croatia uses to control its border in the Municipality of Donji Lapac.

When it comes to the analysis of forest transport infrastructure, thereby taking into account that the strategic planning of forest roads, according to [34], implies the level of the forest management area, and the tactical planning, the level of planning in the management unit, forest roads leading to strategic surveillance points are, in this paper, understood as tactically designed routes. On the other hand, conceptual roads that would lead to tactical points of visibility are, in this paper, understood as operationally designed routes. These are individual forest roads that, in addition to surveillance, primarily serve to make accessible a smaller area within the forest complex. Consequently, routes designed on the tactical level of planning are, in this paper, referred to as tactical (multi-purpose) forest roads, while those designed on the operational level are referred to as operational (multi-purpose) forest roads.

Segments of the horizontal and longitudinal development of tactically designed forest roads were analyzed on the basis of *Glavni projekt šumske ceste Lisac* (main/implementing project design of Lisac Forest Road). The planning of conceptual routes for operational forest roads was carried out by designing a zero-line on the Croatian Basic Map (HOK) at a scale of 1:5000 with an equidistance of 5 m between contour lines using the ArcMap 10.1 program (Esri, Redlands, CA, USA). Careful consideration of laying a zero-line polygon of future routes is a major factor in the planning of the forest road network [36] (p. 2456). The starting point of the conceptual route was positioned on the route of the existing primary forest transport infrastructure, while the endpoint was the tactical control point (the elevation at which it was defined). Based on their distance from each other and the altitude difference between them, the average slope of the zero-line polygon was calculated. Then, the so-called divider segment was determined, that is, "[...] the value that represents the constant distance between the contour lines for a certain slope" [37] (p. 63). In doing so, zero-line polygons of conceptual routes are conditioned by a maximum longitudinal slope of 12% to (exceptionally, also for shorter distances) 14%, which has been defined for such terrain by [38].

4. Research Results

The analysis of visibility from strategic surveillance points showed that the Lisac strategic point in the Visočica–Lisac border management unit is defined on the basis of situational awareness. Despite the fact that the sizes of the set parameters on the actual camera are not known, for the purposes of this research, the simulation of sophisticated video surveillance from this point assumed the setting of the parameters of the thermal imaging camera for detection, recognition, and identification of vehicles at maximum values for selected heights of 4 m (which is the height of a forest truck, trailer, or truck unit) and 8 m (twice the height of a truck, trailer, or truck unit). Consequently, it was concluded that, from this point, it is possible to control the state border and half of the management unit. Whether the video surveillance parameters are defined at a height of 4 m or 8 m, detection and recognition of vehicles from this point are possible in equal quality. As expected, the visibility field for vehicle identification is lower in both examples, as the maximum values of the parameters are also lower (Figure 2). This leads to the conclusion that, in this part of the management unit, Croatia has applied prevention by deterring illegal entry into the country.

However, when it comes to the Gologuz strategic point, the analysis shows negative results for the possibility of state border surveillance. By setting the visibility parameters to maximum height values of 4 m and 8 m, it was observed that it is not possible to control either the state border or the management unit from Gologuz Peak (Figure 3). Furthermore, it is important to point out that, from this point, the surveillance of the area of Visočica itself is limited. This is due to the greater circuity of the relief in this part of the management unit on the one hand and the partial overgrowth of the peak on the other. However, the analysis found that, from this point, the place of Donji Lapac, its access roads, and the hamlets on the west side can be monitored. This leads to the conclusion that Gologuz Peak is suitable for surveillance deep into the territory of Croatia.

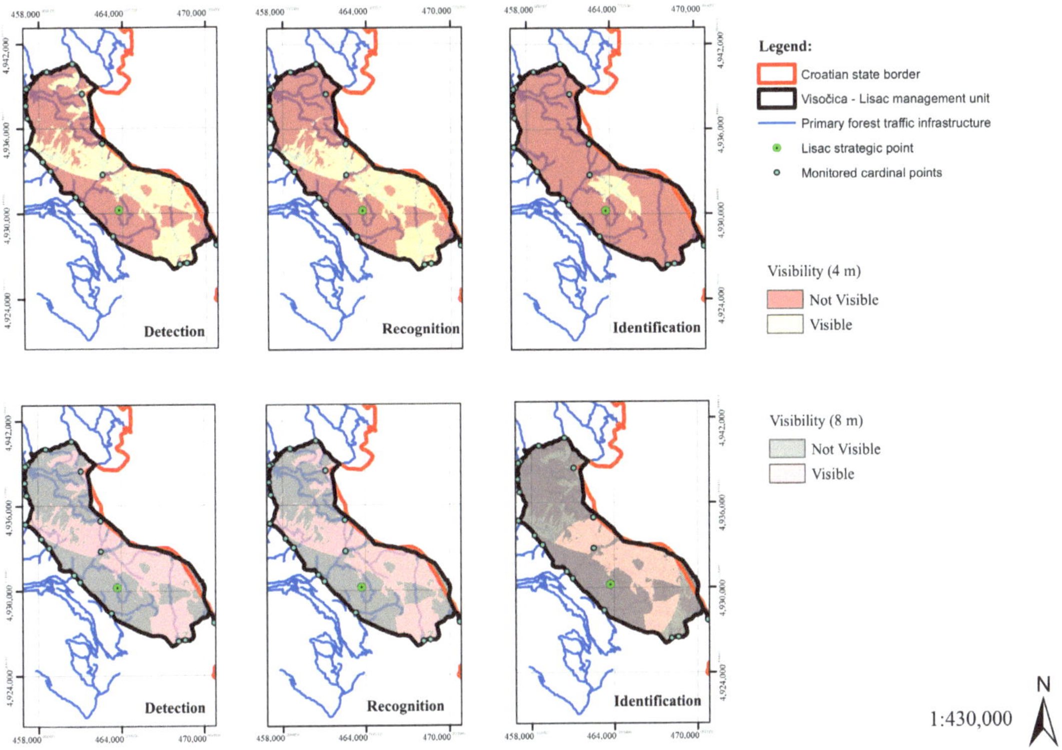

Figure 2. Visibility from the Lisac strategic point by simulating the installation of a thermal imaging camera at a height of 4 m and 8 m. Source: Prepared by the authors.

Based on the above, it can be concluded that height does not have as much influence on the size of the monitored area as does the circuity of the terrain. Namely, the area of Lisac is less hilly compared to Visočica. Therefore, the Lisac Peak dominates the area, so the visibility from that point is better. On Visočica, the terrain is more circuitous, so Gologuz Peak is surrounded by a number of other elevations, which, consequently, interfere with the width of the view from that point. In addition to the above, the overgrowth of the peak should also be taken into account. While clear-cutting was carried out on Lisac, Gologuz is partially overgrown.

An analysis of tactical forest roads (Figure 4) showed that the structure of primary forest roads is actually side forest roads. In [39], it is stated that it starts as a road branch from the previously existing Borićevac–Lisac forest road. From 730 m above sea level to the very peak (997 m above sea level), the road overcomes an altitude difference of 267 m with an average longitudinal slope of 8%. The route ends at chainage 33 + 77.25 hm. This road passes through compartments/sub-compartments 57a, 57b, 57d, 59a, 60a, and 61b. The road to Gologuz Peak is 17 + 20.00 hm long. Its beginning is at 850 m above sea level, and the road continuously climbs to 1022.4 m above sea level, overcoming an altitude difference of 172.40 m. Based on the above data, it is possible to calculate the average slope of the horizontal route, which is 10.02%. However, ocular assessment determined (because the main/implementing project design of Lisac forest roads no longer exists) that certain parts of the route were also built with a higher slope.

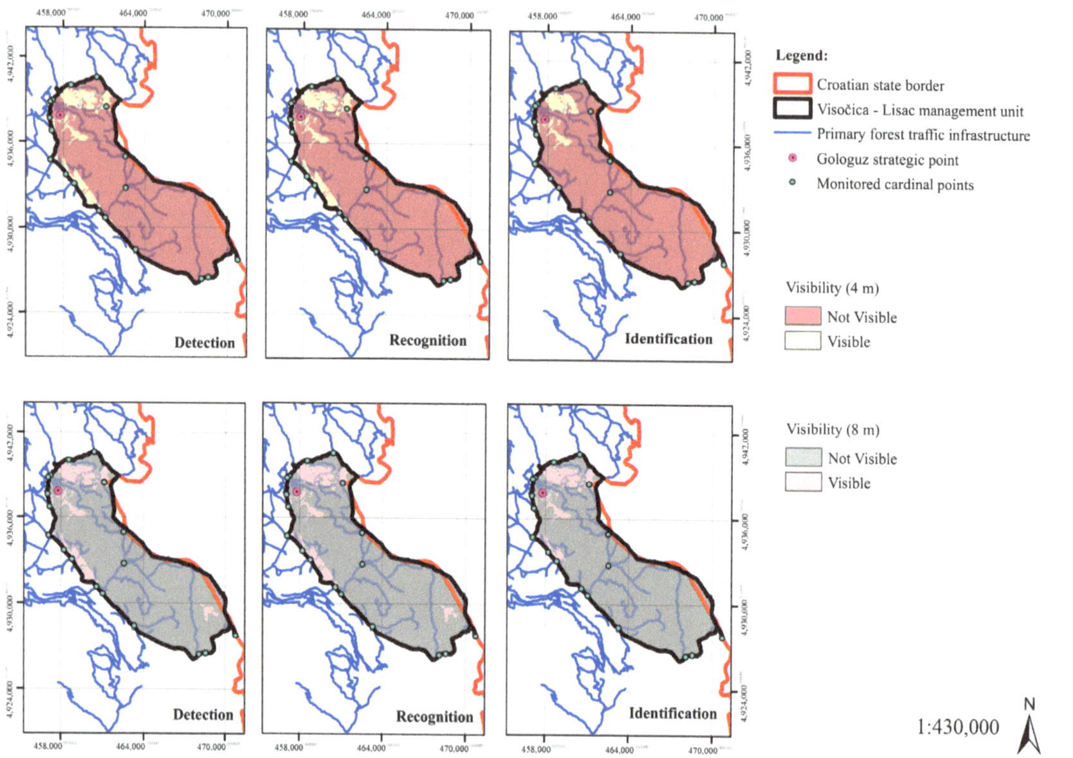

Figure 3. Visibility from the Gologuz strategic point by simulating the installation of a thermal imaging camera at a height of 4 m and 8 m. Source: Prepared by the authors.

When it comes to tactical surveillance points, taking into account their previously mentioned priority of forest protection, the research starts with the assumption that the position of such points depends on the position of cardinal points that are intended to be monitored in the road network. In this research, cardinal points are considered to be road intersections and the points of entry of each category of road into the area of the management unit. It should also be noted that they were all considered equally valuable.

Since tactical points are not significant for national security, it is considered that defining them is exclusively under the competence of foresters. Accordingly, the assumption is that a simpler, not sophisticated, video surveillance system will be installed at these locations. Therefore, a visibility analysis was conducted for the selected locations in such a manner that the stationary video surveillance system was set to half the maximum values from its technical specifications for the height of 8 m. The assumption is that a low-performance thermal imaging camera should be placed at a higher height in order to be able to additionally monitor a certain area.

The research results show that, from the proposed position for Tactical Point 1 (Pod Barićev Peak), it is possible to view (i.e., detect, recognize, and identify the vehicle) with four of the 14 marked cardinal points. Even if three cardinal points in the section dividing the management unit are excluded, it is concluded that the position of the tactical point is not satisfactory with regard to the surveillance of forest roads in the Visočica area. However, it is important to note that, from this position, it is possible to control most of the entry points to this part of the management unit (Figure 5).

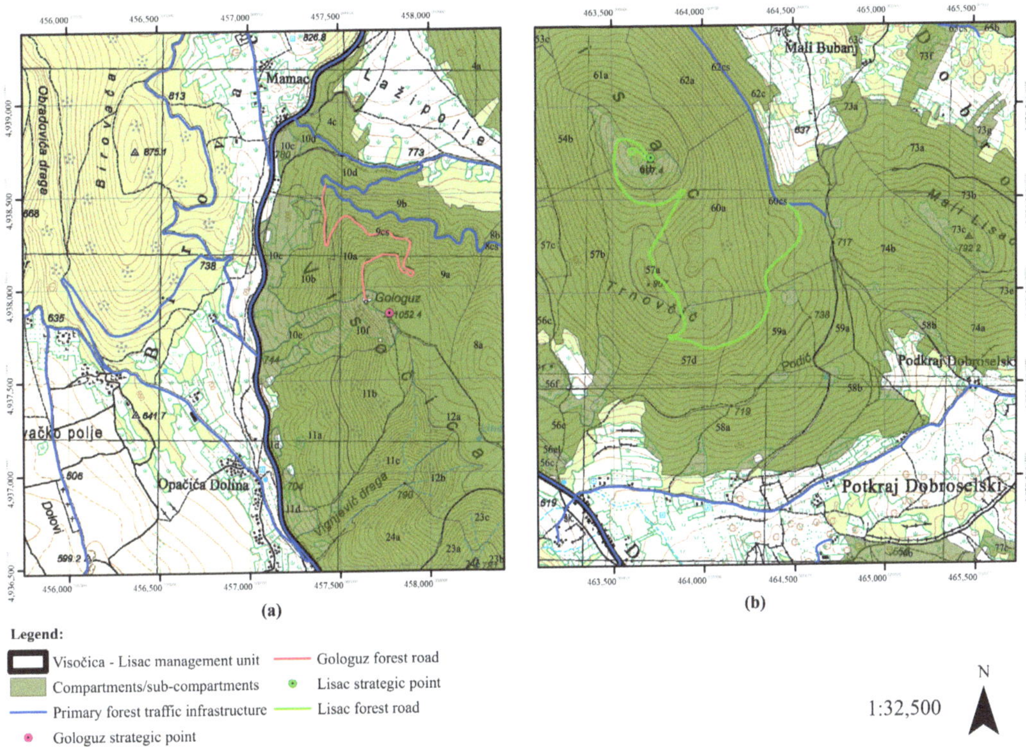

Figure 4. (**a**) Tactical forest road Gologuz and (**b**) Tactical forest road Lisac. Source: Prepared by the authors.

In the Lisac area, from the location of Tactical Point 2 (Špija Peak), of the marked eight cardinal points, it is possible to monitor four of them. It is important to mention that, from this location, it is possible to control most of the entry points to the Lisac area. Furthermore, from this point, it is possible to control the main forest road that divides the management unit in the north–south direction. In addition, visible is most of the section of the main forest road (in that area), which divides the management unit in the east–west direction (Figure 6).

This leads to the general conclusion that defining the position of the tactical point requires precise determination of the cardinal points to be monitored. In this case, it can be said that, from both tactical points, it is possible to monitor most of the entry/exit points to the management unit but not the key intersections within the road network. As the number of tactical points increases, the surveillance area increases proportionally, as [14] (p. 8) also wrote about.

When planning the conceptual routes of operational forest roads, three zero-line polygons were analyzed in the research. Given the position of Tactical Point 1, it has been established that it is convenient to approach this location from two directions (Figure 7a). Consequently, the length of the first conceptual route of the forest road (approaching the point from the south) is 722 m. From 790 m above sea level, it continuously climbs to Pod Barićev Peak (869.75 m above sea level), whereby it overcomes a height difference of 79.75 m. The average slope of the horizontal laying of the route is 10.09%, with a maximum of 12%. The calculated divider segment is 46.13 m. This route would mostly pass through the compartment/sub-compartment 5a and would slightly include the compartment/sub-compartment 4a. The second conceptual route, which approaches this tactical point from the north, is 1095 m long. It overcomes an altitude difference of 74.75 m, i.e., it extends from 795 m above sea level to 869.75 m above sea level. The average slope of the horizontal laying of the route is 6.83%, with a maximum of 12%. The divider segment is 61.29 m. There are two points on this route where there is a change in slope. The first is located at an altitude of 825 m, and the second at 865 m. The forest road route thus planned would pass through four compartments/sub-compartments (3a, 4a, 4b, and 5a).

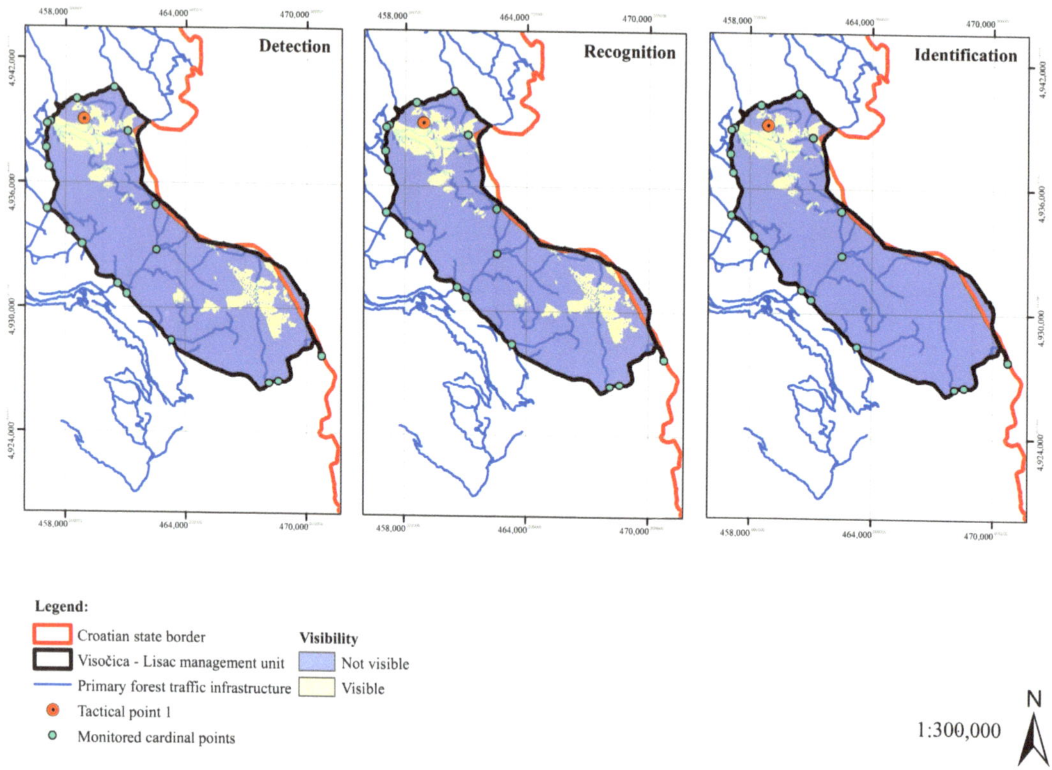

Figure 5. Visibility from Tactical Point 1 by simulating the installation of a thermal imaging camera at a height of 8 m. Source: Prepared by the authors.

Tactical Point 2 would be best approached from the north, along a 505-m-long route (Figure 7b). Its beginning would be at 770 m above sea level, and the end would be at Špija Peak at an altitude of 804.33 m. Consequently, this road would overcome an altitude difference of 34.33 m with an average slope of 6.80%, also not exceeding 12%. The calculated divider segment for this route is 26.51 m. The road would pass through three compartments/sub-compartments (37a, 37c, and 38a).

The results of the research on the planning of conceptual routes for operational multi-purpose forest roads have shown that they can be of a dual character. On the example of zero-line polygons for tactical surveillance points, it can be noted that such roads can be shorter or longer, depending on the position of the tactical point with regard to the existing road network and/or the quality of the accessibility of the part of the management unit where such a point is located. Therefore, it can be concluded that shorter routes should be planned in those parts of the management unit that are well accessible. Longer routes to the tactical surveillance point should be planned in inaccessible or poorly accessible parts of the forest complex.

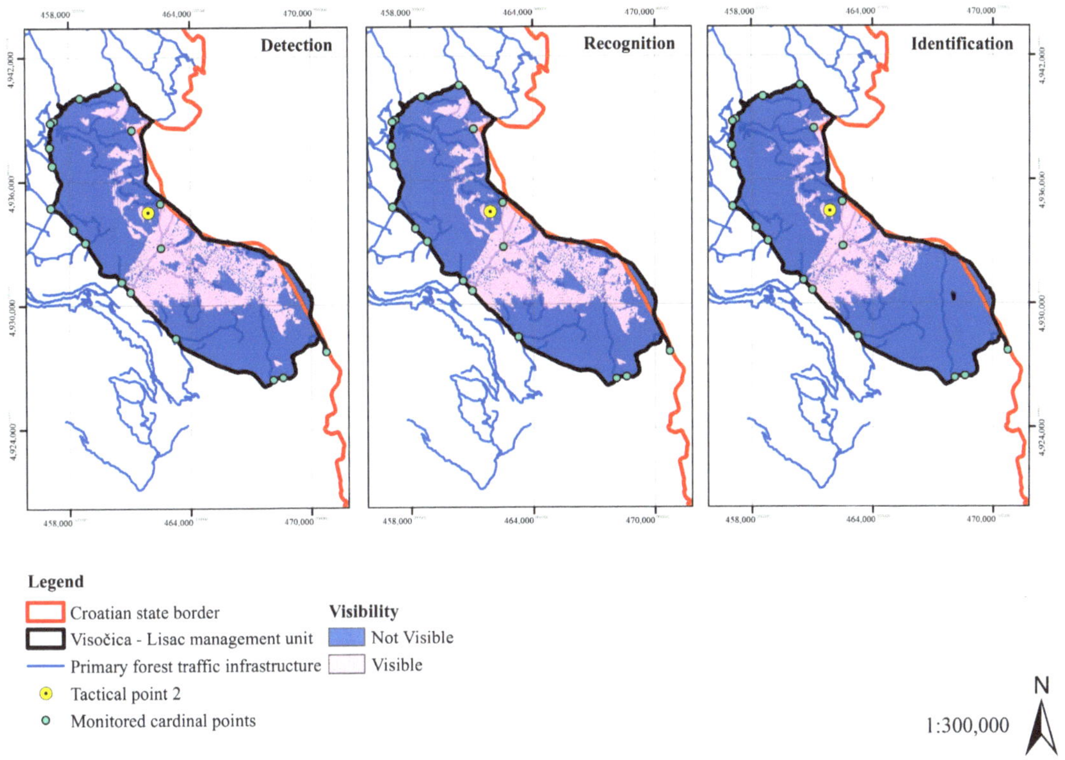

Figure 6. Visibility from Tactical Point 2 by simulating the installation of a thermal imaging camera at a height of 8 m. Source: Prepared by the authors.

In addition, during the analysis of the zero-line polygon in the Visočica area, it was observed that the total length of the route can consist of two parts. Therefore, for the purpose of monitoring the management unit, a shorter part of the route, which leads faster to the location of the tactical point, should be built first. Depending on its position, the assumption is that this part of the route will mostly pass through a smaller number of compartments/sub-compartments. For the purpose of forest management, a longer part of the total route of the operational road may be built first, assuming that it makes a larger number of compartments/sub-compartments more accessible. Until the construction of the full profile of the route, this part can be treated as a commercial forest road with a turning point at the end. However, further up to the elevation defined as the surveillance point, it is only possible to build a skid trail due to the possibility of using a larger longitudinal slope. In this manner, the surveillance point would be ready for activation at any time.

Based on all of the above, it is confirmed by the thesis of [40] (p. 24) that the optimal density of roads within a management unit cannot be discussed in general but that it is adapted to the micro-relief conditions of each management unit separately. However, it has been concluded that it is possible to plan two types of multi-purpose forest roads in the border management unit of a mountain area. The first type consists of short, tactical, or operational forest roads. The analysis determined that, in this management unit, such sections are between 500 and 800 m long, so it can be concluded that the forest roads in this example should be up to 1000 m long, depending on the appearance of the relief and the skill of the designer. In general, short multi-purpose forest roads have the characteristics of classic forest roads, as they separate from the main road towards the surveillance point and, by doing so, make a smaller part of the forest complex accessible (Figure 8a).

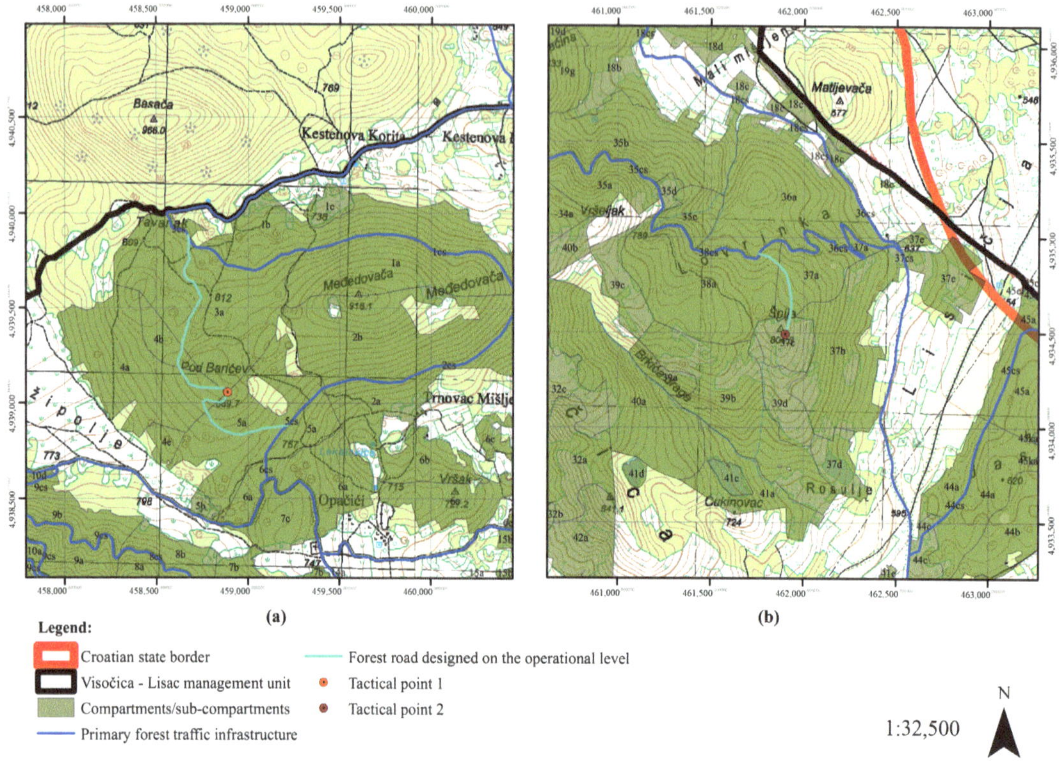

Figure 7. (**a**) Forest road designed on the operational level in Visočica area and (**b**) Forest road designed on the operational level in Lisac area. Source: Prepared by the authors.

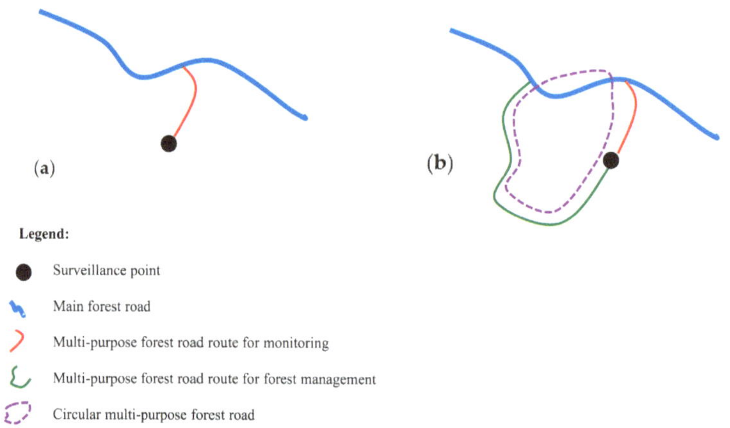

Figure 8. (**a**) Classic multi-purpose forest road and (**b**) multi-purpose circular forest road. Source: Prepared by the authors.

The second type of multi-purpose forest roads are so-called circular roads (Figure 8b). The research found that operational forest roads belong more often to this group, but where possible, this is also feasible with tactical surveillance roads. For example, the assumption is that the Lisac (tactical) forest road is part of a route of such nature. The argument in favor of this is the fact that it currently passes through six compartments/sub-compartments. Assuming that a part of the circular route is built from the east, this road would make two additional compartments/sub-compartments accessible. The shorter route was probably not built first because the strategic point is positioned on a poorly accessible part of the management unit, so the benefits for forest management were also taken into account.

5. Discussion and Conclusions

Taking into account the integrity of the territory within the former Yugoslavia, some forest roads from Croatia continue to those in Bosnia and Herzegovina today. Therefore, the proposal for the establishment of a surveillance system in the mountainous border areas of Croatia is to control the road network in the management units closer to the state border. Based on the analysis of the existing, but also the possibilities for improving the future primary forest transport infrastructure in the Visočica–Lisac border management unit, the following has been concluded about the previously set research questions:

(1) By simulating the surveillance of a management unit with two peaks at opposite ends, *the state border and the territory of the management unit can be controlled if the strategic surveillance point is positioned at the highest peaks of the management unit.* The analysis of the possibility of surveillance from Lisac Peak (the highest elevation in the south of the management unit) showed that, from this strategic point, it is possible to view the state border and half of the area of the management unit. From Gologuz Peak (which is not the highest elevation in the north), a satisfactory level of state border surveillance and of that part of the management unit is not achievable. However, from this point, it is possible to control deep into the Municipality of Donji Lapac.

(2) The analysis has confirmed that *surveillance of forest roads within the management unit depends on defining the position of the monitored cardinal points and terrain relief.* The research found that a simpler video surveillance system would also be suitable for forest protection. By controlling the road network, not only criminal activities in the field of forestry would be prevented, but it would also contribute to strengthening national security in those border management units for which setting up a sophisticated video surveillance system is currently not envisaged. The only prerequisite for this is to draw attention to the position of the monitored cardinal points when defining the position of the tactical point in space. Namely, such a point is best placed on the highest elevation around which the largest number of cardinal points to be monitored are grouped.

(3) It is confirmed that *the multi-purpose forest road designed on a tactical level affects the accessibility of the management unit, although this is not its primary purpose.* Analyzing the layout of the route of the Lisac (tactical) forest road, it can be noted that forest management was taken into account, although the road was built at the request of the Ministry of the Interior for state border surveillance. Namely, six compartments/sub-compartments, which had not been managed previously, were made accessible by its construction. Thus, the mean distance of skidding wood was reduced, as were the costs of forest exploitation. It is also worth mentioning that making accessible the compartments/sub-compartments with protective forests enabled better implementation of fire protection.

(4) The analysis has confirmed that *the length of the multi-purpose forest road designed on the operational level affects the size of the area made accessible in a management unit.* The primary purpose of operating multi-purpose forest roads is to serve forest management; therefore, when planning them, one should always take into account the phytocoenological map of the management unit and inaccessible compartments/sub-compartments through which it would be favorable for the route to pass. The analysis found that, in the Visočica area, it would be desirable for the multi-purpose forest road to be longer because, in this manner, more compartments/sub-compartments

of the commercial beech forest would be made accessible, i.e., in phytocoenological terms, beech forest with autumn moor grass (Seslerio autumnalis-Fagetum sylvaticae (Horvat) M. Wraber ex Borhidi 1963) and mountain beech forest with large red dead nettle (Lamio orvalae-Fagetum (Horvat 1938) Borhidi 1963). However, for surveillance purposes, the route should be as short as possible, as the goal is to reach the tactical point in the fastest possible time. Consequently, in order to meet both requirements, circular roads should be planned, consisting of two routes: shorter (for surveillance) and longer (for forest management). On the other hand, the results of the analysis determined that the operational forest road in the Lisac area should be shorter, as it is defined in the well-accessible area of that part of the management unit. Consequently, its entire route can be used for both surveillance and forest management purposes.

The conducted research shows that the issue of green border control requires an interdisciplinary approach to the problem, which means that the forestry profession, in the process of establishing surveillance over forested areas along the state border, must not be ignored. Furthermore, in order to maintain a continuum in controlling the green state border, it has been concluded that the application of a two-level surveillance system is required in border management units. The higher level refers to the issue of national security. For this purpose, it is necessary to determine the position of one or more strategic surveillance points and to plan the construction of tactical multi-purpose forest roads that would lead to them. The lower level is intended to protect forests from activities primarily related to illegal tree felling, theft of wood assortments, poaching, and arson. For this purpose, it is proposed to determine the tactical surveillance points and plan the route of operational multi-purpose forest roads leading to them.

In further research on this topic, it is also worth considering the possibility of developing the categorization of primary forest roads for the border mountain area, given the priority of their surveillance for the purpose of forest protection but also for the control of the state's green border. By combining such a categorization of primary forest roads with the proposed two-level surveillance of forested areas, Croatia would have an elaborate system of not only controlling its green border in the mountainous area but could also control forest operations in border management units.

Author Contributions: Conceptualization, D.K.-T. and I.P.; methodology, D.K.-T. and I.P.; software, D.K.-T.; validation, I.P. and M.K.; formal analysis, D.K.-T. and I.P.; investigation, D.K.-T.; resources, M.K.; data curation, I.P.; writing—original draft preparation, D.K.-T.; writing—review and editing, I.P. and M.K.; visualization, D.K.-T.; supervision, I.P.; project administration, M.K.; funding acquisition, D.K.-T. and I.P. All authors have read and agreed to the published version of the manuscript.

Funding: This research received no external funding.

Informed Consent Statement: Not applicable.

Data Availability Statement: Data presented in this study are available upon request from the corresponding author. The data is not publicly available due to the protection of the state border.

Conflicts of Interest: The authors declare no conflict of interest.

References

1. Andrić, A. Nonmilitary Security. Available online: https://wsb.edu.pl/files/pages/634/3.pdf (accessed on 6 January 2023).
2. Beznec, B.; Speer, M.; Mitrović, S.M.M. *Governing the Balkan Route: Macedonia, Serbia and the European Border Regime*; Rosa Luxemburg Stiftung: Beograd, Serbia, 2016.
3. Bukowski, P. Frontex activities on the Western Balkan route during the migrant crisis (2015–...). *Pol. Rev. Int. Eur. Law* **2019**, *8*, 97–111. [CrossRef]
4. Çalişkan, E. Planning of Forest Road Network and Analysis in Mountains Area. *Life Sci. J.* **2013**, *10*, 2456–2465.
5. Çelik, G.; Sabuncuoğlu, I. Simulation modelling and analysis of a border security system. *Eur. J. Oper. Res.* **2007**, *180*, 1394–1410. [CrossRef]

6. Chaudhary, M.; Vinay, P.; Kumari, N. Identification Vehicle Movement Detection in Forest Area using MFCC and KNN. In Proceedings of the International Conference on System Modeling & Advancement in Research Trends, Moradabad, India, 23–24 November 2018.
7. Clayton, A. *Warfare in Woods and Forests*; Indiana University Press: Bloomington, Indiana, 2012.
8. Cocco, E. Where is the European frontier? The Balkan migration crisis and its impact on relations between the EU and the Western Balkans. *Eur. View* **2017**, *16*, 293–302. [CrossRef]
9. Debuyser, C. Infomigrants. Available online: https://www.infomigrants.net/en/post/15833/bosnia--croatia-the-game-a-catandmouse-chase-between-migrants-and-border-police (accessed on 3 March 2021).
10. Demir, M.; Hasdemir, M. Functional Planning Criterion of Forest Road Netowrk Systems According to Recent Forestry Development and Suggestion in Turkey. *Am. J. Environ. Sci.* **2005**, *1*, 22–28. [CrossRef]
11. Dobler, G. The green, the grey and the blue: A typology of cross-border trade in Africa. *J. Mod. Afr. Stud.* **2016**, *54*, 145–169. [CrossRef]
12. Düvell, F. Transit Migration: A Blurred and Politicised Concept. *Popul. Space Place* **2012**, *18*, 415–427. [CrossRef]
13. Edmonda. Migrations' changing scenario: The new Balkan Route and the European Union. *Riv. Di Studi Politici Internazionali* **2018**, *85*, 189–206.
14. Gumus, S. Constitution of the forest road evaluation form for Turkish forestry. *Afr. J. Biotechnol.* **2009**, *8*, 5389–5394.
15. Hayati, E.; Majnounian, B.; Abdi, E.; Sessions, J.; Makhdoum, M. An expert-based approach to forest road network planning by combining Delphi and spatial multi-criteria evaluation. *Environ. Monit. Assess.* **2013**, *185*, 1767–1776. [CrossRef]
16. Henning, V. *Jungle Warfare: According to the Armed Forces Theory of Fundamental Capabilities*; Officer Program 10–13; Swedish National Defense College: Stockholm, Sweden, 2013.
17. Hosseini, S.A.O.; Moghadasi, P.; Fallah, A. Forest Road Network Design based on Multipurpose Forestry Management in Hyrcanian Forest. *J. Environ. Sci. Manag.* **2019**, *22*, 13–20. [CrossRef]
18. Hrvatske šume. *Glavni Projekt Šumske Ceste Lisac*; Hrvatske šume d. o. o.: Zagreb, Croatia, 2020.
19. Hrvatske šume. *Osnova gospodarenja G. J. «Visočica—Lisac»*; Hrvatske šume d. o. o.: Zagreb, Croatia, 2018.
20. Hulaj, A.; Shehu, A.; Bajrami, X. Application of Wireless Multimedia Sensor Networks for Green Borderline Surveillance. In Proceedings of the 27th DAAAM International Symposium on Intelligent Manufacturing and Automation, Mostar, Bosnia and Herzegovina, 26–29 October 2016.
21. Kolb, V. *Borbena Dejstva u Šumi*; Mala vojna biblioteka: Beograd, Yugoslavia, 1951.
22. Koszteczky, B.; Vakulya, G.; Simon, G. Forest intrusion detection system with sensor network. In Proceedings of the IEEE International Instrumentation and Measurement Technology Conference, Pisa, Italy, 11–14 May 2015.
23. McIntyre, D.L.; Weeks, J.R. Environmental Impacts of Illegal Immigration on the Cleveland National Forest in California. *Prof. Geogr.* **2002**, *54*, 392–405. [CrossRef]
24. Mporas, I.; Perikos, I.; Kelefouras, V.; Paraskevas, M. Illegal Logging Detection Based on Acoustic Surveillance of Forest. *Appl. Sci.* **2020**, *10*, 7379. [CrossRef]
25. Murphy, G. The use of photographic detection system for counting and classifying vehicles on forest service roads. *Proc. SPIE* **1973**, *37*, 15–28. [CrossRef]
26. Najafi, A.; Sobhani, H.; Saeed, A.; Makhdom, M.; Mohajer, M.M. Planning and Assessment of Alternative Forest Road and Skidding Networks. *Croat. J. For. Eng.* **2008**, *29*, 63–73.
27. Newman, D. The lines that continue to separate us: Borders in our 'borderless' world. *Prog. Hum. Geogr.* **2006**, *30*, 143–161. [CrossRef]
28. Oddone, E. Npr.org. Available online: https://www.npr.org/2021/03/27/976648642/for-migrants-in-bosnia-the-game-is-a-perilous-journey-to-a-better-life (accessed on 3 March 2021).
29. Papa, I.; Pentek, T.; Janeš, D.; Valinčić, E.; Đuka, A. Studija primarnog otvaranja šuma gospodarske jedinice Crno jezero—Marković rudine Šumarije Otočac. *Nova Meh. Šumarstva* **2019**, *40*, 59–70. [CrossRef]
30. Pentek, T. *Forest Opening—PowerPoint*; Faculty of Forestry and Wood Technology of the University of Zagreb: Zagreb, Croatia, 2018.
31. Pentek, T.; Pičman, D.; Nevečerel, H.; Lepoglavec, K.; Papa, I.; Potočnik, I. Primarno otvaranje šuma različitih reljefnih područja Republike Hrvatske. *Croat. J. For. Eng.* **2011**, *32*, 401–416.
32. Potočnik, I. Categorization of forest roads on the basis of their multiple use and social needs. Науковий вісник **2002**, *12*, 66–73.
33. Rossmo, D.K.; Thurman, Q.C.; Jamieson, J.; Dand Egan, K. Geographic patterns and profiling of illegal crossings of the southern US border. *Secur. J.* **2008**, *21*, 29–57. [CrossRef]
34. Said, E.W. *Orijentalizam*; XX VEK: Beograd, Serbia, 2008.
35. Shoemaker, R.; Sandbrink, R.; van Voorthuijsen, G. Intelligent Route Surveillance. In *Unattended Ground, Sea, and Air Sensor Technologies and Applications XI, Orlando, FL, USA, 9–12 April 2007*; Carapezza, E.M., Ed.; SPIE—International Society for Optics and Photonics: Bellingham, WA, USA, 2009.
36. Simonenkova, A.; Simonenkov, M.; Bacherikov, I. Optimization of forest road network layout problem. *IOP Conf. Ser. Mater. Sci. Eng.* **2019**, *817*, 012032. [CrossRef]
37. Stewart, H.E.; Ostericher, I.; Goke, C.; De León, J. Surveilling Surveillance: Countermapping Undocumented Migration in the USA-Mexico Borderlands. *J. Contemp. Archaeol.* **2016**, *3*, 159–174. [CrossRef]

38. Sun, Z.; Wang, P.; Vuran, M.C.; Al-Rodhaan, M.A.; Al-Dhelaan, A.M.; Akyildiz, I.F. BorderSense: Border patrol through advanced wireless sensor networks. *Ad Hoc Netw.* **2011**, *9*, 468–477. [CrossRef]
39. Vallet, É.; David, C.P. The (Re)Building of the Wall in International Relations. *J. Borderl. Stud.* **2012**, *27*, 111–119. [CrossRef]
40. van Houtum, H.; van Naerssen, T. Bordering, ordering and othering. *Tijdschr. Voor Econ. En Soc. Geogr.* **2002**, *93*, 125–136. [CrossRef]

sustainability

MDPI

Article

The Impact of Group- and Single-Tree-Selection Cuttings on Runoff and Sediment Yield in Mixed Broadleaved Forests, Northern Iran

Hassan Samdaliri [1], Meghdad Jourgholami [1], Ali Salajegheh [2], Hadi Sohrabi [1], Rachele Venanzi [3,*], Rodolfo Picchio [3] and Angela Lo Monaco [3]

1 Department of Forestry and Forest Economics, Faculty of Natural Resources, University of Tehran, Karaj 14176-43184, Iran; hsamdaliri@ut.ac.ir (H.S.); mjgholami@ut.ac.ir (M.J.); hadi.sohrabi@ut.ac.ir (H.S.)

2 Department of Arid and Mountainous Regions Reclamation, Faculty of Natural Resources, University of Tehran, Karaj 31587-77871, Iran; salajegh@ut.ac.ir

3 Department of Agricultural and Forest Sciences, Tuscia University, 01100 Viterbo, Italy; r.picchio@unitus.it (R.P.); lomonaco@unitus.it (A.L.M.)

* Correspondence: venanzi@unitus.it

Abstract: Silvicultural treatment and the forest harvesting operations using different methods can lead to an increase in the production of runoff and sediment by changing the canopy and soil surface where they are conducted. In order to investigate this issue, sampling plots were established in the Namkhaneh district of the Kheyrud forest with three replications for every treatment: control stand and tree harvesting systems using single-selection cuttings and group-selection cuttings. The amount of runoff and sediment was collected and estimated from precipitation over a period of one year. Also, some soil physical properties such as bulk density, penetration resistance, sand, silt, and clay content, soil moisture, and soil organic matter were measured. The results showed that tree harvesting systems has a significant effect on runoff, the runoff coefficient, and sediment but the season (growing season and fall) and the combined effect of tree harvesting systems and the season have no significant effect on the runoff coefficient and sediment. The mean runoffs of each rainfall event for the control, single-tree, and group-selection treatments were 5.67, 8.42, and 10.28 mm, respectively, and the sediment amounts were 3.42, 6.70, and 11.82 gr/m^2, respectively. Furthermore, the total annual erosion amounts of the control, selection, and grouping treatments were 0.427, 0.838, and 2.178 t/ha, respectively. The bulk density, penetration resistance, and percentage of sand and silt were positively related and the percentages of clay and organic matter were negatively related with the amount of runoff and sediment. In the method of individual selection cuttings, the damage to the forest in terms of the amount of runoff and soil erosion was less than for the group-selection cuttings. Forest harvesting by the selection method (single-selection and group-selection) has caused different changes in the vegetation canopy. The final summary of our results could be the advice to predominantly use the single-selection method in high-slope stands.

Keywords: forest harvesting; single- and group-selection method; canopy cover; annual erosion; runoff and sediment

Citation: Samdaliri, H.; Jourgholami, M.; Salajegheh, A.; Sohrabi, H.; Venanzi, R.; Picchio, R.; Lo Monaco, A. The Impact of Group- and Single-Tree-Selection Cuttings on Runoff and Sediment Yield in Mixed Broadleaved Forests, Northern Iran. *Sustainability* **2024**, *16*, 1830. https://doi.org/10.3390/su16051830

Academic Editor: Eben N. Broadbent

Received: 19 January 2024
Revised: 14 February 2024
Accepted: 20 February 2024
Published: 23 February 2024

1. Introduction

Forest ecosystems play an effective role in regulating and maintaining the water balance in watersheds, and the ecosystem's function of water supply and regulation by trees includes regulating seasonal water flows, providing water for various uses, and water purifying and storing. In forests, due to humus and organic matter in the soil, the amount of water absorption and infiltration is high [1]. The area of Hyrcanian forests in northern Iran is about 1.9 million ha, of which an area equal to 1.07 million ha has been exploited in the past decades. Disturbance in a forest's hydrological function leads to an increase in the soil erosion intensity, the loss of topsoil, and a decrease in its water holding capacity [2–10]. Undisturbed forest areas usually have the least runoff and sediment production among all ecosystems due to the presence of tree and herbaceous cover, a litter layer, and tree remains. In this regard, the soil erosion rate in undisturbed forests has been reported in a range from 0.02 to 1.2 t/ha [11]. Forest harvesting operations which cut trees and remove vegetation from forest floor lead to an increase in the volume of surface runoff [12].

One of the key goals of sustainable forestry and agricultural management is to minimize soil erosion. Accordingly, the soil properties and vegetation are two important factors which influence soil erosion processes [6,7]. The study of the role of vegetation and soil physical properties on runoff and sediment yields in different land cover types (forest, monoculture plantation, and agroforestry) in China by Zhu et al. [13] showed that the runoffs in land with monoculture plantations are 33.2 and 2.6 times greater than those of forest and the agroforestry fields, respectively. Furthermore, the sedimentation rates in forest, monoculture plantation, and agroforestry systems are 0.041, 11.54, and 2.73 t/ha, respectively. This shows that the conversion of forests to other types of land cover leads to negative hydrological consequences (excessive runoff and sedimentation). The amounts of runoff and sedimentation are two important parameters related to soil erosion, which is considered as an important environmental problem related to natural hazards and forest management operations [14,15]. Vegetation evaporates part of the rain that it receives on the one hand and slowly transfers another portion to the earth's surface. At the same time, creating a barrier to the movement of water on the ground surface, it increases the penetration of water into the soil [5]. Therefore, land cover change has a profound effect on the water cycle. The reduction of vegetation cover due to forest harvesting generally increases the average surface runoff volume and overall water performance [16]. The removal of a certain number of trees in the stand as a consequence of exploitation induces a change in the crown layer of forest cover, reducing the amount of interception and increasing evaporation. It also increases the amount of throughfall on the ground [17,18]. In deciduous forest stands, trees experience two distinct periods, leafy and leafless periods, which have significant effects on hydrological characteristics such as interception, followed by runoff and sediment [19]. The average runoffs in plots with low, medium, and high traffic intensity in the leafless period were 95.5, 54.2, and 21.7% higher than the runoff values in the leafy period, respectively. The average yields of sediment in low-, medium-, and high-traffic treatments in the leafless period were 7.1, 5.1, and 3.3 times higher than the amount of sediment in the leafy period [19].

Soil erosion is affected by many factors, including the rainfall intensity, soil type, vegetation, soil moisture, and slope [20]. The dry bulk density, moisture content, and total porosity were affected considerably on skid trails by the slope and traffic frequency [8,10]. In harvested areas, the presence of foliage and litter on the forest floor is very important to prevent splash erosion, and somewhat to reduce runoff and particle movement to downslope [21]. In areas with steep slopes, the strength of the soil is weak, and heavy rains and high temperatures make the erosive soils more sensitive to the effects of machinery and the disturbance of vegetation [22–24]. The research of Akbarimehr and Naghdi [25] shows that the main variable affecting runoff and soil erosion is the slope. Bahadur [26] also stated that the highest amounts of runoff and soil loss come from plots with a steep slope. Runoff generally increases in forest watersheds in a short time after forest clear cutting, but it decreases in the long term due to the evapotranspiration increase with tree regrowth in the stand [27]. Etehadi Abari et al. [28] studied the runoff and sediment changes following changes in some soil properties due to forest harvesting operations and they concluded that changes in the crown and plant cover have a significant effect on the amount of runoff and sediment. Furthermore, the percentage of clay and the soil bulk density have a significant positive correlation with the amount of runoff [1,5]. The percentage of sand, the pH, and organic matter have a significant negative correlation with the amount of runoff. It was also found that the variables of soil moisture percentage, sand percentage, and soil organic matter have a negative relationship, and the clay percentage and soil bulk density have a significant positive correlation with the amount of sediment [28]. Soil compaction by changes in the soil properties is an important factor in soil losses. The amounts of soil sediment under the influence of harvesting operations in areas with a steep slope and for the control are estimated at 2.56 and 0.13 tons per ha per year, respectively [29].

Nowadays, sustainable forest management aims to reduce environmental degradation at all levels. In the Hyrcanian forests, due to the implementation of single-selection and group-selection harvesting systems as well as the creation of small and large patches caused by tree harvesting, it is necessary to evaluate changes in vegetation, soil properties, runoff, and sediment. The purpose of this is to reduce the effects of disturbance factors and improve the conditions of the forests. The objective of this study was to determine the effect of forest cover on sediment yield and runoff rates. In particular, the effects of the single-tree- and group-selection methods were compared. Furthermore, since the studied forest is a deciduous forest, the effect of seasonal coverage due to the presence or absence of leaves was taken into consideration. The aim was, therefore, to have a better understanding of the factors that have a significant effect on sediment and runoff according to the hypothesis that forestry methods (single- and group-selection) and relative logging activities have an influence.

2. Materials and Methods

2.1. Site Description

This study was conducted in compartment 207 of the Namkhaneh district in the Kheyrud forest (Forest Research Station of University of Tehran). The studied area in the Hyrcanian forests is a remnant of the third ice age and belongs to the Upper Jurassic period. These forests have a high biodiversity and, on the other hand, are exposed to the use of local people and commercial harvesting. The annual rainfall in this area is 1380 mm. The rainiest months of the year are October and November, with averages of 273.6 mm, and July is the least rainy month of the year with 47.5 mm of rainfall. The average annual temperature is 16.1 °C. Parcel area is 49.7 ha with north-west aspect and the altitude is between 710 and 900 m above sea level. The soil type is brown forest, and the forest type is beech-hornbeam. In the Namkhaneh district, the average stem density and average growing stock of stands were 251 tree ha^{-1} and 510.6 m^3 ha^{-1}, respectively. The type of harvesting was done in the form of the selective method (single-tree selection method and group-selection method). Marked trees were felled and processed using a chainsaw, and the timber was extracted from stump to landing by a TAF E655 wheeled skidder. The operations of cutting and logging trees for both treatments were carried out in 2018.

The natural rainfall of the region (not using the rain simulator) was used to determine the amount of runoff and sediment due to the application of the forestry method of selection cutting by groups or by single trees. The studied treatments included undisturbed forest (control), forest with single-tree harvest (single-selection), and forest with more than one tree harvest (group-selection). Also, considering the coverage of tree leaves during the research, the season was considered as another variable. The season factor was divided into two parts: growing season (with leaves, maximum herbaceous cover and litter) and non-growing season (no leaves, limited herbaceous cover and litter). The durations of the growing and non-growing seasons in Hyrcanian forests are 7 and 5 months, respectively. The characteristics of the undisturbed area and the two harvested areas with single-selection and group-selection methods are shown in Table 1.

Table 1. General characteristics of the Namkhaneh district under the investigated treatments.

Location	Features	Average Patch Area (m^2)	Average of Cover by Season				Soil Texture
			Tree Cover (%)		Grass Cover (%)		
			Fall	Growing	Fall	Growing	
Undisturbed (Control)	No disturbance, unharvested, existence of livestock in the field	0	30	85	4	26	Loamy clay
Single-selection cuttings	Low disturbance, effects of water conversion and transportation, existence of livestock in the field	100	10	50	12	65	Loamy clay
Group-selection cuttings	High disturbance, effects of water conversion and transportation, existence of livestock in the field	400	50	15	23	88	Loamy silt

2.2. Experimental Design

In order to determine the sampling location, areas with the same conditions (geographic direction, altitude, forest type) were selected and the plots of each treatment were randomly installed in areas with a 30% slope. Plots with dimensions of 2 m^2 with longer length in the slope direction and with 3 repetitions of the three treatments were designed [1,12,30]. The plots were enclosed using 25 cm-wide wooden planks so that the runoff would not leak out from the plots, allowing them to represent the actual amount of runoff. At the ends of these plots, a pipe is installed to direct the runoff flowing through the plots to the collection reservoir (Figure 1). Data collection was performed for one year, after each rainfall event from 26 September 2018 to 28 September 2019.

Figure 1. Study area and design of plots for measuring runoff and sediment.

2.2.1. Data Collection and Laboratory Analysis

During the research period, after each occurrence of rainfall, the amount of runoff per plot was measured by recording the volume of water in the reservoirs. After the rain stopped, several liters of water were collected from each plot and transported to the laboratory to measure the sediment concentration (g/L) using the water discharge method [31]. In this method, one liter of water and sediment sample is poured into the container, it is kept in a stationary state for 48 h, and then the water is slowly separated from the sediments. The sediments remaining in the container were washed, poured into pre-prepared and weighed foils, and dried in the oven at a temperature of 105 °C for 24 h [2]. By weighing the samples of containers (plates) with sediment and subtracting the weight of the original containers (plates), the sediment sample weight is calculated in g/L.

Then, by performing the necessary calculations, the total concentration of sediments in each rainfall was estimated at the scale of plots (m^2) and stand (ha).

To measure the amount of rainfall, six rain gauges were installed in the undisturbed area and in the harvesting areas for single-selection and grouping-selection methods. To determine the amount of runoff, the amount of water leaving the plots that was stored in the container during each precipitation event was measured. The runoff height was calculated by dividing the runoff volume by the plot area. By dividing the runoff height by the rainfall height, the runoff coefficient was calculated.

2.2.2. Soil Properties

Soil samples were taken from the 0–10 cm layer using metal cylinders with a length of 10 cm and a diameter of 5 cm in all plots. In order to reduce the error, three soil samples were randomly taken in each plot, the samples were mixed together and, at the end, a soil sample was collected from each plot. Soil samples were dried at 105 °C for 24 h to calculate the soil moisture content and dry bulk density. The bulk density was calculated using Equation (1):

$$BD = \frac{WD}{VC} \tag{1}$$

where BD is the dry bulk density (g cm^{-3}), WD is the weight of the dry soil (g), and VC is the volume of the cylinder (cm^3).

Penetration resistance in the field was measured using a manual penetrometer (Eijkelkamp, 06.01.SA penetrometer, Giesbeek, The Netherlands) and soil texture was measured by hydrometric method in the laboratory. Soil penetration resistance in each plot was measured at the location of soil samples (3 sample locations) by applying vertical force to the penetrometer.

2.3. *Statistical Analyses*

The Kolmogorov–Smirnov test was used to test the normal distribution of the data. For homogeneity of variance test, Levene's test was used. The comparison of runoff rate, runoff coefficient, and sediment yield among treatments was done by two-way analysis of variance (ANOVA). If two-way analysis of variance between treatments was significant, Duncan's multiple range test was used to compare means. Pearson correlation test was also used to investigate the relationship between soil physical properties with runoff and sedimentation. Polynomial regression model was used to predict the relationship between runoff and sediment with rainfall. All statistical tests were performed using the SPSS 20 software package.

3. Results

3.1. *Precipitation Events*

Over the course of one year, 28 precipitation events were recorded, the highest amount of precipitation was recorded for event number 2 (250 mm) and the lowest amount was 7 mm on 13 December 2018 (Figure 2). In order to avoid errors in our data analysis, precipitation event number 2 was deleted due to excessive precipitation and an insufficient capacity of the collection container. Additionally, events number 22 and 23, in July 2019, were removed due to cracking of the ground due to lack of moisture, extreme water penetration into the ground, and the number of sunny days before the rain (very low runoff production). The total amounts of precipitation during the 25 recorded events in the field without harvesting (undisturbed control), with single-selection, and with group-selection were 871, 970, and 1057 mm per year, respectively (Figure 2).

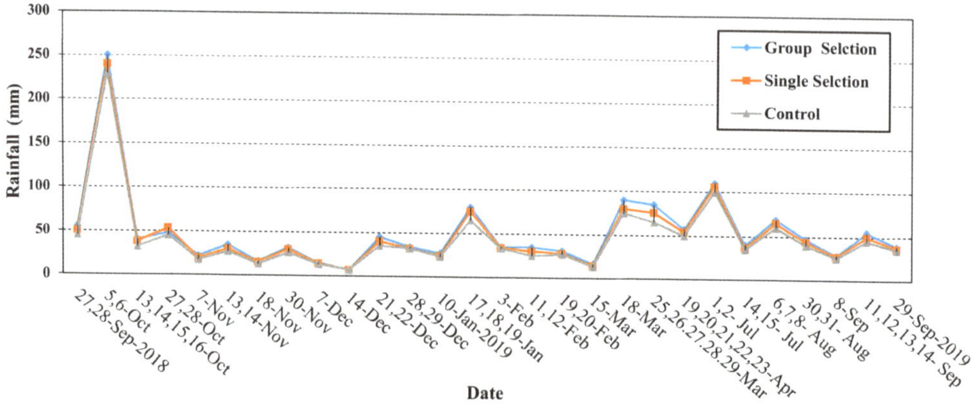

Figure 2. The amount of precipitation in each rainfall event in different treatments.

3.2. Effects of Stand Management and Season

The results of Table 2 showed that the lowest and highest average values of the bulk density, penetration resistance, percentage of sand, and silt were obtained in the undisturbed area and the area with group-selection harvesting. On the other hand, the percentages of clay, soil moisture, and soil organic matter were the highest and lowest in the undisturbed area and the area with group-selection harvesting, respectively. Also, the values of soil physical properties were significantly different between the undisturbed area and the areas with single-selection and group-selection harvesting (Table 2).

Table 2. Average (±standard error) soil physical properties in each of the study areas (treatments).

Soil Properties	Control	Single-Selection	Group-Selection
Bulk density (g/cm^{-3})	1.2 ± 0.08 ^{b}	1.24 ± 0.10 ^{ab}	1.3 ± 0.09 ^{a}
Penetration resistance (MPa)	1.37 ± 0.07 ^{c}	1.43 ± 0.11 ^{b}	1.5 ± 0.12 ^{a}
Sand (%)	19 ± 3 ^{c}	22 ± 4.27 ^{b}	25 ± 2.47 ^{a}
Silt (%)	44 ± 5.96 ^{b}	48 ± 5.45 ^{ab}	51 ± 6.49 ^{a}
Clay (%)	27 ± 4.87 ^{ab}	30 ± 3.50 ^{a}	24 ± 6.06 ^{b}
Water Content (%)	27 ± 11.03 ^{b}	31 ± 7.85 ^{a}	28 ± 4.45 ^{b}
Organic Matter (%)	7.64 ± 0.82 ^{a}	5.24 ± 0.25 ^{b}	3.2 ± 0.44 ^{c}

Note: Different letters in a row indicate significant differences among soil physical property values ($p < 0.05$).

According to the results of Table 3, the bulk density, penetration resistance, and percentage of sand have a positive significant relationship, and the percentages of clay and organic matter have a significant negative relationship with the amount of runoff and sediment. The percentage of silt is not correlated with the runoff rate, while it has a significant positive correlation with the amount of sediment. The water content (moisture percentage) was not correlated with the amount of runoff and sediment (Table 3).

The results of the variance analysis of the studied treatments on the amount of runoff, runoff coefficient, and sediment showed that the tree harvesting system, the seasonal changes, and their interaction effects have a significant effect on the runoff and sediment, while only the effect of seasonal changes was significant on the runoff coefficient (Table 4).

Regression analysis showed that the runoff and sediment changes increased linearly in relation to the rainfall changes for the areas with different harvesting systems (Figures 3 and 4a). The increase in the runoff and sediment in relation to rainfall in the area with the group-selection harvesting is more intense compared to the other areas. Comparing the average runoff from rainfall events using Duncan's test showed that the lowest amounts of runoff and sediment were obtained in the control area (undisturbed) and the highest amounts were obtained in the area with the group-selection harvesting (Figures 3 and 4a). Also, the amounts of runoff and sediment in all three areas have significant differences.

Table 3. Pearson correlation results between soil physical properties with runoff and sediment.

Soil Properties	Runoff		Sediment	
	R	*p* Value	R	*p* Value
Bulk density	0.68 *	0.043	0.72 *	0.028
Penetration resistance	0.82 **	0.006	0.82 **	0.006
Sand	0.71 *	0.031	0.68 *	0.045
Silt	0.61 ns	0.082	0.67 *	0.048
Clay	−0.78 **	0.012	−0.82 **	0.007
Water Content	−0.59 ns	0.092	−0.56 ns	0.116
Organic Matter	−0.84 **	0.005	−0.78 *	0.012

Note: *: $p < 0.05$; **: $p < 0.01$; ns: not significant.

Table 4. Variance analysis of the effect of the studied treatments on the amount of runoff, runoff coefficient, and sediment.

Variable	Source of Change	df	MS	*F* Value	*p* Value
Runoff	Tree harvesting system	2	103.520	12.441	**0.000**
	Change of season	1	138.390	17.036	**0.000**
	their interaction	2	4.854	0.598	**0.012**
Runoff coefficient	Tree harvesting system	2	0.069	3.835	0.113
	Change of season	1	0.069	3.835	**0.026**
	their interaction	2	0.003	0.141	0.869
Sediment	Tree harvesting system	2	425.854	10.214	**0.005**
	Change of season	1	49.640	5.826	**0.000**
	their interaction	2	11.639	0.762	**0.024**

p-values (<0.01 and <0.05) are given in bold.

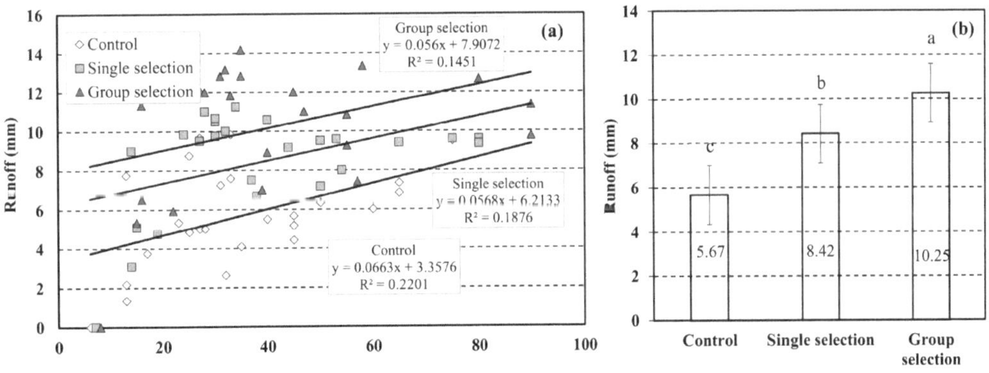

Figure 3. Relationship between rainfall and runoff in plots with different treatments (**a**), comparison of mean runoff in different treatments using Duncan's test (**b**). Different lowercase letters indicate significant differences by Duncan's test in B.

As the height of the runoff increases, the amount of sediment increases. The increase in the sediment rate due to the increase in runoff in the area with group-selection harvesting is the highest and, in the control area (undisturbed area), it is the lowest. For low runoff values, the amount of sediment has less differences between the studied treatments, and with increasing runoff, the differences in sediment values increase significantly (Figure 5). The total annual runoff values in the current study for the control treatments, single-selection and group-selection harvesting were 148, 219, and 267 mm per year, respectively. Also, the amount of runoff in the growing season for all treatments was less than that for the fall season (Figure 6a). Also, the amount of erosion in the growing season for all treatments is less than that for the fall season (Figure 6b).

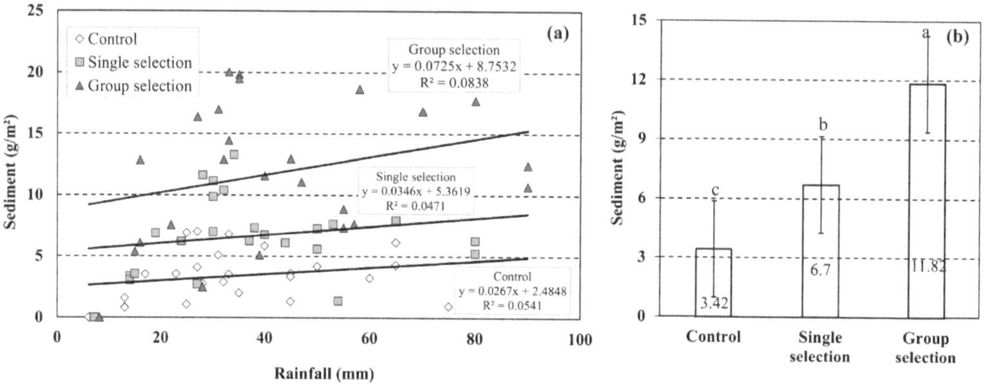

Figure 4. Relationship between rainfall and sediment in plots with different treatments (**a**), comparison of mean sediment in different treatments using Duncan's test (**b**). Different lowercase letters indicate significant differences by Duncan's test in B.

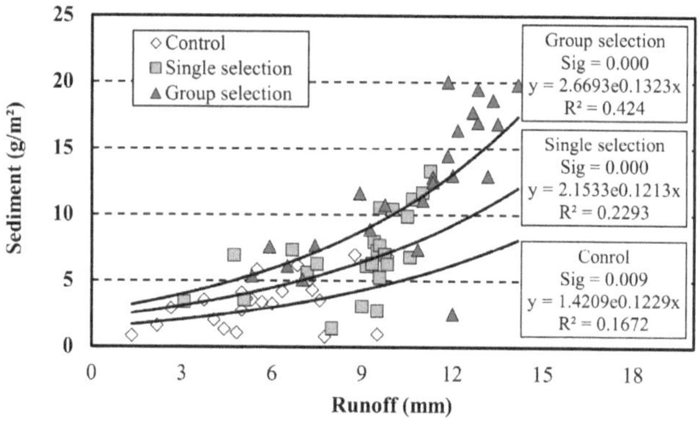

Figure 5. Relationship between runoff versus sediment in areas with different harvesting systems.

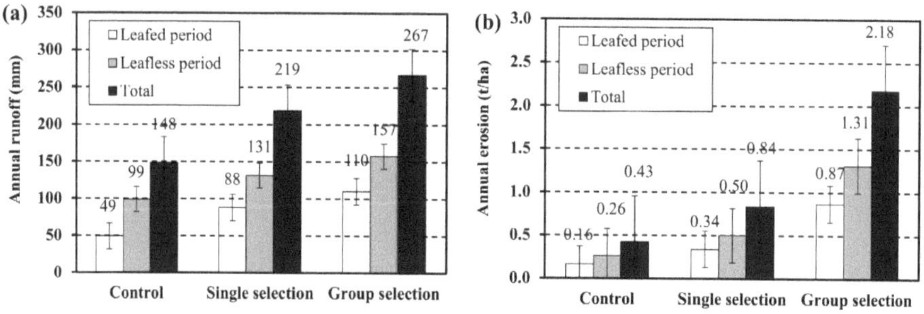

Figure 6. Annual runoff (**a**) and annual erosion (**b**) of rainfall in different treatments in the growing season (leafy period of trees) and autumn season (leafless period of trees).

4. Discussion

4.1. Precipitation Events

Change in forest hydrological characteristics are largely due to the effects of management and executive operations and, finally, the change in the type of land cover, because they can affect many components of the hydrological cycle [32,33]. Different factors such as the soil characteristics, vegetation type, and intensity and duration of rainfall affect the amount of infiltrating water in the soil or surface runoff [34,35]. With the increase in the amount of rainfall in each event, the amount of runoff and sediment increases, and the amount of runoff in each studied treatment was different due to the difference in the amount of rain that reached the soil and forest floor cover. According to the results, in the control area and the area with single-selection harvesting, the amounts of rainfall that reached the soil were 26% and 8% less, respectively, than in the area with group-selection harvesting. In areas with group-selection harvesting, due to lower densities of the canopy and litter, the amount of precipitation reaching the soil is higher, which increases the runoff and sediment rate. With the operation of tree harvesting and the resulting reduced forest cover, raindrops directly hit the soil; thus, the runoff volume also increases [27,36]. The curve of rainfall and runoff for the different treatments in Figure 3 is linear, and should be parabolic naturally [36]. This issue could be due to the fact that the rainfall intensity, which is an important factor in the amount of runoff, was not measured. The lack of measurement of the rainfall intensity and the lack of simultaneity of its measurement due to the distance of the plots in each of the three study areas represent the limitations of this study. Also, the rainfall of the recorded events varied from 1 h to 5 days.

4.2. Effects of Stand Management and Season

The soil physical properties (bulk density, penetration resistance, percentage of sand, silt, and clay, moisture content, and soil organic matter) are effective in determining the amount of soil erosion. Also, many studies showed that the amount of water infiltration is influenced by the soil physical properties [37]. The bulk density is known as an important factor determining the amount of soil loss due to its direct relationship with the penetration resistance [29].

Soil compaction is a factor in reducing properties such as the soil porosity, water infiltration, ventilation, and gas exchange [38,39]. In this study, the highest amount of runoff was observed in the group-selection harvesting treatment, in which the bulk density is the highest. This study is consistent with the research of Cleophas et al. [29], who stated that, with an increase in bulk density, which has a direct and positive relationship with penetration resistance [40,41], less water infiltrates into the soil. Also, with the increase in bulk density, the sediment rate increased, which is consistent with the research of Lotfalian et al. [42]. In the area with group-selection harvesting, due to the greater traffic of logging machines, and the litter layer with a low thickness and low moisture (open canopy and more light reaching the forest floor), the bulk density and penetration resistance increased, while the infiltration decreased [4,43,44], which ultimately led to an increase in runoff and sediment. In the control area (unharvested), the soil had a lower bulk density due to the presence of organic matter and more porosity, and as a result, the water penetration rate was higher [45,46].

Harvesting operations reduce the ability of water to penetrate into forest soils [47,48]. By increasing the percentage of sand in the soil, the amount of water infiltration increases, while increasing the percentage of clay and silt causes a decrease in water infiltration [23]. However, the results of this research are inconsistent with this assumption. In fact, the discrepancy may have been caused by the influence of other factors such as organic matter, the slope, and soil resistance to water infiltration. The increase in soil organic matter causes more water infiltration, and as a result, the rate of infiltration is higher and the amount of runoff and sediment is lower, which is consistent with the findings of the research by Etehadi Abari et al. [28].

Creating patches of different sizes in the forest affects the percentage of crown and herbaceous cover. Changing the vegetation cover (crown and herbaceous) causes a change in the number of raindrops reaching the soil, and ultimately the amount of runoff and sediment [35]. The results of this study showed that changes in the vegetation cover have a significant effect on the amount of runoff and sediment, which is consistent with the results of previous studies [28,29,48]. The relationship between the rainfall and sediment in the plots and the comparison of the mean sediment for the different treatments clearly show that, in the forest with group-selection harvesting of trees, due to the lower amount of canopy cover and vegetation, the average runoff was higher and the amount of sediment was greater. In the area with the group-selection harvesting, the amount of canopy cover was the lowest compared to the single-selection and control treatments; consequently, the amount of rainfall reaching the forest soil was greater [48] and, finally, the amount of runoff and sediment increased [49,50].

In the control area (unharvested), the vegetation cover causes part of the raindrops to be absorbed and evaporated by the canopy and not hit the soil surface [29], so the erosive power caused by the raindrops is reduced [48]. Furthermore, the presence of a greater litter layer on the forest floor leads to an increase in infiltration and a decrease in runoff [51,52]. Various studies have pointed out the role of organic materials in improving the physico-chemical and biological characteristics of forests and reducing the amount of runoff and erosion. For example, Salehi et al.'s research [53] showed that there is a positive relationship between the amount of soil organic matter and the weighted average diameter of soil grains, and that organic matter can improve the stability of soil grains and reduce the amount of sediment and soil erosion. Siegrist et al. [54] stated that organic matter in the soil increases the porosity and water-holding capacity of the soil and causes water to penetrate into the soil and decrease the volume of runoff. The leafiness and leaflessness of trees are also related to the amount of rain that reaches the soil [50]. Canopy interception affects the water level in forest ecosystems, and some of the rain is evaporated, returning to the atmosphere due to interception [6,55]. Leaf loss during the autumn period reduces the amount of interception, the canopy maintenance capacity, and the ratio of evaporation to the rain intensity during the rainy period [19,27]. Our study shows that the annual amount of runoff and erosion in the growing season, which is accompanied by the presence of leaves and herbaceous cover, was lower than in the fall season.

In the treatment with group-selection harvesting, with the increase in the amount of rainfall reaching the soil, the amount of sediment increased in direct relation with the runoff. In this treatment (group-selection harvesting), due to the open canopy, the erosive power caused by raindrops is greater, and also, due to the small quantity of litter on the floor, the amount of infiltration is lower; as a result, the amount of runoff and sediment is greater than for other treatments. The amount of sediment in this research is higher than that in the research of Etehadi Abari et al. [28]. One of the main reasons could be the difference in slope between our study and that of Etehadi Abari et al. [28]. The average slope in the present study was 30%, while in the study of Etehadi Abari et al. [28], the slope was 15%. The tendency for soil erosion is greater on slopes with higher grades [56,57]. Our findings showed that, with the increase in the harvesting rate and the consequent greater canopy opening, on average, the runoff and sediment increased, as reported also in Jourgholami et al. [19]. The rates of soil erosion in the control area, single-selection, and group-selection treatments were 0.864, 1.743, and 3.076 tons per ha per year, respectively, while, in the study by Cleophas et al. [29], in the unharvested and harvested area, were estimated at 0.13 and 2.56 tons per ha per year, respectively. These differences are due to the different slope and amount of annual rainfall.

5. Conclusions

Forest harvesting by selection cutting methods (single-selection and group-selection) has caused different changes in vegetation canopies. According to the results of this study, in the control area and the area with single-selection harvesting, the amount of rainfall that reached the soil was less than in the area with group-selection harvesting. In the area where group-selection harvesting was carried out, due to lower amounts of canopy and litter, the amount of rainfall reaching the soil was higher, which increases the runoff and sediment rate. The soil physical properties are effective in the amount assessment of soil erosion. In the area with group-selection harvesting, due to more traffic of logging machines and increases in soil compaction, the amount of litter layer with a low thickness and low moisture (open canopy and more light reaching the forest floor), the bulk density, and the penetration resistance increased, while the infiltration decreased. Finally, this led to an increase in runoff and sediment. The leafiness and leaflessness of trees also affect the amount of rain reaching the soil surface and soil erosion. The annual amount of runoff and erosion in the growing season, which is accompanied by the presence of leaves and herbaceous cover, was less than in the fall season. Therefore, it is suggested to use the single-selection method on high slopes for forest harvesting. In this research, the studied areas were almost the same in terms of topography and other effective factors, but due to the influence of the slope and its direct and positive relationship with the amount of runoff and sediment, and in order to obtain more accurate results, it is deemed necessary to investigate further by conducting studies in different topographical conditions. Considering the existence of livestock in forest areas and the long-term natural regeneration of degraded areas caused by the group-selection method, it is necessary to accelerate the improvement of these areas by planting seedlings.

Author Contributions: Conceptualization, H.S. (Hassan Samdaliri), M.J., A.S., H.S. (Hadi Sohrabi) and R.P.; Data curation, H.S. (Hassan Samdaliri), A.S., H.S. (Hadi Sohrabi) and A.L.M.; Formal analysis, H.S. (Hassan Samdaliri), A.S. and H.S. (Hadi Sohrabi); Investigation, H.S. (Hassan Samdaliri), A.S. and H.S. (Hadi Sohrabi); Methodology, H.S. (Hassan Samdaliri), M.J., A.S., H.S. (Hadi Sohrabi), A.L.M., R.V. and R.P.; Software, H.S. (Hassan Samdaliri); Supervision, M.J. and R.P.; Validation, M.J., A.S., H.S. (Hadi Sohrabi), A.L.M., R.V. and R.P.; Writing—original draft, H.S. (Hassan Samdaliri), M.J., A.S., H.S. (Hadi Sohrabi), A.L.M., R.V. and R.P.; Writing—review & editing, M.J., A.L.M., R.V. and R.P. All authors have read and agreed to the published version of the manuscript.

Funding: This research received no external funding.

Data Availability Statement: Data are available from the corresponding author upon reasonable request.

Acknowledgments: This research was carried out within the framework of the Ministry for Education, University, and Research (MIUR) initiative "Department of Excellence" (Law 232/2016) DAFNE Project 2023-27 "Digital, Intelligent, Green and Sustainable (acronym: D.I.Ver.So)".

Conflicts of Interest: The authors declare no conflicts of interest.

References

1. Ebeling, C.; Lang, F.; Gaertig, T. Structural recovery in three selected forest soils after compaction by forest machines in Lower Saxony, Germany. *For. Ecol. Manag.* **2016**, *359*, 74–82. [CrossRef]
2. Bagheri, I.; Naghdi, R.; Moradmand Jalali, A. Evaluation of factors affecting water erosion along skid trails (case study; Shafarood Forest, Northern Iran). *Casp. J. Environ. Sci.* **2013**, *11*, 151–160.
3. Sohrabi, H.; Jourgholami, M.; Labelle, E.R. The effect of forest floor on soil microbial and enzyme indices after forest harvesting operations in Hyrcanian deciduous forests. *Eur. J. For. Res.* **2022**, *141*, 1013–1027. [CrossRef]
4. López-Vicente, M.; Sun, X.; Onda, Y.; Kato, H.; Gomi, T.; Hiraoka, M. Effect of tree thinning and skidding trails on hydrological connectivity in two Japanese forest catchments. *Geomorphology* **2017**, *292*, 104–114. [CrossRef]
5. Picchio, R.; Jourgholami, M.; Zenner, E.K. Effects of Forest Harvesting on Water and Sediment Yields: A Review Toward Better Mitigation and Rehabilitation Strategies. *Curr. For. Rep.* **2021**, *7*, 214–219. [CrossRef]
6. Mohammad, A.G.; Adam, M.A. The impact of vegetative cover type on runoff and soil erosion under different land uses. *Catena* **2010**, *81*, 97–103. [CrossRef]
7. Latterini, F.; Venanzi, R.; Picchio, R.; Jagodziński, A.M. Short-term physicochemical and biological impacts on soil after forest logging in Mediterranean broadleaf forests: 15 years of field studies summarized by a data synthesis under the meta-analytic framework. *Forestry* **2023**, *96*, 547–560. [CrossRef]

8. DeArmond, D.; Ferraz, J.B.S.; Lovera, L.H.; Souza, C.A.S.; de Corrêa, C.; Spanner, G.C.; Lima, A.J.N.; dos Santos, J.; Higuchi, N. Impacts to soil properties still evident 27 years after abandonment in Amazonian log landings. *For. Ecol. Manag.* **2022**, *510*, 120105. [CrossRef]
9. Stednick, J.D. Monitoring the effects of timber harvest on annual water yield. *J. Hydrol.* **1996**, *176*, 79–95. [CrossRef]
10. Ampoorter, E.; Schrijver, A.; Nevel, L.; Hermy, M.; Verheyen, K. Impact of mechanized harvesting on compaction of sandy and clayey forest soils: Results of a meta-analysis. *Ann. For. Sci.* **2012**, *69*, 533–542. [CrossRef]
11. Wagenbrenner, J.; MacDonald, L.; Rough, D. Effectiveness of three post-fire rehabilitation treatments in the Colorado Front Range. *Hydrol. Process.* **2006**, *20*, 2989–3006. [CrossRef]
12. Poirier, V.; Paré, D.; Boiffin, J.; Munson, A.D. Combined influence of fire and salvage logging on carbon and nitrogen storage in boreal forest soil profiles. *For. Ecol. Manag.* **2014**, *326*, 133–141. [CrossRef]
13. Zhu, X.; Liu, W.; Jiang, X.J.; Wang, P.; Li, W. Effects of land-use changes on runoff and sediment yield: Implications for soil conservation and forest management in Xishuangbanna, Southwest China. *Land Degrad. Dev.* **2018**, *29*, 2962–2974. [CrossRef]
14. Borrelli, P.; Paustian, K.; Panagos, P.; Jones, A.; Schütt, B.; Lugato, E. Effect of good agricultural and environmental conditions on erosion and soil organic carbon balance: A national case study. *Land Use Policy* **2016**, *50*, 408–421. [CrossRef]
15. Gessesse, B.; Bewket, W.; Bräuning, A. Determinants of farmers' tree-planting investment decisions as a degraded landscape management strategy in the central highlands of Ethiopia. *Solid Earth* **2016**, *7*, 639–650. [CrossRef]
16. Suryatmojo, H.; Masamitsu, F.; Mizuyama, T. Effects of elective logging methods on runoff characteristics in paired small headwater catchment. *Procedia Environ. Sci.* **2013**, *17*, 221–229. [CrossRef]
17. Cristan, R.; Aust, W.M.; Bolding, M.C.; Barrett, S.M.; Munsell, J.F.; Schilling, E. Effectiveness of forestry best management practices in the United States: Literature review. *For. Ecol. Manag.* **2016**, *360*, 133–151. [CrossRef]
18. Malvar, M.C.; Silva, F.C.; Prats, S.A.; Vieira, D.C.; Coelho, C.O.; Keizer, J.J. Short-term effects of post-fire salvage logging on runoff and soil erosion. *For. Ecol. Manag.* **2017**, *400*, 555–567. [CrossRef]
19. Jourgholami, M.; Fathi, K.; Labelle, E.R. Effects of foliage and traffic intensity on runoff and sediment in skid trails after trafficking in a deciduous forest. *Eur. J. For. Res.* **2018**, *137*, 223–235. [CrossRef]
20. Guo, M.; Zhang, T.; Li, Z.; Xu, G. Investigation of runoff and sediment yields under different crop and tillage conditions by field artificial rainfall experiments. *Water* **2019**, *11*, 1019. [CrossRef]
21. Hartanto, H.; Prabhu, R.; Widayat, A.S.; Asdak, C. Factors affecting runoff and soil erosion: Plot-level soil loss monitoring for assessing sustainability of forest management. *For. Ecol. Manag.* **2003**, *180*, 361–374. [CrossRef]
22. Zachar, D. *Soil Erosion*; North-Holland Inc.: New York, NY, USA, 1982; p. 547, ISBN 9780444997258.
23. Ross, S.M.; Dykes, A. Soil conditions, erosion and nutrient loss on steep slopes under mixed dipterocarp forest in Brunei Darussalam. In *Monographiae Biologicae*; Edwards, D.S., Booth, W.E., Choy, S.C., Eds.; Springer: Berlin/Heidelberg, Germany, 1996; Volume 74. [CrossRef]
24. Bathurst, J.C.; Iroumé, A. Quantitative generalizations for catchment sediment yield following forest logging. *Water Resour. Res.* **2014**, *50*, 8383–8402. [CrossRef]
25. Akbarimehr, M.; Naghdi, R. Assessing the relationship of slope and runoff volume on skid trails (Case study: Nav 3 district). *J. For. Sci.* **2012**, *58*, 357–362. [CrossRef]
26. Bahadur, K.C.K. Spatio-temporal patterns of agricultural expansion and its effect on watershed degradation: A case from the mountains of Nepal. *Environ. Earth Sci.* **2012**, *65*, 2063–2077. [CrossRef]
27. Ide, J.I.; Finér, L.; Laurén, A.; Piirainen, S.; Launiainen, S. Effects of clear-cutting on annual and seasonal runoff from a boreal forest catchment in eastern Finland. *For. Ecol. Manag.* **2013**, *304*, 482–491. [CrossRef]
28. Etehadi Abari, M.; Majnounian, B.; Malekian, A.; Jourgholami, M. Runoff and sediment variations due to change in some soil properties following forest harvesting (Case study: Kheyrud Forest). *Iran. J. For.* **2018**, *10*, 267–278.
29. Cleophas, F.; Musta, B.; How, P.M.; Bidin, K. Runoff and soil erosion in selectively-logged over forest, Danum Valley Sabah, Malaysia. *Trans. Sci. Technol.* **2017**, *4*, 449–459.
30. Sensoy, H.; Kara, Ö. Slope shape effect on runoff and soil erosion under natural rainfall conditions. *iForest* **2014**, *7*, 110. [CrossRef]
31. Walling, D.E.; Collins, A.L.; Sichingabula, H.M.; Leeks, G.J.L. Integrated assessment of catchment suspended sediment budgets: A Zambian example. *Land Degrad. Dev.* **2001**, *12*, 387–415. [CrossRef]
32. Vergani, C.; Giadrossich, F.; Buckley, P.; Conedera, M.; Pividori, M.; Salbitano, F.; Rauch, H.S.; Lovreglio, R.; Schwarz, M. Root reinforcement dynamics of European coppice woodlands and their effect on shallow landslides: A review. *Earth-Sci. Rev.* **2017**, *167*, 88–102. [CrossRef]
33. Giadrossich, F.; Schwarz, M.; Cohen, D.; Cislaghi, A.; Vergani, C.; Hubble, T.; Phillips, C.; Stokes, A. Methods to measure the mechanical behaviour of tree roots: A review. *Ecol. Eng.* **2017**, *109*, 256–271. [CrossRef]
34. Vergani, C.; Schwarz, M.; Soldati, M.; Corda, A.; Giadrossich, F.; Chiaradia, E.A.; Morando, P.; Bassanelli, C. Root reinforcement dynamics in subalpine spruce forests following timber harvest: A case study in Canton Schwyz, Switzerland. *Catena* **2016**, *143*, 275–288. [CrossRef]
35. Du, E.; Link, T.E.; Wei, L.; Marshall, J.D. Evaluating hydrologic effects of spatial and temporal patterns of forest canopy change using numerical modelling. *Hydrol. Process.* **2016**, *30*, 217–231. [CrossRef]

36. Wagenbrenner, J.W.; MacDonald, L.H.; Coats, R.N.; Robichaud, P.R.; Brown, R.E. Effects of post-fire salvage logging and a skid trail treatment on ground cover, soils, and sediment production in the interior western United States. *For. Ecol. Manag.* **2015**, *335*, 176–193. [CrossRef]
37. Zemke, J.J. Runoff and soil erosion assessment on forest roads using a small scale rainfall simulator. *Hydrology* **2016**, *3*, 25. [CrossRef]
38. Fernández, C.; Vega, J.A. Effects of mulching and post-fire salvage logging on soil erosion and vegetative regrowth in NW Spain. *For. Ecol. Manag.* **2016**, *375*, 46–54. [CrossRef]
39. Prats, S.A.; Wagenbrenner, J.W.; Martins, M.A.S.; Malvar, M.C.; Keizer, J.J. Mid-term and scaling effects of forest residue mulching on post-fire runoff and soil erosion. *Sci. Total Environ.* **2016**, *573*, 1242–1254. [CrossRef]
40. Vaz, C.M.; Bassoi, L.H.; Hopmans, J.W. Contribution of water content and bulk density to field soil penetration resistance as measured by a combined cone penetrometer–TDR probe. *Soil Tillage Res.* **2001**, *60*, 35–42. [CrossRef]
41. Ampoorter, E.; Van Nevel, L.; De Vos, B.; Hermy, M.; Verheyen, K. Assessing the effects of initial soil characteristics, machine mass and traffic intensity on forest soil compaction. *For. Ecol. Manag.* **2010**, *260*, 1664–1676. [CrossRef]
42. Lotfalian, M.; Babadi, T.Y.; Akbari, H. Impacts of soil stabilization treatments on reducing soil loss and runoff in cutslope of forest roads in Hyrcanian forests. *Catena* **2019**, *172*, 158–162. [CrossRef]
43. Ezzati, S.; Najafi, A.; Rab, M.A.; Zenner, E.K. Recovery of soil bulk density, porosity and rutting from ground skidding over a 20-year period after timber harvesting in Iran. *Silva Fenn.* **2012**, *46*, 521–538. [CrossRef]
44. Sohrabi, H.; Jourgholami, M.; Tavankar, F.; Venanzi, R.; Picchio, R. Post-Harvest Evaluation of Soil Physical Properties and Natural Regeneration Growth in Steep-Slope Terrains. *Forests* **2019**, *10*, 1034. [CrossRef]
45. Webb, R.H. Recovery of severely compacted soils in the Mojave Desert, California, USA. *Arid. Land Res. Manag.* **2002**, *16*, 291–305. [CrossRef]
46. Rab, M.A. Recovery of soil physical properties from compaction and soil profile disturbance caused by logging of native forest in Victorian Central Highlands, Australia. *For. Ecol. Manag.* **2004**, *191*, 329–340. [CrossRef]
47. Ziegler, A.D.; Negishi, J.N.; Sidle, R.C.; Noguchi, S.; Nik, A.R. Impacts of logging disturbance on hillslope saturated hydraulic conductivity in a tropical forest in Peninsular Malaysia. *Catena* **2006**, *67*, 89–104. [CrossRef]
48. Nunes, A.N.; De Almeida, A.C.; Coelho, C.O. Impacts of land use and cover type on runoff and soil erosion in a marginal area of Portugal. *Appl. Geogr.* **2011**, *31*, 687–699. [CrossRef]
49. Zhang, L.; Wang, J.; Bai, Z.; Lv, C. Effects of vegetation on runoff and soil erosion on reclaimed land in an opencast coal-mine dump in a loess area. *Catena* **2015**, *128*, 44–53. [CrossRef]
50. Li, X.; Xiao, Q.; Niu, J.; Dymond, S.; van Doorn, N.S.; Yu, X.; Xie, B.; Lv, X.; Zhang, K.; Li, J. Process-based rainfall interception by small trees in Northern China: The effect of rainfall traits and crown structure characteristics. *Agric. For. Meteorol.* **2016**, *218*, 65–73. [CrossRef]
51. Tsiko, C.T.; Makurira, H.; Gerrits, A.M.J.; Savenije, H.H.G. Measuring forest floor and canopy interception in a savannah ecosystem. *Phys. Chem. Earth Parts A/B/C* **2012**, *47*, 122–127. [CrossRef]
52. Shah, N.W.; Baillie, B.R.; Bishop, K.; Ferraz, S.; Högbom, L.; Nettles, J. The effects of forest management on water quality. *For. Ecol. Manag.* **2022**, *522*, 120397. [CrossRef]
53. Salehi, A.; Zahedi Amiri, G. Study of physical and chemical soil properties variations using principal component analysis method in the forest, North of Iran. *Casp. J. Environ. Sci* **2005**, *3*, 131–137.
54. Sieegrist, S.; Schaub, D.; Pfiffner, L.; Mäder, P. Does organic agriculture reduce soil erodibility? The results of a long-term field study on loess in Switzerland. *Agric. Ecosyst. Environ.* **1998**, *69*, 253–264. [CrossRef]
55. Dung, B.X.; Gomi, T.; Miyata, S.; Sidle, R.C.; Kosugi, K.; Onda, Y. Runoff responses to forest thinning at plot and catchment scales in a headwater catchment draining Japanese cypress forest. *J. Hydrol.* **2012**, *444–445*, 51–62. [CrossRef]
56. Thanapakpawin, P.; Richey, J.; Thomas, D.; Rodda, S.; Campbell, B.; Logsdon, M. Effects of landuse change on the hydrologic regime of the Mae Chaem river basin, NW Thailand. *J. Hydrol.* **2007**, *334*, 215–230. [CrossRef]
57. Kinnell, P.I. A review of the design and operation of runoff and soil loss plots. *Catena* **2016**, *145*, 257–265. [CrossRef]

sustainability

Article

Effects of Successive Planting of *Eucalyptus* Plantations on Tree Growth and Soil Quality

Qiongling Dai, Tianhui Wang, Penglian Wei * and Yunlin Fu

Guangxi Key Laboratory of Forest Ecology and Conservation, College of Forestry, Guangxi University, Nanning 530004, China; 2009392003@st.gxu.edu.cn (Q.D.); 2009392028@st.gxu.edu.cn (T.W.); fuyunlin@gxu.edu.cn (Y.F.)

* Correspondence: weipenglian@gxu.edu.cn

Abstract: The ultra-short-cycle successive planting of *Eucalyptus* plantations has caused environmental and social problems, and changing the rotation cycle is a very good option to solve this issue. However, the effects of successive planting on *Eucalyptus* growth and soil quality after changing the cultivation period are unclear. This study evaluated the effects of successive *Eucalyptus* planting on growth, soil nutrients, and bacterial and fungal community structure with an eight-year cultivation period. *Eucalyptus* plantations with different succession generations (first, second and third generation) were selected, and tree height and diameter at breast height were measured. Ten indicators of soil nutrients in different soil layers (0–20 cm and 20–40 cm) were measured, and soil bacteria and fungi were sequenced in high throughput. Results show that there is an upward trend in tree growth after three successive generations, reaching the highest timber yield in the third-generation plantation. Soil nutrients also showed changes, in the 0–20 cm soil layer, with decreased TN, NH_4^+-N, NO_3^--N and AK and increased AP, particularly for OM and TP content. In the 20–40 cm soil layer, the content of NH_4^+-N and NO_3^--N increased slightly and the soil's OM, TP, and TK content increased significantly. The diversity of bacterial and fungal communities in different soil layers increased significantly, and the community structure composition changed. Bacterial and fungal community structures were mainly driven by pH, NH_4^+-N, TP and AP factors and by OM, NH_4^+-N, TP and TK factors, respectively. Thus, successive plantings of *Eucalyptus* plantations with a cultivation period of eight years is conducive to the growth of trees. Some nutrients of the soil were returned, and the soil microbial diversity increased. Successive planting has brought efficiency and economic benefits while maintaining the soil's fertility.

Keywords: successive planting; cultivation period; *Eucalyptus* plantation; sustainable development; tree growth; soil nutrients; soil fungal and bacterial communities

Citation: Dai, Q.; Wang, T.; Wei, P.; Fu, Y. Effects of Successive Planting of *Eucalyptus* Plantations on Tree Growth and Soil Quality. *Sustainability* **2023**, *15*, 6746. https://doi.org/10.3390/su15086746

Academic Editors: Ivica Papa and Andreja Đuka

Received: 13 March 2023
Revised: 11 April 2023
Accepted: 13 April 2023
Published: 17 April 2023

1. Introduction

Eucalyptus has been widely planted in southern China due to its fast growth, adaptability, high yield, and excellent pulping properties. By 2018, the planting area had exceeded 5.4 million hm^2 [1]. To address market demands and timber shortages, short-cycle multi-generation successive planting has been widely applied in *Eucalyptus* plantation management. This management model allows trees to sprout and regrow on their own stumps after felling, without replanting. In China, the cultivation period is usually between 5 to 7 years, or even 4 to 5 years, generally shorter than in other countries where these periods typically range from 6 to 14 years [1,2], due to limited forest land resources and timber scarcity [3].

Eucalyptus plantation cultivation has significantly contributed to the timber supply security in China [4]. Planting fast-growing species helps in restoring degraded forestland quickly and generates employment opportunities in afforestation, forest management, the wood pulp industry, and fiberboard production. Short cycle rotation offers more employment opportunities than traditional, long cycle rotation and increases the income of forestry farmers, which in turn stabilizes the prosperity of the rural population. However, this commercial management model has created ecological and environmental issues [5,6]. This form of land use conversion often exhausts soil nutrients and decreases soil quality [7], which consequently influences the sustainability of plantations.

Multi-generational succession in *Eucalyptus* plantation management has attracted considerable attention regarding its impact on tree growth, soil nutrients, and microorganisms. Soil nutrients play a crucial role in sustaining *Eucalyptus* plantation growth. The nature of the soil, its fertility and the conditions of cultivation and management determine the growth state of the stand. Moreover, the decay of trees through the litter material constantly drives new nutrient replenishment back into the forest soil, and soil nutrients are constantly updated [8,9]. Research has shown that successive plantations of *Eucalyptus* with a cultivation period of 5 years had a significant negative impact on soil nutrients, leading to a decrease in total nitrogen (TN) and available phosphorus (AP) [10]. Furthermore, as soil acidification pH increased with the increase in successive planting generations, soil organic matter (OM) exhibited a significant decreasing trend [11].

Soil microbes are essential for maintaining soil function and ecosystem sustainability through the regulation of nutrient cycling via the decomposition of litter and organic matter [12,13]. The growth, activities, and structures of forest soil microbial communities are influenced by various factors, including the quality and quantity of litter and organic matter inputs, soil nutrient effectiveness, and anthropogenic disturbances [14,15]. Multi-generational succession of plantations has caused significant impacts to soil bacterial and fungal community diversity, richness and metabolic activities in *Eucalyptus* plantations [11,16]. Studies have found that soil fungal and bacterial community diversity significantly decreased in *Eucalyptus* plantations with a 5-year cultivation period, leading to a negative impact on soil multi-functionality and microbial communities [11,17].

The operation mode of short-cycle, multi-generational successive plantations of *Eucalyptus* plantations is highly controversial [5]. Addressing soil quality issues associated with short-cycle successive plantations is necessary for the sustainable development of *Eucalyptus* plantations. Extending the age of *Eucalyptus* plantations and maintaining a near-natural state of operation can improve soil quality, but this may not meet market demand and may result in job losses. For some foresters and *Eucalyptus* investors, fund recovery is especially important. Balancing ecological problems and practical production needs is crucial when extending the cultivation period. Therefore, the cultivation period can only be reasonably extended from the existing ultra-short cultivation period. However, current studies have focused mainly on the 4–5-year ultra-short cultivation period, and it is still uncertain whether successive plantation is beneficial to tree growth and soil quality beyond the 5–7-year cultivation period.

Based on the above background, in this study, the first, second, and third generations of 8-year-old *Eucalyptus* plantations in the Guangxi state-owned Dongmen Forest Farm was used as the research object, in consideration of the fact that 8 years is not very short compared with the prevailing rotation cycle, but it is also not a particularly long period for some foresters who are under pressure to attain a cash flow. Tree growth and soil nutrient indicators in two soil layers were assessed and soil bacteria and fungi were identified with high-throughput sequencing to determine the effects of successive planting on tree growth, soil nutrients, and bacterial–fungal community structure and diversity. Our hypotheses were that (1) successive *Eucalyptus* plantations with an 8-year cultivation period had effects on stand growth, soil physical and chemical properties, and soil bacterial and fungal structure; (2) with successive generations, tree growth did not decline, some soil nutrients were replenished, and soil bacterial and fungal diversity and richness increased; and (3) changes in soil bacterial and fungal community structure were linked to changes in one or more of the soil's physical and chemical properties. The study aimed to determine the effects of successive planting on tree growth, soil nutrients, and bacterial–fungal community structure and diversity of *Eucalyptus* plantations with an 8-year cultivation period, and to establish the relationship between soil nutrients and microbial community diversity and structure. It also aimed to promote the sustainable development of *Eucalyptus* plantation forests.

2. Materials and Methods

2.1. Study Area

The experimental site was located in the Guangxi state-owned Dongmen Forest Farm in Dongmen Town, Fusui County, Guangxi, and the test area was about 5 ha. The distance between the three stands of first generation plantation, second generation plantation and third generation plantation did not exceed 1 km, and the plantation conditions, planting density and forestry measures were the same. The initial planting density was 1665 trees/ha. All three stands were fertilized with 622.5 kg/ha of compound fertilizer (composition: total nutrients $\geq$ 30%, N-P_2O_5-K_2O: 15-6-9, organic matter $\geq$ 15%) per year until 2020. After 2020, annual application of 1245 kg/ha organic fertilizer (composition: effective live bacteria $\geq$ 200 million/g, organic matter $\geq$ 40%, nitrogen, phosphorus and potassium $\geq$ 8%, plus amino acids, humic acid and trace elements) was added. The soil of the three stands is a typical brick red loam red soil in the Dongmen area, with a pH value between 4.5 and 5.6. The topography is low hills, with an elevation of around 100–130 m and a slope of 8–12°. The average annual temperature is 21.2–22.3 degrees Celsius, with extreme maximum temperature of 38–41 degrees Celsius and extreme minimum temperature of −4–1.9 degrees Celsius. The annual frost-free period is 346 days, the annual rainfall is 1000–1300 mm, and the relative humidity is 74–83%.

2.2. Site Setup and Stand Growth Survey

The experiment was selected from three different stands of the first, second, and third generation plantations of the fast-growing *Eucalyptus* DH32-26 asexual line at the age of 8 years. The first-generation stands were planted in 2014, the second and third generation stands were obtained from felled stump sprouts after felling of the previous generation stand. Three standard sample squares with an area of 20 m × 20 m were selected from each of the three generations, for a total of nine sample squares. The height of the trees in each sample square was measured with a height meter and recorded as *H*. The diameter at breast height of the trees in each sample square was measured with a diameter at breast height ruler at 1.3 m and recorded as *D*. The wood volume was estimated according to the Guangxi *Eucalyptus* binary material accumulation table. Visual inspection and recording of *Eucalyptus* leaves in each sample plot revealed no obvious disease or pest infestation.

2.3. Soil Sampling

Three standard sample squares (9 in total) with an area of 20 m × 20 m were selected from each of the three generations, and a near-center point was selected within each sample square. Six sampling points were established at a distance of 5 m from the near-center point along the 0°, 60°, 120°, 180°, 240°, and 300° direction lines. After removing fresh and semi-decomposed litter residues from the soil surface, the original soil was collected in the 0–20 cm and 20–40 cm soil layers. Careful measures were taken to avoid soil clods and disturbance from external soil during the digging process. Soil from the six sampling points in each sample plot were mixed, and 10–25 g was collected and stored on dry ice in sterile centrifuge tubes for soil bacterial and fungal DNA extraction and sequencing. The remaining soil samples were packed in square aluminum boxes and transported back to the laboratory for sieving and storage at 4 °C to facilitate testing of soil physical and chemical properties.

2.4. Soil Nutrient Analysis

Soil pH was determined by mixing the 10 g air-dried soil sample and deionized water (1:2.5 soil:water ratio) and, after the soil solution had clarified, the supernatant was tested [18]. Organic matter (OM) was quantified using the potassium dichromate-sulfate colorimetric method [19]. The total nitrogen (TN) and the soil's ammonium nitrogen (NH_4^+-N) and nitrate nitrogen (NO_3^--N) were determined by the continuous flow analyzer method. Total phosphorus (TP) was determined by the $HClO_4$–H_2SO_4 method. Available phosphorus (AP) was determined by the double acid leaching–molybdenum antimony anti-colorimetric method [20,21]. Total potassium (TK) and available potassium (AK) were determined by flame spectrophotometric method [22].

2.5. Soil Bacterial and Fungal DNA Extraction and Sequencing

The DNA was extracted from the soil subsamples stored at −80 degrees Celsius with the TGuide S96 Magnetic Soil /Stool DNA Kit (Tiangen Biotech (Beijing, China) Co., Ltd., Beijing, China) according to manufacturer's instructions. The DNA concentration of the samples was measured with the Qubit dsDNA HS Assay Kit and Qubit 4.0 Fluorometer (Invitrogen, Thermo Fisher Scientific, Portland, OR, USA). The bacterial 16S rRNA V3–V4 region was amplified using the primer pair 338F (5′-ACTCCTACGGGAGGCAGCAG-3′) and 806R (5′-GGACTACHVGGGTWTCTAAT-3′). The fungal rDNA ITS2 region was amplified using PCR with the primers ITS2F (5′-GCATCGATGAAGAACGCAGC-3′) and ITS2R (5′-TCCTCCGCTTATTGATATGC-3′). After amplification, PCR products were detected by electrophoresis using agarose of 1.8% (manufacturer: Boomerang Fuxin Technology (Beijing, China) Co., Ltd., Beijing, China). The total of PCR amplicons were purified with Agencourt AMPure XP Beads (Beckman Coulter, Indianapolis, IN, USA) and quantified using the Qubit dsDNA HS Assay Kit and Qubit 4.0 Fluorometer. After the individual quantification steps, amplicons were pooled in equal amounts to construct sequencing libraries, and the libraries that passed quality control were sequenced using an Illumina novaseq 6000 (Illumina, Santiago, CA, USA). The raw data were quality filtered using Trimmomatic (version 0.33). Primers were identified and removed using Cutadapt (version 1.9.1), followed by the splicing of double-end reads and the removal of the chimera (UCHIME, version 8.1) using USEARCH (version 10) to provide high quality sequences. The amplicon sequence variants (ASVs) were obtained by denoising the sequences using the dada2 method included in QIIME2 software, and finally the feature sequences were annotated taxonomically using a plain Bayesian classifier based on the ASVs sequence information with Silva.138 as the reference database.

2.6. Data Processing and Statistical Analysis

The differences between tree growth data and soil physical and chemical properties were analyzed using one-way ANOVA (LSD) in Excel and SPSS 26.0 (SPSS Inc., Chicago, IL, USA) software ($p < 0.05$ or $p < 0.01$). The results were then presented using histograms and box line plots created in Origin2022. Soil fungal and bacterial AVS Venn diagrams were produced using a highly customizable Venn and Euler diagram package in the R language, the visualization parameters were as follows: alpha = 0.50, fontfamily = 'serif', margin = 0.2. The inter-group variability of soil fungal and bacterial communities was compared using PCA analysis, and PCA plots were drawn using Origin 2022 based on the Bray–Curtis algorithm. Pearson correlation analysis was performed using SPSS 26.0 (SPSS Inc., Chicago, IL, USA) to examine the correlation between soil fungal and bacterial diversity and soil physicochemical properties. The relationship between soil physicochemical properties and fungal and bacterial community structure was determined using RDA analysis, RDA was conducted using CANOCO software for Windows (version 5.0). The bacterial and fungal community diversity index (Chao1) and richness index (Shannon) were analyzed by software QIIME2 (https://qiime2.org/, Date accessed: 1 October 2022).

3. Results

3.1. Tree Growth Status

The growth indexes of the trees were significantly different among different generations and showed an increasing trend with the addition of generations, as depicted in Figure 1. The average tree height of the first, second and third generation plantations were

24.63 m, 24.49 m and 28.53 m, respectively. The average tree height of the third generation plantation was significantly higher than that of first and second generation plantations ($p < 0.05$), with an increase of 15.87% and 16.53%, respectively. The mean diameter at breast height of the third generation plantation was significantly larger than that of the first generation plantation ($p < 0.05$), with an increase of 10.94%. The average individual-tree volume of the third generation plantation was 1.43 times that of the first generation plantation and 1.30 times that of the second generation plantation ($p < 0.05$).

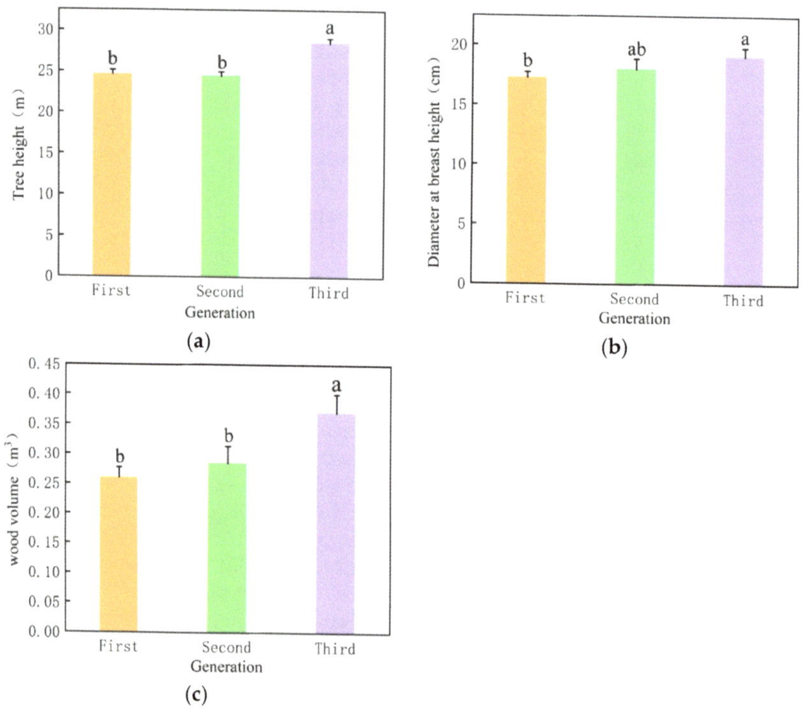

Figure 1. Average tree height (**a**), diameter at breast height (**b**), and individual-tree volume (**c**) in *Eucalyptus* plantations of different succession generations. Different lowercase letters indicate significant differences between successive planting generations ($p < 0.05$).

3.2. Changes in Soil Nutrients

Differences in soil properties were observed among generations ($p < 0.05$, Table 1). In the 0–20 cm soil layer, pH did not change significantly ($p < 0.05$) and OM content (27.91–39.30 $g \cdot kg^{-1}$) showed a significant upward trend ($p < 0.05$). TN content decreased, with the third generation plantation TN content (0.72 $g \cdot kg^{-1}$) significantly lower than that of the first generation plantation (1.06 $g \cdot kg^{-1}$). NH_4^+-N content was significantly lower than that of the first generation plantation (17.62 $mg \cdot kg^{-1}$) in both the second and third generation plantations ($p < 0.05$). NO_3^--N content was the lowest at 13.82 $mg \cdot kg^{-1}$ in the third generation plantation, significantly lower than that in the first and second generation plantations (15.21 $mg \cdot kg^{-1}$, 15.94 $mg \cdot kg^{-1}$) ($p < 0.05$). The TP content increased, and the first generation plantation of 0.35 $g \cdot kg^{-1}$ was significantly lower than that of the second and third generation forests (p <0.05). AP content showed a slight decrease and then an increase. The highest TK content was 2.62 $g \cdot kg^{-1}$ in the second generation plantation. The soil AK content (22.78–33.28 $g \cdot kg^{-1}$) decreases significantly in each generation. In the 20–40 cm soil layer, the contents of pH, TN and AK decreased slightly, but the content of NH_4^+-N, NO_3^--

N and TK increased. The soil OM, TP and AP content increased significantly ($p < 0.05$) and reached the maximum value in the third generation plantation.

Table 1. Effect of different successive planting generations on soil physicochemical properties of *Eucalyptus* plantations.

Indicator	Soil Layer/cm	First Generation	Second Generation	Third Generation
pH	0–20	4.76 ± 0.50 a	4.52 ± 0.10 a	4.51 ± 0.11 a
	20–40	4.62 ± 0.06 a	4.57 ± 0.06 a	4.50 ± 0.11 a
OM ($g \cdot kg^{-1}$)	0–20	27.91 ± 1.92 b	28.85 ± 1.08 b	39.30 ± 0.79 a
	20–40	26.60 ± 2.10 b	27.70 ± 1.26 b	36.96 ± 0.97 a
TN ($g \cdot kg^{-1}$)	0–20	1.06 ± 0.16 a	0.85 ± 0.01 ab	0.72 ± 0.04 b
	20–40	0.89 ± 0.09 a	0.85 ± 0.06 a	0.79 ± 0.20 a
NH_4^+-N ($mg \cdot kg^{-1}$)	0–20	17.62 ± 0.81 a	14.20 ± 0.42 b	14.44 ± 0.49 b
	20–40	14.64 ± 0.66 a	14.77 ± 0.84 a	16.31 ± 1.04 a
NO_3^--N ($mg \cdot kg^{-1}$)	0–20	15.21 ± 0.0.48 a	15.94 ± 0.42 a	13.82 ± 0.20 b
	20–40	14.54 ± 0.26 a	15.31 ± 0.59 a	14.85 ± 0.59 a
TP ($g \cdot kg^{-1}$)	0–20	0.35 ± 0.01 b	0.70 ± 0.49 a	0.73 ± 0.30 a
	20–40	0.38 ± 0.05 b	0.61 ± 0.01 a	0.66 ± 0.04 a
AP ($mg \cdot kg^{-1}$)	0–20	1.07 ± 0.05 a	0.98 ± 0.05 a	1.29 ± 0.15 a
	20–40	1.03 ± 0.01 b	1.06 ± 0.01 ab	1.10 ± 0.03 a
TK ($g \cdot kg^{-1}$)	0–20	2.47 ± 0.09 a	2.62 ± 0.49 a	2.19 ± 0.20 a
	20–40	2.70 ± 0.18 a	2.32 ± 0.53 a	3.18 ± 0.57 a
AK ($mg \cdot kg^{-1}$)	0–20	33.28 ± 1.03 a	27.29 ± 1.34 b	22.78 ± 1.14 c
	20–40	24.12 ± 1.96 a	22.44 ± 1.21 a	19.56 ± 1.96 a

All results in the table are the mean ± standard error. OM: organic matter, TN: total nitrogen, NH_4^+-N: ammoniacal nitrogen, NO_3^--N: nitrate nitrogen, TP: total phosphorus, AP: available phosphorus, TK: total potassium, AK: available potassium. Different lowercase letters indicate significant differences ($p < 0.05$).

3.3. Diversity and Structure of Soil Bacterial and Fungal Communities

The number of amplicon sequence variants (ASVs) measured for bacteria in the 0–20 cm soil layer of the plantation was 910, 992, and 993, for the first, second and third generation plantations, respectively, and the number of ASVs measured in the 20–40 cm soil layer was 923, 978, and 960, respectively, which increased with the addition of successive planting generations (Figure 2). The number of specific ASVs of bacteria was greater than the common number among the three generations, indicating that there were large differences in bacterial communities among them. The number of ASVs of fungi in the 0–20 cm soil layer was 323, 489, and 471, for the first, second and third generation plantations, respectively, and the number of ASVs in the 20–40 cm soil layer was 352, 545, and 381, respectively, all of which were the smallest in the first generation plantation. The number of fungal-specific ASVs was greater than the common number among the three generations, indicating that there were large differences in the fungal communities among them.

PCA analysis of soil bacterial and fungal community structure in different successive plantation generations is presented in Figure 3. The sample points for soil fungal and bacterial groups in the first, second, and third generation plantations were located in the same quadrant, suggesting that the composition of these three generations was quite similar within their respective groups. Conversely, sample points between the groups were scattered and distributed in different quadrants, indicating significant differences between the groups in terms of the structure of the bacterial and fungal communities across the three generations.

Table 2 demonstrates that the structural richness and diversity of soil bacterial and fungal communities increased significantly with successive planting generations. The Chao1 indices of soil bacteria and fungi increased with successive planting generations in the 0–20 cm soil layer and were significantly lower in the first generation plantation compared with the second and third generation plantations ($p < 0.05$). In the 20–40 cm soil layer, the soil bacterial Chao1 index was significantly higher in the second and third generation plantations compared with the first generation plantation ($p < 0.05$), while the soil fungal Chao1 index was highest in the second generation plantation and significantly lower in the first and third generation plantation compared with the second generation plantation ($p < 0.05$). The Shannon index of soil bacteria in both the 0–20 cm and 20–40 cm soil layers increased with successive planting generations, with the largest index observed in the second generation plantation, and a significantly lower index in the first generation plantation compared with the second and third generation plantations ($p < 0.05$). The Shannon index of soil fungi in both the 0–20 cm and 20–40 cm soil layers initially increased and then decreased with successive planting generations. In the 0–20 cm soil layer, the first generation plantation had a significantly lower index compared with the second and third generation plantations ($p < 0.05$), while in the 20–40 cm soil layer, the second generation plantation had a significantly larger index than the first and third generation plantations ($p < 0.05$).

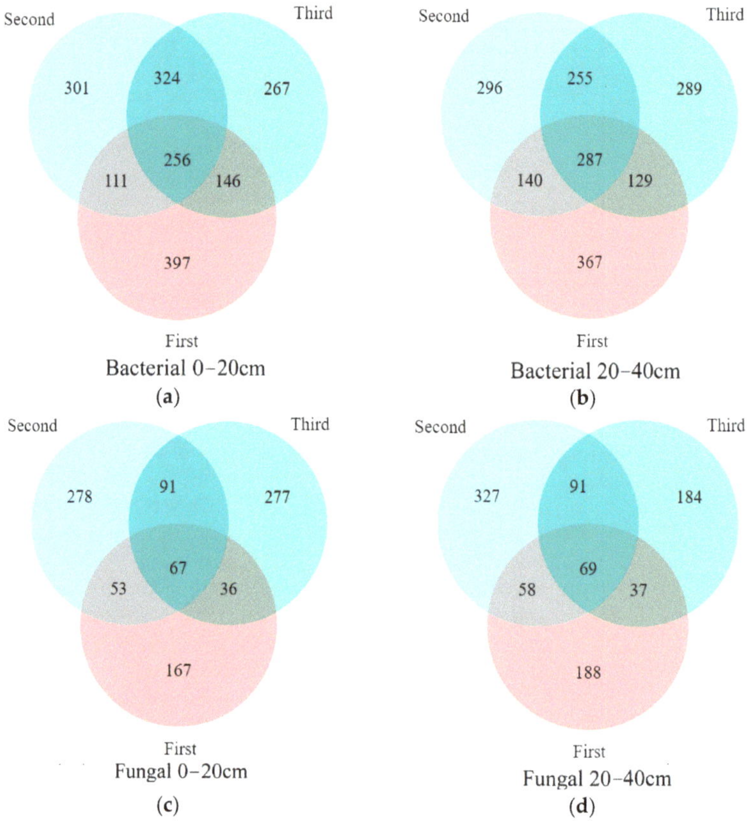

Figure 2. Venn diagram of ASV of soil bacteria (**a**,**b**) and fungi (**c**,**d**) in different successive planting generations.

Bacterial and fungal communities showed significant variations at the phylum level across different generations of successive plantation. Figure 4 reveals that the relative abundance of *Proteobacteria*, the dominant bacterial phylum, differed significantly between soil layers in successive planting generations ($p < 0.05$). Moreover, the relative abundance of *Acidobacteriota*, another dominant bacterial phylum, showed a highly significant difference ($p < 0.01$). Similarly, *Chloroflexi*, *Gemmatimonadota*, *Verrucomicrobiota*, *Myxococcota*, and other bacterial phyla displayed significant differences ($p < 0.05$). As for fungi, the relative abundance of *Ascomycota* increased significantly in the 20–40 cm soil layer with successive planting generations ($p < 0.05$). Furthermore, the dominant phylum *Basidiomycota* showed a significant difference between the two soil layers ($p < 0.05$), while *Chytridiomycota*, *Rozellomycota*, and *Glomeromycota* also exhibited significant differences among generations ($p < 0.05$).

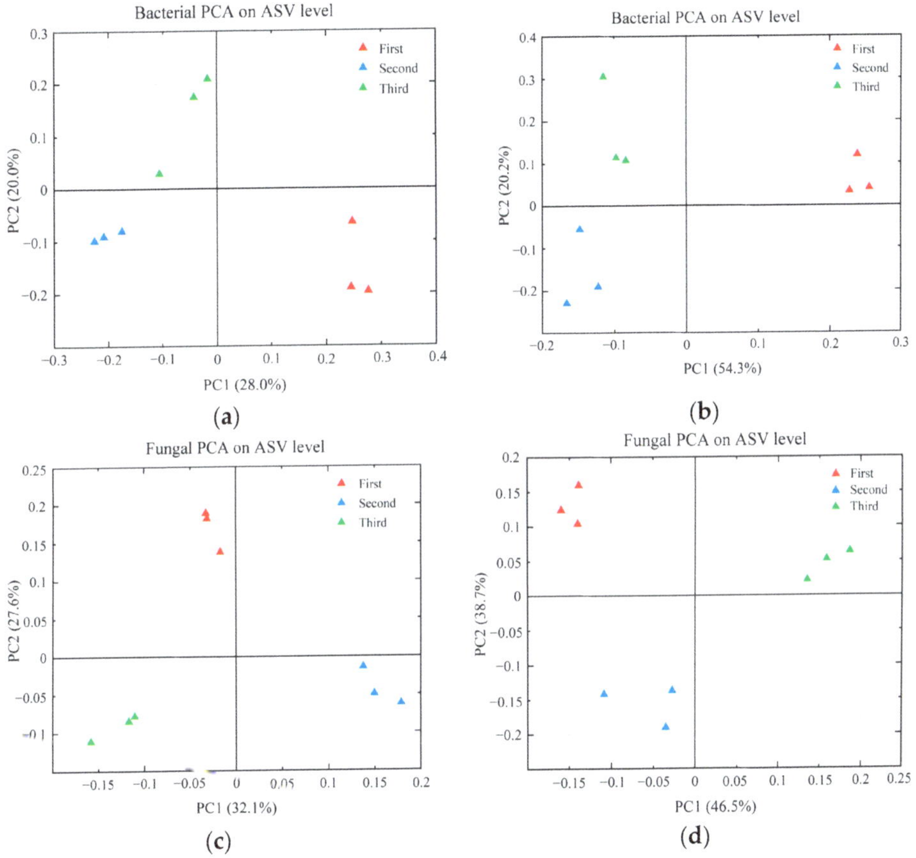

Figure 3. Principal component analysis of soil bacteria (**a**,**b**) and fungi (**c**,**d**) at ASV level in different successive generations of *Eucalyptus* plantations.

Table 2. Analysis of bacterial and fungal community structure diversity in different generations of *Eucalyptus* plantation.

Diversity	Index	Soil Layer/cm	First Generation	Second Generation	Third Generation
Bacterial	Chao1	0–20	417.67 ± 8.69 c	534.62 ± 11.59 a	487.73 ± 7.83 b
		20–40	436.58 ± 11.51 b	535.61 ± 15.32 a	507.33 ± 17.34 a
	Shannon	0–20	7.17 ± 0.21 B	8.27 ± 0.04 A	8.05 ± 0.07 A
		20–40	6.96 ± 0.02 B	8.21 ± 0.03 A	8.00 ± 0.09 A
Fungal	Chao1	0–20	142.78 ± 7.01 b	205.33 ± 8.86 a	205.50 ± 7.18 a
		20–40	163.25 ± 12.25 b	222.44 ± 11.33 a	172.61 ± 4.40 b
	Shannon	0–20	3.50 ± 0.30 b	5.40 ± 0.24 a	4.85 ± 0.19 a
		20–40	3.56 ± 0.31 b	6.19 ± 0.68 a	4.19 ± 0.36 b

All results in the table are the mean ± standard error. Different lowercase letters indicate significant differences ($p < 0.05$) and different capital letters indicate highly significant differences ($p < 0.01$).

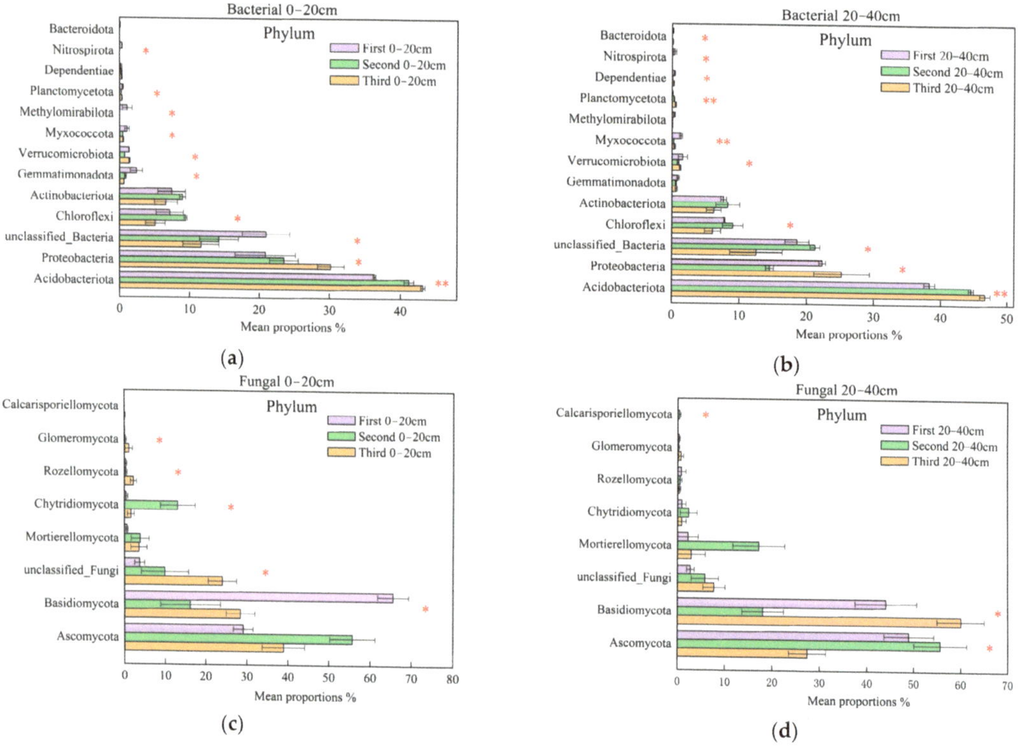

Figure 4. Community structure of bacteria (**a**,**b**) and fungi (**c**,**d**) at the phylum level. Column length indicates the mean relative abundance of the corresponding species, * indicates the statistical difference between treatments ($p < 0.05$), and ** indicates the statistical difference between treatments ($p < 0.01$).

3.4. Relationship between Soil Nutrients and the Diversity and Structure of Bacterial and Fungal Communities

Soil nutrients were significantly correlated with bacterial and fungal community diversity. As shown in Table 3, in the 0–20 cm soil layer, NH_4^+-N had a significant negative correlation ($-0.726 < r < -0.948$) with Chao1 and Shannon indices for both bacterial and fungal community diversity ($p < 0.01$). TP had a significant positive correlation ($0.766 < r < 0.896$) with Chao1 and Shannon indices for both bacterial and fungal community diversity ($p < 0.01$). AP showed significant positive correlation ($p < 0.05$) with bacterial Chao1 index in the 20–40 cm soil layer.

The RDA results show that, in the 0–20 cm soil layer (Figure 5a), the bacterial community structure was mainly driven by pH, TN, TP and NH_4^+-N factors. The first axis (RDA1) explains 56.59% of the total variation of bacterial phyla, the second axis (RDA2) explains 19.11% of the total variation of bacterial phyla, and both axes explain 75.70% of the total bacterial variation. The bacterial dominant phylum *Acidobacteriota* showed a strong positive correlation with TP and a strong negative correlation with TN, NO_3^--N, and pH. The bacterial dominant phylum *Proteobacteria* showed positive correlations with OM and AP. The bacterial dominant phylum *Chloroflexi* showed a positive correlation with NO_3^--N. *Gemmatimonadota*, *Methylomirabilota*, *Planctomycetota*, *Nitrospirota*, and *Bacteroidota* showed strong correlations with pH, TN, NO_3^--N, and AK.

Table 3. The Pearson correlation analysis of soil physicochemical properties and bacterial and fungal community diversity, n = 9.

Indicator	Soil Layer/cm	Bacterial		Fungal	
		Chao1	Shannon	Chao1	Shannon
pH	0–20	−0.643	−0.614	−0.524	−0.629
	20–40	−0.591	−0.201	−0.258	−0.223
OM ($g \cdot kg^{-1}$)	0–20	0.076	0.229	0.439	0.321
	20–40	0.182	0.275	−0.067	−0.317
TN ($g \cdot kg^{-1}$)	0–20	−0.423	−0.380	−0.637	−0.492
	20–40	−0.046	−0.286	−0.001	−0.215
NH_4^+-N ($mg \cdot kg^{-1}$)	0–20	−0.847 **	−0.948 **	−0.726 *	−0.883 **
	20–40	0.026	0.232	−0.013	−0.372
NO_3^--N ($mg \cdot kg^{-1}$)	0–20	0.180	−0.126	−0.129	−0.044
	20–40	0.038	0.400	0.356	0.033
TP ($g \cdot kg^{-1}$)	0–20	0.881 **	0.896 **	0.766 *	0.838 **
	20–40	0.356	0.570	0.268	0.383
AP ($mg \cdot kg^{-1}$)	0–20	−0.074	0.115	0.120	−0.222
	20–40	0.680 *	0.508	−0.033	0.238
TK ($g \cdot kg^{-1}$)	0–20	−0.076	0.061	0.074	0.035
	20–40	−0.208	−0.066	0.212	−0.202
AK ($mg \cdot kg^{-1}$)	0–20	−0.450	−0.416	−0.224	−0.570
	20–40	−0.248	−0.153	0.084	−0.435

** Significant correlation at 0.01 (bilateral), * correlation at 0.05 (bilateral).

In the 20–40 cm soil layer (Figure 5b), bacterial community structure was mainly driven by pH, OM, AP and AK factors. The first axis (RDA1) explains 69.27% of the variation and has the strongest positive correlation with pH. The second axis (RDA2) explains 14.64% of the variation and has the strongest positive correlation with AP. Both axes explain 83.91% of the variation in environmental factors and bacterial community structure. The bacterial dominant phylum *Acidobacteriota* was positively correlated with TP and negatively correlated with pH, AK, and TN. The bacterial dominant phylum *Chloroflexi* was strongly positively correlated with NO_3^--N.

The results of the RDA between the fungal community structure and soil physicochemical properties in the 0–20 cm soil layer are shown in Figure 5c. The first axis (RDA1) explains 59.42% of the variation, the second axis (RDA2) explains 17.54% of the variation, with the strongest positive correlations with TP. Both axes explain 76.96% of the variation in environmental factors and bacterial community structure. TN, NH_4^+-N, TP and TK are the main drivers of fungal community structure, mainly manifested in the strong positive correlation between NH_4^+-N and *Basidiomycota* and *Calcarisporiellomycota*, and a strong negative correlation with *Mortierellomycota*, *Mucoromycota*, and TP phosphorus. TK was positively correlated with *Olpidiomycota* and *Chytridiomycota*.

In the 20–40 cm soil layer (Figure 5d), the fungal community structure was mainly driven by OM, TP, TK, and AK factors. The first axis (RDA1) explains 73.17% of the variation, with positive contributions from most physical and chemical property factors, the second axis (RDA2) explains 14.66% of the variation, with a positive correlation with TP. Both axes together explain 87.83% of the variation in environmental factors and bacterial community structure. *Mortierellomycota*, *Chytridiomycota*, and *Olpidiomycota* were positively correlated with AK, NO_3^--N, and pH, while they were negatively correlated with TP, AP, and TK. *Ascomycota*, *Mucoromycota*, *Calcarisporiellomycota*, and *Rozellomycota* were negatively correlated with OM, NH_4^+-N, and TN.

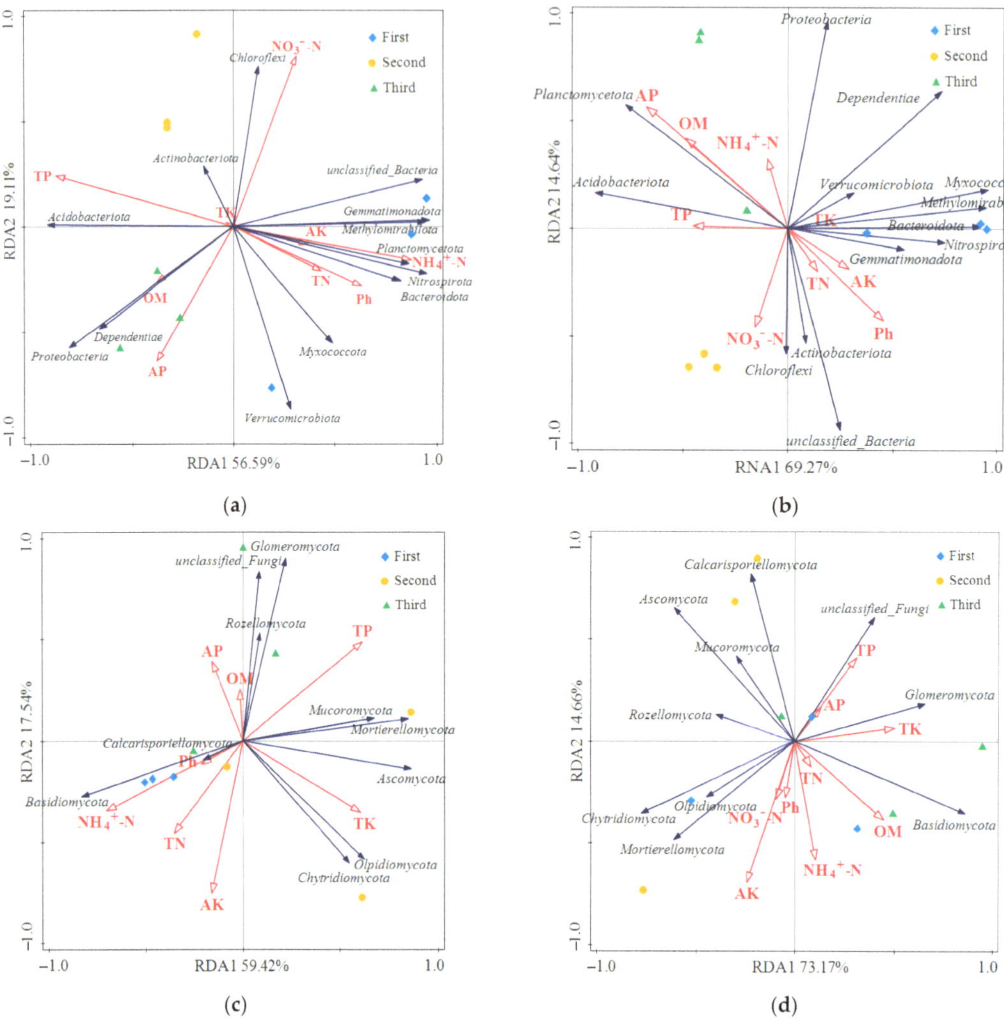

Figure 5. RDA analysis of soil physical and chemical properties and bacterial and fungal community structure composition, (**a**,**c**) are the redundancy analyses of bacteria and fungi in soil at 0–20 cm respectively, and (**b**,**d**) are the redundancy analyses of bacteria and fungi in the soil at 20–40 cm respectively.

4. Discussion

4.1. Variation in Growth of Successive Plantations of Eucalyptus

Eucalyptus plantations with an eight-year cultivation period showed a significant increase in growth index with an increase of successive planting generations ($p < 0.05$). The third generation of plantation had the highest average tree height, diameter at breast height, and wood volume per tree. Compared with the first generation of plantation, there was an increase of 15.87%, 10.94%, and 42.58% in these parameters, respectively. This indicates that, with a cultivation period of eight years, the successive plantation increased growth of *Eucalyptus* plantations.

The significant increase in stand growth at the study site can be attributed to several factors. Firstly, the Guangxi state-owned Dongmen Forest Farm has excellent maintenance and management practices for *Eucalyptus* plantations and uses high-quality strains for planting. Secondly, the second and third generation plantations have well-developed root systems due to budding from the previous generation, which reduces their demand for soil nutrients. *Eucalyptus* is also known for its fast growth rate, and the growth indexes of the third generation plantation are exceptional under adequate nutrient supply. Thirdly, as successive planting generations are added, the content of OM, TP, and AP increases. OM is a significant source of plant nutrients that retains water and absorbs cations [23], and it can also serve as food for soil microorganisms, which enhances soil activity and physical properties. This further promotes forest tree growth [24]. TP and AP are essential macronutrients required for plant growth and metabolism, and they are also critical components of the photosynthesis process [25,26]. Most activities, such as plant growth, respiration and reproduction, depend on the phosphorus level in the soil [27], so the rise of TP and AP content also promotes the growth of trees.

4.2. Changes in Soil Nutrients in Successive Plantations of Eucalyptus

The study found that, after an eight-year period of cultivating *Eucalyptus*, the negative impact of successive planting on soil nutrients decreased and some nutrients were returned. In the 0–20 cm soil layer, N content decreased significantly ($p < 0.05$), indicating that N supplementation is necessary during successive planting. However, OM and TP content significant increased with the increase of successive planting generations ($p < 0.05$). In contrast, in the 20–40 cm soil layer, most indicators showed insignificant changes, with only the soil OM, TP, and AP content increased significantly. This could be because the surface soil layer (0–20 cm) is more affected by external environmental conditions compared with the deeper layer (20–40 cm), which has a more stable soil environment [28]. The increase in some nutrient content in both soil layers can be attributed to the eight-year cultivation period allowing for some repair of the damage caused by full reclamation methods. Additionally, as the age of the plantation trees increased, the biodiversity of the understory also increased [29,30]. As a result, the forest's soil nutrient content was replenished [29,31].

4.3. Diversity and Structure Change of Bacterial and Fungal Communities in Continuous Eucalyptus Plantation

The diversity of soil bacteria and fungi in *Eucalyptus* plantations differed significantly among successive planting generations ($p < 0.05$). The Chao1 index for bacteria was highest in both soil layers of the second generation plantation, increasing by 28.00% and 22.91% compared with the first generation plantation. The bacterial Shannon index was also highest in the second generation plantation. Additionally, fungal community richness and diversity were best in the second generation plantation and lowest in the first generation plantation. PH, NH_4^+-N, TP and AP factors were found to mainly drive bacterial community structure, while NH_4^+-N, TP, TK, and OM factors drove fungal community structure, as determined by RDA results. Correlation analysis showed that changes in soil bacterial and fungal community diversity were significantly related to NH_4^+-N, TP, and AP content ($p < 0.05$), with NH_4^+-N showing mainly negative correlation, indicating a lesser diversity index in higher NH_4^+-N soils. The diversity and richness of soil bacterial and fungal communities in different successive planting generations of *Eucalyptus* plantations may be affected by microclimate changes, litter production, secretions, and root symbionts, as well as root-associated microorganisms [32,33]. The reasonable cultivation period may also indirectly contribute to an increase in understory vegetation diversity and soil faunal diversity [34].

The study found significant changes in the structure of soil bacterial and fungal communities in *Eucalyptus* plantations across different succession generations ($p < 0.05$). *Acidobacteriota*, *Proteobacteria*, and *Chloroflexi* showed a significant increase in relative abundance with successive planting generations. Moreover, *Acidobacteriota* was positively correlated with TP, *Proteobacteria* with OM and AP, and *Chloroflexi* with NO_3^--N, which are the primary drivers of changes in bacterial community structure. *Basidiomycota* and

Ascomycota are the two main fungal groups in forest soils [35–37], with *Basidiomycota* being capable of completely decomposing lignin in soil litter into water and carbon dioxide [38]. The relative abundance of *Ascomycota* increased slightly and then decreased significantly with increasing generations in the 20–40 cm soil layer. The relative abundance of *Ascomycota* was significantly lower in the third generation plantation than in the second generation plantation. The relative abundance of the dominant phylum *Basidiomycota* decreased significantly with the addition of successive planting generations, driven mainly by $NH_4{}^+$-N, TN, and pH.

5. Conclusions

In conclusion, there were significant differences in tree growth, soil physical and chemical properties, and soil fungal bacterial structure between successive plantation generations of *Eucalyptus* plantations with an eight-year cultivation period. The results of the study show that the growth of the stands became better with the addition of successive generations. The diversity and richness of soil bacterial and fungal communities were significantly enhanced. The content of some nutrients, such as OM, TP, and AP, also increased with the addition of successive planting generations. Consequently, successive plantings of *Eucalyptus* plantations during the eight years of cultivation period is reasonable, and it improves the productivity of *Eucalyptus* plantations. It brings efficient economic benefits and alleviates soil quality problems caused by short-cycle successive planting of *Eucalyptus* and further solves social and environmental problems. This demonstrates that the use of more scientific and quantitative plantation management measures can maintain the trend of increasing productivity and soil fertility of eucalyptus plantations even after multiple generations of successive plantings. This study provides an empirical basis for the multi-generational succession management of *Eucalyptus* in Guangxi and promotes the sustainable development of the Guangxi *Eucalyptus* plantation forest industry.

Author Contributions: Conceptualization, P.W. and Q.D.; methodology, Q.D.; software, Q.D.; validation, P.W., Q.D. and T.W.; formal analysis, Y.F.; investigation, Q.D.; resources, P.W.; data curation, Q.D.; writing—original draft preparation, Q.D.; writing—review and editing, P.W.; visualization, Q.D.; supervision, P.W.; project administration, P.W.; funding acquisition, P.W. All authors have read and agreed to the published version of the manuscript.

Funding: This research was funded by Guangxi Science and Technology Base and Talent Special Project: Research on New Varieties of *Eucalyptus* and Efficient Ecological Cultivation Technology 2022AC12010.

Data Availability Statement: All graphs and data obtained or generated during the investigation appear in the published article.

Conflicts of Interest: The authors declared that they have no conflicts of interest with this work. We declare that we do not have any commercial or associative interest that represents a conflict of interest in connection with the work submitted.

References

1. Xu, Y.; Li, C.; Zhu, Y.; Wang, Z.; Zhu, W.; Wu, L.; Du, A. The shifts in soil microbial community and association network induced by successive planting of *Eucalyptus* plantations. *For. Ecol. Manag.* **2022**, *505*, 119877. [CrossRef]
2. Xu, Y.; Du, A.; Wang, Z.; Zhu, W.; Li, C.; Wu, L. Effects of different rotation periods of *Eucalyptus* plantations on soil physiochemical properties, enzyme activities, microbial biomass and microbial community structure and diversity. *For. Ecol. Manag.* **2020**, *456*, 117683. [CrossRef]
3. Li, X.; Yin, X.; Wang, Z.; Fan, W. Litter mass loss and nutrient release influenced by soil fauna of *Betula ermanii* forest floor of the Changbai Mountains, China. *Appl. Soil Ecol.* **2015**, *95*, 15–22. [CrossRef]
4. Xie, Y.; Arnold, R.J.; Wu, Z.; Chen, S.; Du, A.; Luo, J. Advances in eucalypt research in China. *Front. Agric. Sci. Eng.* **2017**, *4*, 380. [CrossRef]
5. Arnold, R.J.; Xie, Y.J.; Luo, J.Z.; Wang, H.R.; Midgley, S.J. A tale of two genera: Exotic *Eucalyptus* and Acacia species in China. 2. Plantation resource development. *Int. For. Rev.* **2020**, *22*, 153–168. [CrossRef]
6. Zohar, Y.; Gafni, A.; Morris, J.; Shalhevet, S. Eucalyptus plantations in Israel: An assessment of economic and environmental viability. *New For.* **2008**, *36*, 135–157. [CrossRef]

7. Wei, Y.C.; Ouyang, Z.Y.; Miao, H.; Zheng, H. Exotic *Pinus carbaea* causes soil quality to deteriorate on former abandoned land compared to an indigenous *Podocarpus* plantation in the tropical forest area of southern China. *J. For. Res.* **2009**, *14*, 221–228. [CrossRef]
8. Lebenya, R.M.; van Huyssteen, C.W.; du Preez, C.C. Change in soil organic carbon and nitrogen stocks eight years after conversion of sub-humid grassland to Pinus and *Eucalyptus* forestry. *Soil Res.* **2018**, *56*, 318. [CrossRef]
9. Ellis, R.C. The respiration of the soil beneath some *Eucalyptus* forest stands as related to the productivity of the stands. *Soil Res.* **1969**, *7*, 349–357. [CrossRef]
10. Zhou, X.; Zhu, H.; Wen, Y.; Goodale, U.M.; Zhu, Y.; Yu, S.; Li, C.; Li, X. Intensive management and declines in soil nutrients lead to serious exotic plant invasion in *Eucalyptus* plantations under successive short-rotation regimes. *Land Degrad. Dev.* **2019**, *31*, 297–310. [CrossRef]
11. Zhu, L.; Wang, X.; Chen, F.; Li, C.; Wu, L. Effects of the successive planting of *Eucalyptus* urophylla on soil bacterial and fungal community structure, diversity, microbial biomass, and enzyme activity. *Land Degrad. Dev.* **2019**, *30*, 636–646. [CrossRef]
12. Gu, S.; Hu, Q.; Cheng, Y.; Bai, L.; Liu, Z.; Xiao, W.; Gong, Z.; Wu, Y.; Feng, K.; Deng, Y.; et al. Application of organic fertilizer improves microbial community diversity and alters microbial network structure in tea (*Camellia sinensis*) plantation soils. *Soil Tillage Res.* **2019**, *195*, 104356. [CrossRef]
13. Mganga, K.Z.; Razavi, B.S.; Kuzyakov, Y. Land use affects soil biochemical properties in Mt. Kilimanjaro region. *Catena* **2016**, *141*, 22–29. [CrossRef]
14. Zheng, H.; Ouyang, Z.Y.; Wang, X.K.; Fang, Z.G.; Zhao, T.Q.; Miao, H. Effects of regenerating forest cover on soil microbial communities: A case study in hilly red soil region, Southern China. *For. Ecol. Manag.* **2005**, *217*, 244–254. [CrossRef]
15. Gao, Q.; Hasselquist, N.J.; Palmroth, S.; Zheng, Z.; You, W. Short-term response of soil respiration to nitrogen fertilization in a subtropical evergreen forest. *Soil Biol. Biochem.* **2014**, *76*, 297–300. [CrossRef]
16. Wu, Z.; Haack, S.E.; Lin, W.; Li, B.; Wu, L.; Fang, C.; Zhang, Z. Soil Microbial Community Structure and Metabolic Activity of Pinus elliottii Plantations across Different Stand Ages in a Subtropical Area. *PLoS ONE* **2015**, *10*, e135354. [CrossRef]
17. Xu, Y.; Ren, S.; Liang, Y.; Du, A.; Li, C.; Wang, Z.; Zhu, W.; Wu, L. Soil nutrient supply and tree species drive changes in soil microbial communities during the transformation of a multi-generation *Eucalyptus* plantation. *Appl. Soil Ecol.* **2021**, *166*, 103991. [CrossRef]
18. Reijonen, I.; Metzler, M.; Hartikainen, H. Impact of soil pH and organic matter on the chemical bioavailability of vanadium species: The underlying basis for risk assessment. *Environ. Pollut.* **2016**, *210*, 371–379. [CrossRef] [PubMed]
19. Walkley, A. An Examination of Methods for Determining Organic Carbon and Nitrogen in Soils. (With One Text-figure.). *J. Agric. Sci.* **1935**, *25*, 598–609. [CrossRef]
20. Murphy, J.; Riley, J.P. A modified single solution method for the determination of phosphate in natural waters. *Anal. Chim. Acta* **1962**, *27*, 678–681. [CrossRef]
21. Olsen, S.R. *Estimation of Available Phosphorus in Soils by Extraction with Sodium Bicarbonate*; Miscellaneous Paper Institute for Agricultural Research: Washington, DC, USA, 1954; p. 19.
22. Lu, K. *Analytical Methods of Soil and Agricultural Chemistry*; China Agriculture Science Press: Beijing, China, 1999; pp. 107–240.
23. Tautua, A.; Bamidele, M.W.; Onigbinde, A.O.; Ere, D. Assessment of some heavy metals and physicochemical properties in surface soils of municipal open waste dumpsite in Yenagoa, Nigeria. *Afr. J. Environ. Sci. Technol.* **2014**, *8*, 41–47. [CrossRef]
24. Brady, N.C. *The Nature and Properties of Soils*; McMillan: New York, NY, USA, 1996; Volume 11, p. 621.
25. Singh, D.P.; Rathore, M.S. Available nutrient status and their relationship with soil properties of Aravalli mountain ranges and Malwa Plateau of Pratapgarh, Rajasthan, India. *Acad. J.* **2013**, *8*, 5096–5103.
26. Jain, S.A.; Jagtap, M.S.; Patel, K.P. Physico-Chemical Characterization of farmland Soil used in some villages of Lunawada Taluka. Dist: Mahisagar (Gujarat) India. *Int. J. Sci. Res. Publ.* **2014**, *4*, 1–5.
27. Sayyed, M. Assessment of Macro and Micronutrients in Soils from Panvel Area, Maharashtra, India. *Univers. J. Environ. Res. Technol.* **2013**, *3*, 72–78.
28. Zhao, H.; Zheng, W.; Zhang, S.; Gao, W.; Fan, Y. Soil microbial community variation with time and soil depth in Eurasian Steppe (Inner Mongolia, China). *Ann. Microbiol.* **2021**, *71*, 21. [CrossRef]
29. Sun, H.; Wang, Q.; Liu, N.; Li, L.; Zhang, C.; Liu, Z.; Zhang, Y. Effects of different leaf litters on the physicochemical properties and bacterial communities in Panax ginseng -growing soil. *Appl. Soil. Ecol.* **2017**, *111*, 17–24. [CrossRef]
30. Tiarks, A.; Elliottsmith, M.; Stagg, R.; Nambiar, E.; Tiarks, A.; Cossalter, C.; Ranger, J. *Loblolly Pine (Pinus Taeda) Plantations in the Semitropical Southeastern United States. Site Management & Productivity in Tropical Plantation Forests: A Progress Report Workshop*; Center for International Forestry Research: Bogor, Indonesia, 2000; pp. 101–103.
31. Mendham, D.S.; Sankaran, K.V.; O'Connell, A.M.; Grove, T.S. Eucalyptus globulus harvest residue management effects on soil carbon and microbial biomass at 1 and 5 years after plantation establishment. *Soil Biol. Biochem.* **2002**, *34*, 1903–1912. [CrossRef]
32. Colombo, F.; Macdonald, C.A.; Jeffries, T.C.; Powell, J.R.; Singh, B.K. Impact of forest management practices on soil bacterial diversity and consequences for soil processes. *Soil Biol. Biochem.* **2016**, *94*, 200–210. [CrossRef]
33. Burton, J.; Chen, C.; Xu, Z.; Ghadiri, H. Soil microbial biomass, activity and community composition in adjacent native and plantation forests of subtropical Australia. *J. Soil Sediments* **2010**, *10*, 1267–1277. [CrossRef]

34. Wu, J.; Liu, Z.; Wang, X.; Sun, Y.; Zhou, L.; Lin, Y.; Fu, S. Effects of understory removal and tree girdling on soil microbial community composition and litter decomposition in two Eucalyptus plantations in South China. *Funct. Ecol.* **2011**, *25*, 921–931. [CrossRef]
35. Qiao, S.S.; Zhou, Y.N.; Chai, B.F.; Jia, T.; Li, C. Characteristics of Fungi Community Structure and Genetic Diversity of Forests in Guandi Mountains. *Huan Jing Ke Xue=Huanjing Kexue* **2017**, *38*, 2502–2512.
36. Ji, L.; Ni, K.; Ma, L.; Chen, Z.; Guo, S. Effect of different fertilizer regimes on the fungal community of acidic tea-garden soil. *Acta Ecol. Sin.* **2018**, *38*, 8158–8166.
37. Chen, Y.; Xu, T.; Veresoglou, S.D.; Hu, H.; Hao, Z.; Hu, Y.; Liu, L.; Deng, Y.; Rillig, M.C.; Chen, B. Plant diversity represents the prevalent determinant of soil fungal community structure across temperate grasslands in northern China. *Soil Biol. Biochem.* **2017**, *110*, 12–21. [CrossRef]
38. Bi, C.X.; Rong, S.Y.; Yang, H.X.; Ning, H.L.; Ming, L.Y.; Zhen, F.S.; Hui, G.Y.; Wei, X. Basidiomycetous laccase gene diversity in two subtropical forest soils. *Chin. J. Appl. Ecol.* **2011**, *22*, 2699–2704.

sustainability

MDPI

Article

Forest Fires, Stakeholders' Activities, and Economic Impact on State-Level Sustainable Forest Management

Stjepan Posavec [1,*], Damir Barčić [1], Dijana Vuletić [2], Višnjica Vučetić [3,†], Ivana Čavlina Tomašević [4] and Špela Pezdevšek Malovrh [5]

1 Faculty of Forestry and Wood Technology, University of Zagreb, 10000 Zagreb, Croatia; dbarcic@sumfak.unizg.hr
2 Croatian Forest Research Institute, 10450 Jastrebarsko, Croatia; dijanav@sumins.hr
3 Independent Researcher, 21450 Hvar, Croatia; visnjica.vucetic@gmail.com
4 Croatian Meteorological and Hydrological Service, 10000 Zagreb, Croatia; ivana.tomasevic@cirus.dhz.hr
5 Department of Forestry and Renewable Forest Resources, Biotechnical Faculty, University of Ljubljana, 1000 Ljubljana, Slovenia; spela.pezdevsekmalovrh@bf.uni-lj.si
* Correspondence: sposavec@sumfak.unizg.hr
† Retired.

Abstract: In Europe, forest fires are a serious and constant threat. They destroy forests and forest land, causing damage, financial loss, and long-lasting impacts on forest ecosystem services. There are several ways to decrease the number of forest fires, including continuous investment in fire prevention measures and the intensive implementation of adaptive sustainable forest management measures, which need additional financial resources. In many cases, forest management activities in karst forests are not implemented in a timely manner and in coordination with other stakeholders. A comprehensive study about the impact of forest fires on different economic activities (tourism and protected areas) is not currently available. In this study, the legislative framework in Croatia was analysed in relation to the fire protection activities and jurisdictions of different institutions. From data collected in the period 2013–2020, the first-age class afforestation costs and growing stock assortment value were calculated, and the non-wood forest functions were estimated. The aforementioned data were further compared to the Fire Weather Index (FWI) and Seasonal Severity Rating (SSR) of fire seasons. The total estimated damage is EUR 326,810,724.72. The research emphasises the need for the implementation of cross-sectoral forest policy measures. The state forest company should allocate more financial resources for biological forest restoration in the future. Forest management practices should implement climate-adaptable silviculture measures to preserve forest and forest land.

Keywords: forest policy; management costs; forest ecosystem services; forest protection; Fire Weather Index

Citation: Posavec, S.; Barčić, D.; Vuletić, D.; Vučetić, V.; Čavlina Tomašević, I.; Pezdevšek Malovrh, Š. Forest Fires, Stakeholders' Activities, and Economic Impact on State-Level Sustainable Forest Management. *Sustainability* **2023**, *15*, 16080. https://doi.org/10.3390/su152216080

Academic Editor: Simon Bell

Received: 13 September 2023
Revised: 9 October 2023
Accepted: 30 October 2023
Published: 18 November 2023

1. Introduction

Current forest-fire management policies are mainly focused on fire suppression and less adapted to climate change. Forest management should incorporate species selection considering flammability, fire resistance, and resilience, and the adoption of silvicultural practices that decrease the fire hazard [1].

Recognising the complexity of the negative impacts of climate change on different sectors, including forestry, the European Union (EU) has emphasised the importance of and the need to develop adaptive sustainable forest management regimes, appropriate institutional frameworks, and supportive measures to combat climate change [2]. The European Green Deal [3] aims to bring into effect a set of policy initiatives proposed by the European Commission to adapt to climate change due to the estimation that 60% of global ecosystem services have deteriorated, contributing to a significant rise in the number of floods and major wildfires on all continents [4]. The Green Deal is an integral part of the

Commission's strategy to implement the United Nations 2030 Agenda and Sustainable Development Goals. The UN Sustainable Development Goals (SDGs) encompass a broad range of issues [5], among which managing climate change was set as a priority (Goal 13—Climate action: Take urgent action to combat climate change and its impacts, and Goal 15—Life on land with sustainable use of terrestrial ecosystems, and sustainably manage forests). The new EU forest strategy for 2030 presents a vision and proposes actions to improve the quantity and quality of EU forests and strengthen their protection, restoration, and resilience. Its aim is to adapt Europe's forests to the new conditions, weather extremes, and high uncertainty brought about by climate change [6]. The aim of the EU biodiversity strategy for 2030 is to put Europe's biodiversity on the path to recovery by 2030 for the benefit of the people, the climate, and the planet [7]. Therefore, a variety of policies and measures for climate change mitigation were developed at international, EU, and national levels (for example, the EU Joint Research Centre communication for 2021; The Paris Protocol, and COP26—a blueprint for tackling global climate change beyond 2020; a Roadmap for moving to a competitive low-carbon economy in 2050; and the 2030 Framework for Climate and Energy Policies, the United Nations Framework Convention on Climate Change). In recent decades, natural disturbances, such as wildfires, windstorms, and insect outbreaks, have severely disrupted large areas of the world's forests at an unprecedented rate [8]. In addition to anthropogenic climate change, southern Europe has experienced strong land cover and land use changes in recent decades, including the progressive abandonment of agricultural land and activities [9]. In Europe, wildfires are one of the most important disturbance agents. The average annual area covered by forest fires was 213,000 ha between 1950 and 2000 [10].

Forest fires are one of the dominant disturbances in the forests of Croatia. Numerous natural disasters (fires, floods, ice storms, and windstorms) during the period 2014–2020 caused significant damage in the forestry sector, requiring substantial investment to preserve and repair these threatened ecosystems [11,12]. Croatia, as a Mediterranean country, is increasingly feeling the effects of climate change. Longer and more intense dry periods create ideal conditions for the development of forest fires, posing an increasing challenge in the organisation and cost of firefighting measures. Fire frequency and burn severity have had a negative impact on the recruitment of pine seedlings in the pine ecosystem forests in the Iberian Peninsula [13]. In Spain (Cordoba), suppression costs increased by between 65.67% and 86.73% in the last decade [14]. The fire environment can be represented as a triangle where the sides represent fuel, topography, and weather [15]. There is a need to analyse flammability and the spread of fire within ecosystems, especially in coastal areas [16]. The European Commission Emergency Response Coordination Centre (ERCC) monitors forest-fire risks and emergencies across Europe [17]. It is supported by national and European monitoring services, such as the European Forest-Fire Information System (EFFIS) [18]. The EFFIS has estimated that on average, about 65,000 fires occur in Europe every year, burning approximately half a million hectares of wildland and forest areas. Most of the burnt area (over 85%) is in the European Mediterranean region. In 2020, fires of over 30 ha affected 20 EU Member States, burning 339,489 ha in total, which is a slight increase from the year before [19]. Therefore, in 2019, the European Commission adopted [20] a goal to strengthen European civil protection, by enhancing prevention and preparedness actions to alleviate the effect of wildfires in the extended European region. Moreover, Europe is developing a pan-European approach to wildfire risk assessment with different variables regarding danger and vulnerability [21]. In 2021, the European Commission adopted a new EU Strategy on Adaptation to Climate Change [22]; due to climate change, droughts cause cascading effects and increase the volume of dry fuel available for forest fires. Forest fires create additional silviculture work and damage to forests, which affects the long-term management and business plans of companies managing these forests, making integral forest management even more challenging. In the Republic of Croatia, more than 250,000 ha of forest has been burnt over the past decade (2010–2020) [23]. Annual burnt areas have increased over the past 60 years, particularly during extremely dry years such as

2017 [24,25]. According to the Seventh National Report of Croatia and the Third Two-Year Report of the Republic of Croatia According to the United Nations Framework Convention on Climate Change [26], it is necessary to consider the long-term resilience and natural capacity for ecosystem adaptation caused by the unprecedented combination of climatic variables, extreme weather, and natural occurrences (fires and breakages by ice, snow, and wind). The Mediterranean region is a major climate change hotspot [27], due to warming of approximately 1.3 °C, which is above the global average [28]. Croatia, as a Central European and Mediterranean country, is increasingly experiencing the consequences of climate change. Longer and more intense dry periods (Figure 1) create ideal conditions for the ignition and spread of forest fires, which results in increasing challenges in the organisation and costs of firefighting measures.

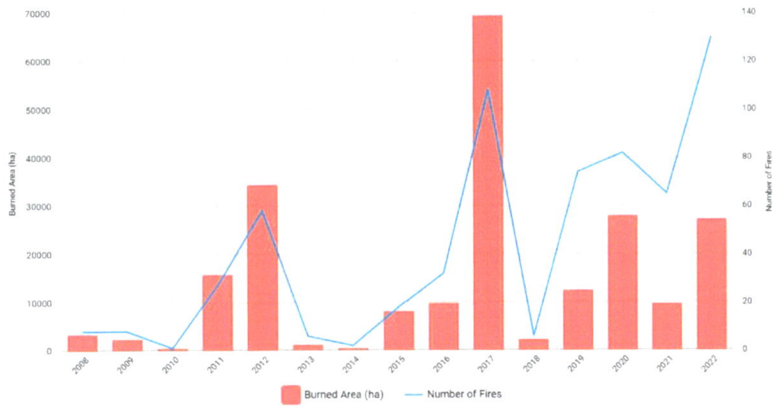

Figure 1. Distribution of burned areas and number of fires in Croatia in the period 2008–2022 (Source: EFFIS—Statistics Portal (https://effis.jrc.ec.europa.eu/apps/effis.statistics/, accessed on 16 March 2023) [29]).

The Republic of Croatia has been participating in the EU Civil Protection Mechanism since 2009, emphasising the cooperation of Mediterranean countries in the field of firefighting, where the state administration body is responsible for firefighting activities and coordinating assistance to other countries in case of open fires on their territory when they seek international assistance through the EU Monitoring and Information Centre in Brussels. For the purposes of the MEDforex project, the total monetary value of Croatia's forests was calculated, and the results showed that approximately 15% of these forests are Mediterranean forests that are most vulnerable to fire [30]. The importance of the socio-economic impact of forest ecosystems in Croatia was further outlined by Krznar [31], Vuletić et al. [32], and Posavec [33]. Forest ecosystem degradation after fire significantly impacts the loss of a range of products and services previously provided by those ecosystems. Degraded habitats demand increased management costs and long-term investments in the revitalisation and protection of forests after a fire. Unfortunately, future climate change [34] will also increase the fire risk, particularly in Mediterranean forests. Decision makers and policy makers in forestry should closely follow international initiatives and ensure implementation in a range of sectors on the ground.

The aim of this study was to use content analysis to review existing legislative frameworks related to the financing of forest-fire protection activities and the jurisdictions of decision makers in Croatia. Based on the data regarding the number of forest fires and total burnt areas in the period 2013–2020, damage to the first-age class, growing stock, and non-wood forest functions for the state forest company were calculated.

The purpose of this research was therefore as follows: (a) to identify policy regulating areas regarding forest-fire protection stakeholders, with a review of the Forest Law and other related laws which define fire protection measures and activities; (b) to explore the characteristics and obligations of different stakeholders which are prescribed in the implementation of special fire protection measures; (c) to calculate the damages and economic impact of forest fires on state forest company management; and (d) to compare extreme weather conditions and the potential meteorological risk of forest fires using the Fire Weather Index (FWI) and Seasonal Severity Rating (SSR) of fire seasons in the period 2013–2020.

2. Materials and Methods

2.1. Research Area

The research area encompasses all state forests in the Republic of Croatia, managed by the state-owned company Croatian Forests Ltd. (Hrvatske šume d.o.o., Zagreb, Croatia). The total forest area covers 2,759,039.05 ha. State-owned forests cover an area of 2,097,318.16 ha (76% of the total forest area), while private forests cover an area of 661,720 ha (24% of the total forest area). The total growing stock of the state forest company amounts to 315.8 million m^3 [35]. Croatian Forests Ltd. is responsible for the management of most (97%) of the state-owned forests and forest areas. The remaining 3% falls under the control of other public institutions, such as municipalities, the military, etc. The main tree species are beech (38%), common oak (14%), sessile oak (8%), fir (9%), hornbeam (8%), ash (4%), and spruce (2%) [36]. Forest fires are primarily an issue in the Mediterranean region. These areas are primarily covered by pubescent and holm oak forests which are indigenous vegetation, with significant areas of allochthonous forests and plantations of Aleppo pine. The forests and forestlands of the Croatian Mediterranean region cover an area of 662,000 ha, in which sub-mediterranean forests cover 457,000 ha, while EU-Mediterranean forests cover an area of 120,000 ha. In addition, there are 85,000 bare forestlands [37]. These forests are of low commercial value but are highly valuable for the preserving and provisioning of ecosystem services. Due to previous fires and mismanagement, most of the Mediterranean forests are in degraded forest forms (maquis, garrigues) that are highly vulnerable to fire, which in turn causes further degradation.

2.2. Methodology

In methodological terms, the paper is divided into three parts. The first part analyses the legislative framework, while the second part assesses the damage from forest fires in the state forest company. In the third part, the potential meteorological risk of forest fires is analysed for Croatia in the period 2013–2022, particularly for two extremely different weather years, 2014 and 2017.

2.2.1. Forestry Legislation and Fire Protection Regulation

The reviewed documents consist of forestry legislation and fire protection regulations: the Forest Law [38], the Fire Protection Law [39], the Fire Fighting Law [40], and the Law on Agricultural Land [41]. Content analysis [42] was used to identify legal acts and articles which regulate fire protection. The legislation documents were collected and reviewed to define regulation areas, legislation, and measures which include fire protection.

2.2.2. Assessment of Forest-Fire Damages

In the second part of the article, the number of forest fires, burned area (ha), and calculations of the amount of damage on wood volume, amount of damage in first-age class, damage on non-wood forest functions are presented. Data regarding the number of forest fires and burned areas were taken from the business report of the state forest company (Croatian Forests Ltd., Zagreb, Croatia) for the period 2013–2022 [36]. To calculate growing stock values, it is necessary to consider the average quantity of growing stock in rotation, price on the forest road from the company price list of the main forest products, and the exploitation costs from Croatian Forests Ltd.'s price list of calculated, internal, and external sales prices; labour costs; and costs of machinery and vehicles. It is also necessary to determine the number of years in the rotation. The stumpage value is the result of the function of price of the forest products and the cost of their exploitation. A higher market price for wood products has a favourable effect on increasing the value of forests, and vice versa; a lower market price for products will reduce forest value. The compensation calculation (Equation (1)) from the Ordinance on the determination of compensation for transferred and restricted rights to forests and forest lands [43] is used to calculate the value of the growing stock damage caused by forest fires:

$$V_m = c\left(1.0 + \frac{p}{100}\right)^n \quad (1)$$

where V_m is the stand value, c denotes the cost of raising the stand, p is the percentage of the increment, and n is the stand rotation.

The damage to the growing stock is calculated on the basis of the actual value of the mature stand for felling V_m, which is determined in the manner prescribed by the Ordinance. Therefore, for the calculation of V_m, the total amount of wood is determined, which is defined by the forest management plan, and exceptionally with the incremental-income tables. The percentage of assortments is estimated, their quantity and value are calculated, and the costs of transport are deducted. The costs of the first-age class include the costs of stand establishment (seeding or planting), costs of thinning, costs of management, and protection of stands. The production services of Croatian Forests Ltd. have established the standard technology for raising stands of specific management classes. The raising costs per hectare are determined by multiplying the labour price from the works pricelist for the calculation year, with a multiplication coefficient for that work, and the calculated values are then added (Equation (2)).

$$Hkm = (B + V + c) \times 1.0pm - (B + V) - (Da \times 1.0{-}pm{-}a + \ldots) \quad (2)$$

Hkm = stand value;
B = land value;
V = value of administration costs;
c = afforestation costs;
p = interest rate;
Da = income from thinning.

The raising costs per hectare in Mediterranean forest for the Aleppo pine management class include different silviculture costs like raising and establishing forests (afforestation), forest care (clearing and thinning), and forest rejuvenation. The value of non-wood forest function is calculated based on the methodology developed by Prpić and Meštrović [44] by assessing ten factors using a judgement scale and points system. The obtained score is then multiplied by the monetary value of the economic function of the forest. The non-wood forest functions, according to the Ordinance on the determination of compensation for the transferred and restricted rights to forests and forest lands [43] fall within two categories: social (tourism, recreation, health, and aesthetic) or protection and ecological (hydrological, anti-erosion, climate protection, and assimilation). In accordance with the overall score of non-wood forest function in each category, the stand is awarded a point. The value of a point is determined by the Croatian Forests executive board, and the value is currently set at 1 point = 0.13 EUR. The official classification of non-wood forest functions, with the price range for their validation, is provided in the Ordinance [43]. The compensation for reductions of the non-wood forest functions is calculated based on a review of the investor's project documentation, as the difference between the value of the non-wood forest functions before and after works were performed in the forests. The value of the non-wood forest function of newly raised forests or plantations, from Article 5 of this Ordinance, is equal to the cost of raising a new forest or plantation to an age of five years, as calculated using the company standards. The non-wood forest functions, protective forests, and special value forests are evaluated using this range of scores [43]:

a. Protection of soil, roads, and other structures from erosion, torrents, and flooding (1–5);
b. Influence on the water regime and hydropower system (1–4);
c. Influence on soil fertility and agricultural production (1–4);
d. Influence on climate (1–4);
e. Protection and improvement of the human environment (0–3);
f. Oxygen generation and atmospheric filtration (1–3);
g. Recreation, tourism, and health functions (1–4);
h. Influence on fauna and hunting (0–4);
i. Protective forests and special purpose forests (8–10).

2.2.3. Fire Weather Index and Assessment of Seasonal Weather Conditions

Since 1982, the Croatian Meteorological and Hydrological Service (DHMZ) has been performing risk assessments of the ignition and spread of forest fires for the Adriatic area, and for the whole of the Croatian area during the last ten fire seasons. The fire weather risk is based on determination of the Fire Weather Index (FWI) for the 41 meteorological stations using the Canadian Forest Fire Weather Index System (CFFWIS) [45]. The daily results of the risk assessment are delivered to the operational fire centre. The FWI considers the dry and wet effects of past and current weather conditions (air temperature, relative humidity, and wind speed at 12 UTC, as well as the daily amount of precipitation from the previous day at 12 UTC to the current day at 12 UTC) for three types of forest fuel indices: Fine Fuel Moisture Code (FFMC), Duff Moisture Code (DMC), and Drought Code (DC). The Initial Spread Index (ISI) and the suitability index of the burning material, called the Buildup Index (BUI), are calculated from the previous indices. ISI provides a fire spread rating after ignition of a standard type fuel of Canadian pine. If the ISI is greater than 18, then there are favourable meteorological conditions for the development of the worst form of fire, which is called a crown fire. Finally, the numerical value of the FWI is determined from the ISI and BUI. The FWI associated with the BUI is generally the best way to describe fire weather risk classes (Table 1).

Table 1. Fire weather risk classes according to the Fire Weather Index (FWI) associated with the Buildup Index (BUI) and Seasonal Severity Rating (SSR) separately [46].

Fire Risk	FWI	BUI	SSR
Very low	≤4	≤48	–
Low	5–8	49–85	<1
Moderate	9–16	86–118	1–3
High	17–32	119–158	3–7
Very high	≥33	≥159	>7

The most appropriate indicator of seasonal potential fire risk is the Seasonal Severity Rating (SSR). The assessment of severity is obtained by applying the Canadian model of the FWI according to the following relation:

$$DSR = 0.0272\ (FWI)^{1.77} \quad (3)$$

The DSR is the Daily Severity Rating from which the mean Monthly Severity Rating (MSR) or mean Seasonal Severity Rating (SSR) is calculated. As the assessment of severity contains meteorological conditions and the moisture condition of dead forest fuel material, MSR and SSR are used for the climatological-fire presentation of the average condition in an area by month or the fire season from June to September. For the assessment of the monthly, seasonal, or annual weather conditions at the meteorological stations for the current year, the anomalies of the mean daily air temperature (°C) and the amount of precipitation (%) in relation to the mean value of the reference climate period 1961–1990 are analysed in the DHMZ. According to the values of the 2nd, 9th, 25th, 75th, 91st, and 98th percentiles obtained from the theoretical distributions, which approximate the basic characteristics of the data for the reference period, the limits for the climate assessment are determined. The normal distribution is applied for the assessment of temperature conditions, and the square root normal distribution is used for the assessment of the precipitation amount [47].

3. Results

This section presents an analysis of the latest Forestry Law with bylaws and related fire protection regulations, grouped according to the main regulation areas: sustainable development, environmental protection, nature protection, biodiversity conservation, protected areas, fire protection, waste management, forestry, agriculture, and other related laws. The economic impact of forest fires on state forest management is presented through the calculation of first-age class afforestation costs, growing stock assortment value, and estimation of non-wood forest functions. Using the mean Seasonal Severity Rating (SSR), the extreme weather conditions and potential meteorological risk of forest fires in the selected period are analysed.

3.1. Legislative and Policy Framework

In Croatia, fire protection at the state level is regulated by the National Fire Protection Strategy for the period from 2013 to 2022 [48], the National Fire Protection Action Plan [49], the report on the state of fire protection in the Republic of Croatia [50], and the program of activities in the implementation of special fire protection measures of interest to the Republic of Croatia [51]. The Government enacts the Fire Protection Law [39], the Ordinance on the Fire Protection Plan [52], the Ordinance on Forest Fire Protection [53], and other ordinances and regulations in the field of fire protection.

The Law on Forests [38] is the most important legal document regulating various aspects of forest management. The protection of forests from natural disasters is legally regulated in Articles 41 to 45 of the Law on Forests, which stipulate that persons who manage forests are obliged to use a set of measures to protect forests from fire, other natural disasters, harmful organisms, and harmful anthropogenic impacts. The same act prescribes the obligation of constituting and managing a unique information system and registry on forest fires (Article 43), in order to improve the control of forest fires their causes, effects, and prevention. On burned areas of forests and forest land, the purpose cannot be changed for 10 years after the fire (Article 46). According to Article 19, forest owners are responsible for restoring burned areas in a period of two years, if this is not defined in the forest management plan. According to Article 20, counties, cities, and municipalities should plan and provide prescribed preventive silviculture works with the aim of decreasing fire danger and the fast spread of forest fires, and enabling early detection, forest-fire alarms, rapid warnings, and fire suppression.

In the Republic of Croatia, the Ordinance on Forest-Fire Protection [53] prescribes technical, preventive silviculture and other fire protection measures to be implemented by owners and users of forests and forest land, holders of other rights in forests and forest lands, legal entities that manage forests and forest lands on the basis of special regulations (counties, cities, and municipalities) in territories which are forests and forest lands owned by forest owners, in order to reduce the risk of forest fires and enable early detection and reports of forest fires, and timely action in extinguishing forest fires. Table 2 presents the main laws that regulate fire protection management and define financial measures for fire suppression. Measures prescribed in the legislation are marked with a plus sign. Legislation without defined measures and activities is marked with a negative sign.

Table 2. Main and other sector-related laws that define fire protection measures and activities.

Regulation Area	Legislation	Measures and Activities
Sustainable development	Sustainable development strategy for Republic of Croatia [54]	–
Environmental protection	National strategy for environmental protection [55]	+
	Climate change adaptation strategy [56]	+
	Law on Environmental Protection [57]	–
	Law on Mitigation and Elimination of Consequences of Natural Disasters [58]	+
Nature protection	National strategy and action plan for nature protection 2017–2025 [59]	+
	Law on Nature Protection [60]	–
Biodiversity conservation	Biodiversity and landscape conservation strategy and action plan [61]	+
	Convention on Biological Diversity [62]	–
Protected areas	Rulebook on Protected Areas [63]	–
	National strategy for fire protection 2013–2022 [64]	+
Fire protection	Program of activities in the implementation of special fire protection measures of interest to the Republic of Croatia [51] National Fire Protection Action Plan [49] Law on Fire Protection [39] Law on Fire Fighters [40] Rulebook on fire protection plan [64] Law on Civil Protection System [65]	+
Waste management	Law on Sustainable Waste Management [66]	–
Forestry	National forest policy and strategy 2003 [67] Forest Law [38] Ordinance on Forest-Fire Protection [53]	+
Agriculture	Law on Agriculture Land [68]	+
Related laws	Law on Flammable Liquids and Gases [69] Law on Roads [70] Law on Explosive Materials and Production and Trade of Weapons [71] Law on Dangerous Materials Transport [72] Law on Sustainable Waste Management [66] Law on Air Protection [73] Law on Safety and Interoperability of Railroad System [74]	+

Measures prescribed in the legislation are marked with a plus sign. Legislation without defined measures and activities are marked with negative sign.

The Law on Fire Protection [39] prescribes measures and activities addressing the causes of fire, elimination and fire extinguishing, and coping with fire disasters and fire protection financing. The Ministry, as an entity responsible for nature protection, participates in the creation of sustainable development documents and nature protection at the state level. Responsible policy departments at a regional level participate in the provision of nature protection regulations with fire protection sections. The government ensures financial means in the state budget. Also, natural and legal persons, local government units, and regional autonomy ensure financing in their budgets (Article 59). Many forest fires have started on agricultural land. The Law on Agriculture Land [68], in Article 87, defines how to provide fire protection measures and order the implementation of particular measures if agricultural land users or owners are not following directions for fire protection. Landowners have to care about the agricultural land according to the Ordinance on the methodology for monitoring the condition of agricultural land [75]. The Law on Fire Fighters [40] regulates firefighters' organisation in the Republic of Croatia at the national, regional, and local levels, as well as their constitution, obligations, method of management, and financing. Article 13 defines that means collected from the green tax fund defined in the Forest Law should be allocated for the needs of the firefighters in a minimum amount of 20%. Units of local and regional administration regulate the fire protection in their area in accordance with the provisions of the Fire Protection Act and other regulations governing the area of fire protection and according to their own plans, needs, and judgments. Financing of planning and operational costs related to forest -fire protection is realised within the approved funds in the State Budget of the Republic of Croatia, with the largest share for the Ministry of Defence, which includes firefighting aircrafts, six Canadairs CL-415 (Canadair Bombardier Aerospace De Havilland Canada, Calgary, Alberta), seven observation aircrafts, and two transport helicopters. The fee for non-wood forest functions (green tax) is paid in the amount of 0.024% of a total income of more than EUR 1 million for defined taxpayers [38]. Funds collected from the fee are spent on demining areas where there are still mined or suspected mined forest areas and for the restoration and fire protection of forests on islands, coasts, and mountainous areas. In the continental part of Croatia, these funds are mostly spent on the remediation and restoration of decaying forests, forest protection, development of forest management plans, afforestation and restoration of forest stands, reconstruction and conversion of forests, forest protection, and construction of forest infrastructure. All these investments are controlled and approved by experts from the Ministry of Agriculture. Since 2014, the funds of the green tax fee have been paid into a special account of the State Budget, and the Ministry of Agriculture has controlled and distributed them to various users in accordance with the Law on Forests. Annually, about EUR 22 million is collected, which is distributed according to the regulations to the firefighters, demining initiatives, private forest owners, and the state forest company.

The laws listed in Table 2 define the measures, activities, and resources required for fire protection. Competencies and partners in the implementation of measures in practice are specifically listed for law enforcement. The obligations of individual stakeholders are prescribed in detail in the Program of activities in the implementation of special fire protection measures of interest to the Republic of Croatia, which is adopted before the fire season. Of course, most of the responsibilities are assumed by the Croatian Fire Brigade, which is responsible for the execution, coordination, monitoring, harmonisation, and direction of all activities related to the implementation of this Program. Planned, preventive, operational, and supervisory activities are carried out through a program of activities for executors and participants in the preparation of the fire season in order to reduce the risk of fire occurrence and spread. In addition to forest owners and public institutions of national parks, nature parks, and institutions for the management of other protected areas, as the most important stakeholders in the implementation of fire protection, other companies and institutions are also involved in the activities. Because of the risk during the tourist season, tourist companies, the Croatian National Tourist Board, and the Croatian Hotel Employers' Association are implementing plans for the evacuation and care of tourists in case of forest-fire threats. The Ministry of the Sea, Transport, and Infrastructure in the Air Transport Department and the Croatian Air Navigation Services is obliged to draw up plans for firefighting activities (number of aircraft, equipment, and use of drones). The same Ministry in the Department of Maritime Transport will determine the conditions and manner of use of vessels, equipment, and techniques and the transportation plan. Croatian Railways companies, private carriers, and the Ministry of the Sea, Transport, and Infrastructure in the field of railway transport are obliged to plan and implement fire protection measures, with special attention during the fire season. Road management companies, as well as companies involved in the production, transmission, and distribution of electricity, and the relevant ministry should develop a plan for clearing flammable substances from roads and distribution zones. Most of the work consists of clearing side vegetation and illegal waste disposals from roads and clearing shrubs and vegetation below electricity transmission lines to reduce the potential fire risk. The Ministry of the Interior—Police Directorate and the Directorate of Civil Protection, during the tourist season, conducts intensified supervision of critical places that are a potential danger to the occurrence and spread of fire (landfills, whether legal or illegal, high-risk forests, neglected agricultural land, and other risky open areas for the occurrence and spread of fire). The Ministry of the Interior—Directorate of Civil Protection—Croatian Mine Action Centre is obliged to include in the Mine Action Plan priority areas for demining for the purpose of easier access, i.e., firefighting, in agreement with mine-affected counties. The Ministry of Health covers the costs of on-call staff and additional activities. Regarding the other responsible stakeholders and participants and their activities, the main responsible authority is the company Croatian Forests Ltd. Based on the Forest Management Plan 2016–2025 [35], the company will continuously work on the construction and maintenance of observation points, forest roads, fire protection corridors, and video surveillance for fire detection. Together with other stakeholders (county prefects, major, municipal heads, Ministry of Agriculture, Ministry of Defence, firefighters' associations, county commander, regional civil protection offices, Ministry of Interior, Croatian Firefighters Association), the company is committed to providing up-to-date data about forest-fire vulnerability, forest roads, observation points, and protection corridors in digital form, and data about forest values. It must provide a list of the teams with equipment and an engagement plan delivered to the territorial responsible firefighters' association (county) and local civil protection office.

Responsible authorities in protected areas are public institutions of national parks, public institutions of nature parks, public institutions for management with protected areas, and the Ministry of the Interior and State Inspectorate. They have to carry out preventive measures for fire protection defined with threat assessment and a fire protection plan, and after the fire season provide joint analyses with the aim of improving fire protection measures. The responsible inspection service is responsible for the planning, organisation, and timely monitoring of the implementation of fire protection measures in national and nature park institutions.

One of the important authorities is the Croatian Meteorological and Hydrological Service (DHMZ). Together with the Croatian Firefighters Association, Ministry of Interior—Directorate for Civil Protection—Operations Centre, and Ministry of Defence, they have to identify on a daily basis the real fire index threat for the creation and spread of fire in open space for 40 stations in coastal and regional country areas. At the end of the year, the DHMZ will make an annual report with the space distribution of monthly (MSR) and seasonal (SSR) intensity at the national level.

3.2. *Economic Impact of Forest Fires on State Forest Management*

Due to climate change, the company Croatian Forests Ltd. has been investing more funds each year into the restoration of forest stands, as a consequence of various natural disasters (fires, floods, icebreaks, windbreaks, wind throw, and snow damage). The company, with the aim of fire prevention, has invested significant funds in fire protection, as presented in Table 2 [36]. Many works were carried out in the analysed period (i.e., designing forest-fire roads, establishing observation and information services, production and maintenance of observational areas, production and maintenance of forest-fire roads, setup of warning signs, firefighting work, fire conservation, and promotional activities). In 2017, the company spent EUR 9.2 million. During the year 2017, 3168 fires were recorded, which burned 48,543 ha of forests and other land, owned together by the state and private forest owners, and agricultural land. In terms of the number of forest fires, this was a year with an above-average number of fires (328 fires compared to the average of 273 per year in the past 26 years). The burned area of 48,543 ha far exceeded the annual average of 14,300 ha. The most catastrophic wildfire in Croatia occurred in 2017, producing downslope fire runs into the most populated area along the Adriatic coast in Croatia near the city of Split [25]. Overall, fires mostly affected karst areas (99%) and state forests, which were burned to 85% of the area, compared to 7% of burned private forests and 8% of agricultural land. Based on the damage cost calculations by the Croatian state forest company for the period 2013–2022 (EUR), presented in Table 3, carried out in the analysed period, the highest amounts were in the first-age class and growing stock in 2017, while the highest calculated amount of damage from the non-wood forest functions of forests was in 2020. These data correlate with the extreme weather conditions in those years, and the highest Fire Weather Index (FWI).

Table 3. Calculated damage by the Croatian state forest company for the period 2013–2022 (EUR).

Year	First-Age Class (EUR)	Growing Stock Damage (EUR)	Non-Wood Forest Functions (EUR)	Number of Fires	Burned Area (ha)	Fire Protection (Mil. EUR)
2013	n.a.	1,832,227.24	6,912,529.86	n.a.	n.a.	n.a.
2014	n.a.	131,271.57	526,811.78	n.a.	n.a.	n.a.
2015	374,037.70	1,535,041.15	8,597,690.47	164	6594	8.9
2016	752,588.27	888,903.64	3,801,831.69	134	6707	3.8
2017	4,598,221.40	10,518,980.62	94,189,673.50	316	41,171	9.2
2018	518,582.43	141,226.44	2,056,809.13	49	1095	13.3
2019	1,936,421.56	757,499.34	4,056,145.47	118	1639	9.6
2020	544,730.55	5,174,807.54	176,964,693.39	141	20,874	8.0
2021	2,419,138.63	1,318,733.82	75,255,425.04	111	4328	8.3
2022	1,595,859.04	4,059,725.26	167,963.24	238	17,760	10.3
Total	8,724,581.90	20,979,957.53	297,106,185.29	1271	100,168	71.4

In the period 2013 to 2022, there were a total of 1271 fires on forest land owned by the Republic of Croatia. A total of 100,168 ha was burned. The average burned area per fire was 78.8 ha/fire. The state company has spent more than EUR 71 million on fire prevention. Most of the fires occurred in the karst area. After analysing the number of fires and burned areas per year and correlating these data with climatic conditions, the results confirm that the dry and hot years were 2017 and 2020.

3.3. *Extreme Weather Conditions and Potential Meteorological Risk of Forest Fires*

Various definitions classify forest fires as a natural disaster, but their occurrence is associated with another natural disaster, i.e., drought. Thus, one natural disaster causes another, but the circle is not closed. Namely, a forest fire causes great changes to the vegetation, which is destroyed by fire, but drastic changes in the plant cover are also caused by erosion. The effect of fire also leads to chemical–physical changes in the content of the soil. All these sudden changes in a burned area cause changes in the plants during and after the fire. This shows the interaction of the processes taking place in the atmosphere and the soil. The Croatian mid-Adriatic region known as Dalmatia is the area most threatened by forest fires according to the period 1991–2020 [76]. The reason for this is the specific and easily flammable Mediterranean vegetation and long-term dry and hot periods [77]. The potential risk of forest fires is certainly increased by human factors due to the increased number of tourists in the summer. The mean SSR values of a 30-year period in the mid-Adriatic are mostly in the range of 8 to 12. The desiccation of dead fuel material and vegetation increases during long hot and dry periods, thereby increasing the potential risk of forest fires. Heat stress for plants is considered a period of at least ten consecutive days with a maximum daily air temperature (tmax) greater than or equal to 30 °C [78]. For precipitation (P) to have some effect on dead fuel material, each fuel type has a different critical precipitation threshold. Therefore, dry periods for FFMC ($P \leq 0.5$ mm), DMC ($P < 1.5$ mm), and DC ($P < 2.8$ mm) were considered separately. Previous research has shown that dry periods lasting longer than a month cause an extremely high potential risk of the ignition and spread of forest fires [79]. The longest hot and very hot periods were analysed for two temperature thresholds (tmax ≥ 30 °C and tmax ≥ 32 °C) as well as the longest dry periods with regard to the type of fuel material in the period 2013–2020 (Table 4). In the summer of 2015 in the mid-Adriatic, the longest hot period of 47 days (from 1 July to 16 August at the Hvar station) and the longest very hot period of 35 days (from 12 July to 15 August at the Split-Marjan and Ploče stations) were recorded. In the considered 8-year period, 2017 had the longest dry period with regard to FFMC, i.e., 73 consecutive days (from 17 June to 2 September), and the daily amount of precipitation at the Lastovo station was less than or equal to 0.5 mm. During the summers of 2014 and 2017, the maximal numbers of dry consecutive days with respect to DMC were approximately the same (81 and 78 days, respectively) as for DC (102 and 100 days, respectively).

Table 4. Maximal numbers of consecutive days with maximum daily air temperatures $t_{max} \geq 30$ °C and $t_{max} \geq 32$ °C as well as with daily amounts of precipitation $P \leq 0.5$ mm, $P < 1.5$ mm, and $P < 2.8$ mm according to the 41 meteorological stations in Croatia in the period 2013–2020.

Years	2013	2014	2015	2016	2017	2018	2019	2020
				Maximal Number of Consecutive Days				
$t_{max} \geq 30$ °C	45	18	47	24	39	41	32	25
$t_{max} \geq 32$ °C	10	0	35	17	15	20	15	10
$P \leq 0.5$ mm	57	34	51	34	73	37	49	37
$P < 1.5$ mm	81	54	56	41	78	39	51	39
$P < 2.8$ mm	102	55	84	42	100	50	78	50

The spatial distributions of the mean SSR were analysed in order to determine extreme years in the considered period 2013–2022. The highest SSR values were estimated in the fire season of 2017, which reached up to 24–28 in Dalmatia (Figure 2). In contrast, the fire season of 2014 was rated with the lowest SSR values and these were mostly between 0 and 4 throughout Croatia. Therefore, those two extreme fire seasons will be analysed in more detail.

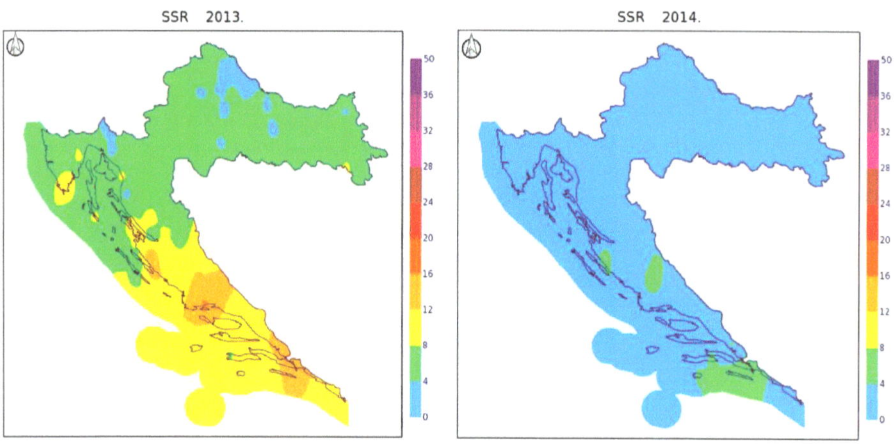

Figure 2. *Cont.*

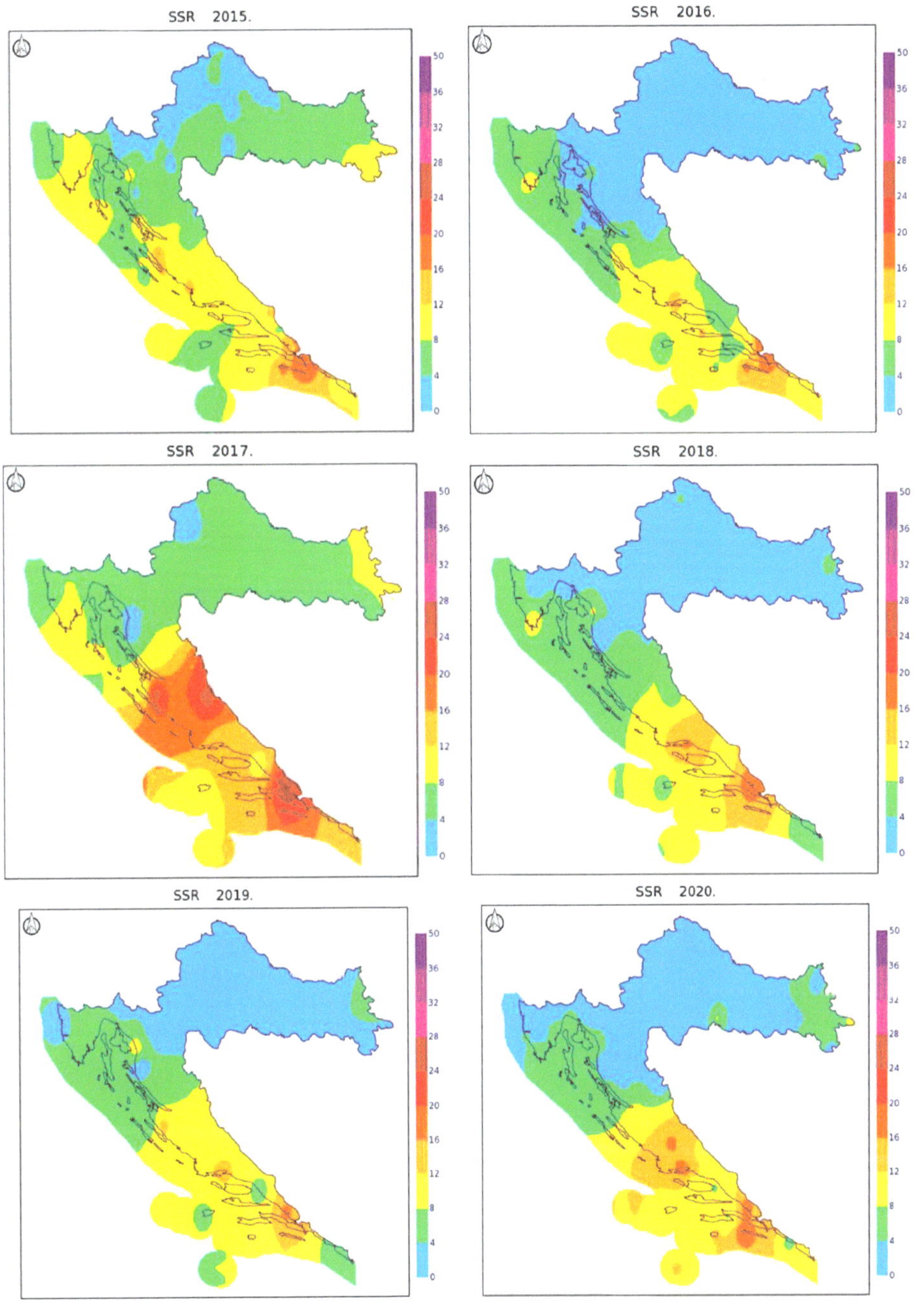

Figure 2. *Cont*.

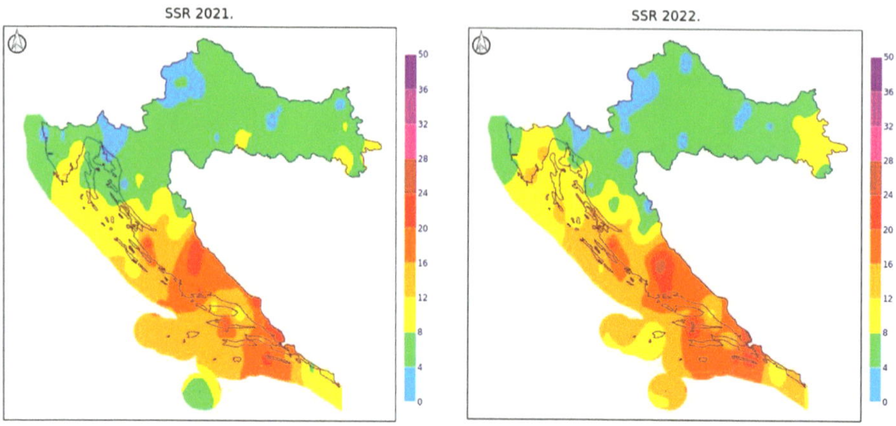

Figure 2. Mean Seasonal Severity Rating (SSR) in Croatia in fire season from June to September in the period 2013–2020. (Source: DHMZ).

The summer of 2017 (June–August) was categorised as extremely warm throughout Croatia according to the temperature conditions of the percentile distribution (Figure 3). Above-average values of air temperature in all three summer months contributed to this [80]. Anomalies in the mean summer air temperature ranged from 2.5 °C to 4 °C. Precipitation amounts in the summer of 2017 were below average throughout the country, with a range from 2% in Dubrovnik to 74% in Zagreb, considering the multi-year summer average. This was partly due to the amount of precipitation being less than or approximately equal to the multi-year average in all three summer months. Therefore, an extremely warm summer and at the same time very and extremely dry conditions in the greater part of Croatia created favourable meteorological conditions for the occurrence and spread of forest fires in the summer months. In contrast, the summer of 2014 in most of Croatia was categorised as warm and very warm according to the temperature conditions of the percentile distribution. Above-average air temperature values in June contributed the most to this, as air temperature values were mostly at the level of the multi-year average in July and August [81]. Anomalies in the mean summer air temperature reached 1.1 °C. Precipitation amounts in the summer of 2014 were above average throughout the country, except in eastern Croatia, where they were within normal limits, with a range from 100% in Osijek to 302% in Zadar, with regard to the multi-year summer average. A very rainy July in almost the entire country and very wet June in Dalmatia contributed the most to this. Thus, the described weather conditions in the summer of 2014 were not favourable for the occurrence of large forest fires, especially not in the Adriatic coast and islands.

In July 2017, in the area of Dalmatia, the greater part of the peninsula of Istria in the northern Adriatic, and part of the interior in the northwest and east of the country, very high values of monthly intensity were assessed (Figure 4). The greatest risk from fire appeared in the Dalmatian hinterland, where MSR values were higher than 35 (Knin 37.0 and Zadar airport 35.6) and at Split airport and Ploče stations, which were twice their average. In the Šibenik, Dubrovnik, and Lastovo areas, the MSR was greater than 20, and in the rest of the Croatia Lastovo areas, the MSR values were mostly up to 12, except at Pula airport (13.3). In northwestern Croatia, MSR values were also 2.5–3.5 times higher than the multi-year average in the period 1981–2010. The very high potential risk of forest fires in almost the entire country, except for a small part in the interior, continued in August 2017. Compared to the multi-year average, extremely warm weather prevailed with simultaneous drought throughout the country. It is to be expected that the MSR values will be above average, which can be clearly seen in Figure 4. In the continental area of Croatia, the potential risk of forest fire was up to three times higher than the average.

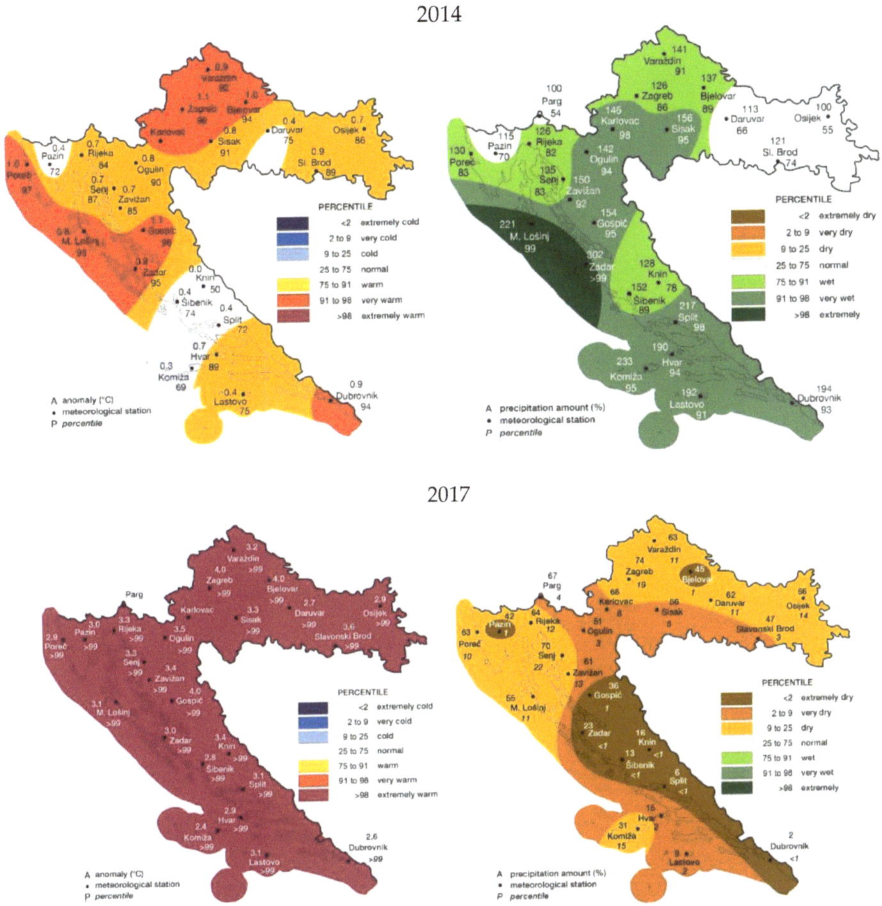

Figure 3. Anomalies in seasonal air temperature (**left**) and seasonal precipitation amount (**right**) for summers from June to August 2014 and 2017 compared to the mean values of the reference climate period 1961–1990. (Source: DHMZ).

Thus, the weather conditions were extremely favourable in July and August 2017 for the ignition and rapid spread of fires, which did occur in the Split, Šibenik, and Zadar areas when catastrophic forest fires raged. The largest wildfire was the Split fire, which firefighters refer to as "the mother of all fires in Croatia". The detailed surface meteorological situation, as well as the vertical structure of the atmosphere during the Split fire in July 2017, were analysed in the paper of Čavlina Tomašević et al. [25].

After the collection of data regarding the total damages in Croatian state forest company for the period 2013–2022 (EUR), we made a correlation with the maximal number of consecutive days with maximum daily air temperature tmax more than 30 °C and tmax more than 32 °C, as well as with the daily amount of precipitation ($P \leq 0.5$ mm, $P < 1.5$ mm, and $P < 2.8$ mm) according to the 41 meteorological stations in Croatia in the same period, as presented in Table 5.

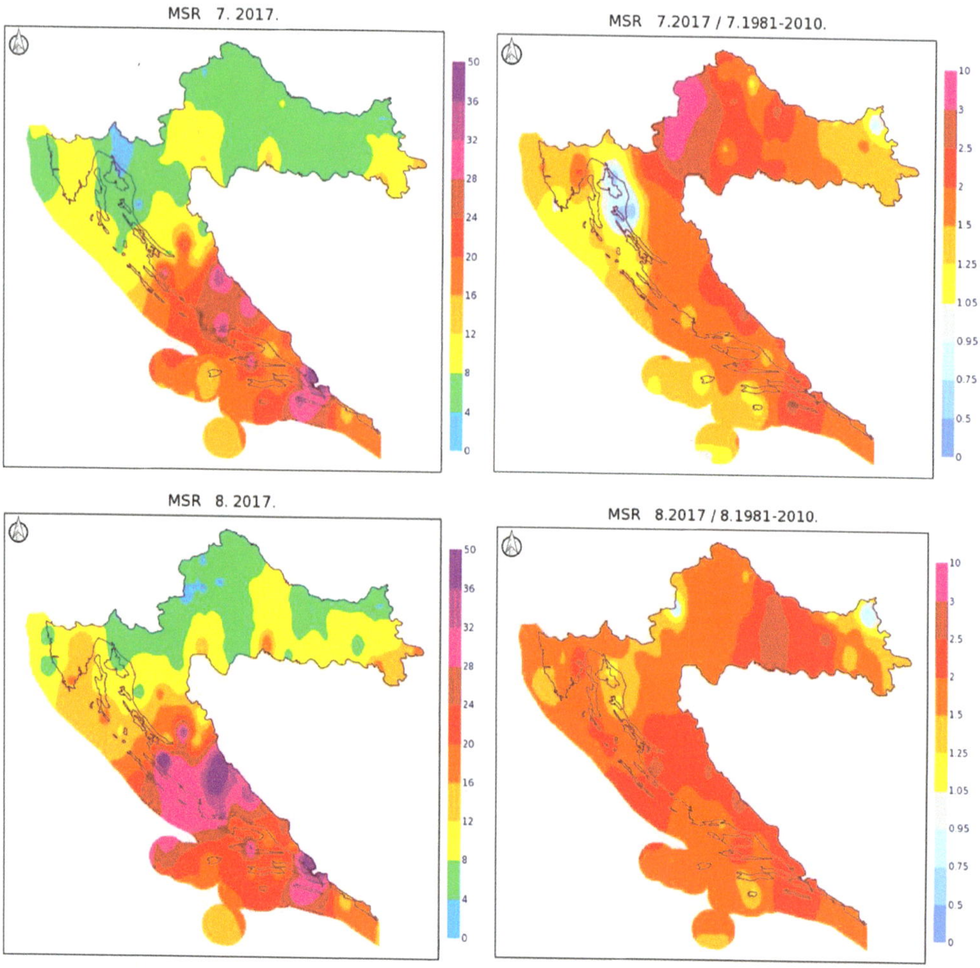

Figure 4. Mean Monthly Severity Rating (MSR) in Croatia in July and August 2017 (**left**) and comparison with the multi-year average in the period 1981–2010 (**right**, source: DHMZ).

Regarding the analysed data, there is a negative correlation between the amount spent on fire protection and total damage amount (R = −0.153). The correlation between a temperature higher than 32 degrees Celsius and total damage is also negative (R = −0.565). The burnt area indicator was created to confirm the highest positive correlation with burned area in the selected period.

Table 5. Correlation of selected variables in the period 2013–2022 (EUR). The values marked with red colour are showing correlations significant at $p < 5\%$.

Variable	Correlations (Fires) Marked Correlations Are Significant at $p < 0.05000$ N = 8 (Casewise Deletion of Missing Data)															
	Means	Std.Dev.	I Age Class (EUR)	Growing Stock Damage (EUR)	Non-Wood Forest Functions (EUR)	Total Damage (EUR)	Number of Fires	Area (ha)	Burnt Area Indicator (ha/fire)	Share of Fires (%)	Sources Fire Protection (Mil. €)	tmax > 30 °C	tmax > 32 °C	P < 0.5 mm	P < 1.5 mm	P < 2.8 mm
I age class (EUR)	1,592,447	1,428,146	1.000	0.712	0.225	0.275	0.689	0.669	0.320	−0.0423	0.043	0.154	−0.360	−0.003	0.380	0.387
Growing stock damage (EUR)	3,049,290	3,484,688	0.712	1.000	0.578	0.624	0.871	0.994	0.852	−0.014	−0.009	−0.013	−0.295	−0.439	−0.091	−0.149
Non-wood forest functions (EUR)	45,636,279	64,696,326	0.225	0.578	1.000	0.998	0.203	0.553	0.819	0.228	−0.159	−0.309	−0.563	−0.177	0.020	−0.025
Total damage	50,278,016	67,151,445	0.275	0.624	0.998	1.000	0.255	0.599	0.840	0.218	−0.153	−0.295	−0.565	−0.194	0.023	−0.024
No. of fres	159	83	0.689	0.871	0.203	0.255	1.000	0.882	0.620	−0.167	−0.119	0.084	−0.010	−0.297	0.007	−0.075
Area (ha)	12,521	13,644	0.669	0.994	0.553	0.599	0.882	1.000	0.867	−0.035	−0.058	−0.055	−0.279	−0.420	−0.099	−0.157
Burnt area indicator (ha/fire)	65	50	0.320	0.852	0.819	0.840	0.620	0.867	1.000	−0.010	−0.187	−0.326	−0.417	−0.370	−0.168	−0.230
Share of fires (%)	79	11	−0.042	−0.014	0.229	0.218	−0.167	−0.035	−0.010	1.000	0.069	0.610	0.414	−0.115	0.114	−0.052
Sources fire protection (Mil. €)	9	3	0.043	−0.009	−0.159	−0.153	−0.119	−0.058	−0.187	0.069	1.000	0.597	0.157	−0.570	−0.389	−0.502
tmax > 30 °C	35	8	0.154	−0.013	−0.309	−0.295	0.084	−0.055	−0.326	0.610	0.597	1.000	0.703	−0.382	−0.078	−0.283
tmax > 32 °C	18	7	−0.360	−0.295	−0.563	−0.565	−0.010	−0.279	−0.417	0.414	0.157	0.703	1.000	−0.296	−0.299	−0.441
P < 0.5 mm	49	18	−0.003	−0.439	−0.177	−0.194	−0.297	−0.420	−0.370	−0.115	−0.570	−0.382	−0.296	1.000	0.863	0.910
P < 1.5 mm	62	28	0.380	−0.091	0.020	0.023	0.007	−0.099	−0.168	0.114	−0.389	−0.078	−0.299	0.863	1.000	0.935
P < 2.8 mm	73	28	0.387	−0.149	−0.025	−0.024	−0.075	−0.157	−0.230	−0.052	−0.502	−0.283	−0.441	0.910	0.93	1.000

4. Discussion

Forest fires have a significant influence on sustainable forest management. After analysing forest-fire protection policy measures and laws, in addition to the state forest company, there are many other stakeholders which participate in forest-fire prevention. Unfortunately, their total costs could not be calculated because of the different (unpublished) database sources (ministries, municipalities, and state companies) used in firefighting operations. There are several important factors for reducing the number of fires and burned (fire-affected) areas, such as the continuation of investment in preventive measures (e.g., maintenance of forest roads), and intensive implementation of silvicultural work. The majority of forests in the Mediterranean karst area in Croatia do not have an important wood production role, so the above-mentioned work often cannot be carried out. These problems do not only concern Croatia, but also other Mediterranean countries like Spain and Greece. According to Pausas and Millian [28], larger numbers of fires and larger affected areas are correlated with land-use changes (rural depopulation increases, land abandonment, and consequently, fuel accumulation) and climate change (extreme droughts and extreme heat). Fire-smart management is essential for forest sustainability in the Mediterranean [82].

In 2017 (the most intensive fire season in the past five years), the accumulated annual burnt area of Portugal, Spain, and Italy alone exceeded 0.8 million ha [83]. Fires caused the loss of 127 lives and estimated costs of EUR 10 billion [84]. The Mediterranean region experienced a heatwave from July to August 2021, with record-breaking temperatures in Italy and Spain in some areas. The widespread dry conditions were conducive to numerous wildfires, in particular in Italy, Greece, and Turkey, while fires of about 7500 ha and 65,000 ha occurred in Portugal and France. The total burnt area in the Mediterranean region in July and August exceeded 800,000 ha. The summer of 2021 was the warmest on record for Europe, with slightly higher temperatures than the previous warmest summers of 2010 and 2018 [85]. The year 2022 was the second worst year in terms of areas burnt since records began in 2006. The cost of damage inflicted by wildfires in 2022 is estimated to be "at least EUR 2 billion" [86]. According to the data from the European Forest-Fire Information System (EFFIS), the area burned in Greece so far in 2023 has surpassed that of the destructive 2021 fire season.

In Croatia, neglected agricultural areas (formerly vineyards and olive groves) and private forests, often allow the rapid and uncontrolled spread of fires. Mediterranean forests should be managed according to the rules of the forestry profession. State forest company implemented projects with the support of EU rural development funds for the conversion of degraded forest stands and forest crops to mixed forests with autochthonous species. However, there is a lot of unused potential (such as non-wood forest functions) in these forests. In this sense, it is necessary to plan the long-term and strategic development of karst forestry, which has not been the case so far. Active and sustainable forest management in the Mediterranean requires the development of competitive new added-value chains on wood, non-wood products, and services. The new EU bio-economy strategy could create opportunities for new markets and innovations [7]. There is a need for strategic, long-term forest development planning, and the valorisation of all forest products and services to obtain more economic value from endangered forests.

Using content analysis, we quantified and analysed the presence, meanings, and relationships of fire protection laws and different stakeholders' obligations and activities. The regulatory, economic, and informational forest policy instruments or their combinations are often used to address challenges in the forestry field [87]. The Croatian government provides a good legal framework aligned with EU directives, which enables different stakeholder tools for fire protection measures. Each of the prescribed measures should have a budget for implementation. Informational forest policy instruments could create awareness among the citizens and private forest owners to consider abandoned forest and agricultural land as the main risk for the spread of forest fires in the state and protected forest areas, which are significant for tourism. The most effective instruments would be economic, such as subsidies for private forest owners or the allocation of more financial sources for fire protection to the state company. This could be realised through different EU financial mechanisms, e.g., ForestEye cross-border project "Protection of nature and environment from forest fires" between Croatia and Bosnia and Herzegovina.

An important factor that cannot be influenced is weather and climate, especially during extremely dry and extremely warm years. Spatial analysis of mean SSR in the period 1991–2020 [88] showed the expansion of areas with high potential risk of forest fires from the Dalmatian islands and the coast towards the hinterland compared to the previous reference climate period (1961–1990). However, the area of moderate to high risk has extended to almost the entire interior of Dalmatia and the northern Adriatic, as well as eastern Croatia and part of northwestern Croatia. Certainly, the most threatened area is still the mid-Adriatic, considering the occurrence of forest fires in the summer months when there are long-lasting dry and hot periods. It follows that areas with an increased potential risk of forest fires are expanding rapidly. In addition to the spatial change, a temporal change is also expected in the prolongation of the fire season from May to October due to climate change [89]. These results fit into the wider picture of the expansion of areas at high risk of forest fires in the Mediterranean and Eastern Europe in the summer months [35]. The meteorological elements that most influence the ignition and spread of fires are solar radiation, air temperature, relative humidity, amount of precipitation, and wind speed and direction, but the vertical structure of the atmosphere is also important [90]. Therefore, in addition to long-lasting hot and dry periods, meteorological analyses during the large forest fires on the Adriatic coast showed that weather changes associated with a cold front passage often trigger fire events due to a sudden increase in wind speed [91]. Contrary to the negative attributes, it is important to stress the exceptional ecological and economic role of pine trees in the Mediterranean [92,93].

The current forest management practice hardly accepts new trends and adaptations to changed conditions. Forest-fire management should be improved by implementing various protection measures, like planning forest-fire access roads, implementing specific silvicultural measures, improving the efficiency of responding to forest fires by investing in fire protection forces, interdisciplinary fire research, and education of the general public, private forest owners, and others about fire risks. There should be continuous improvement in the assessment of the meteorological risk of forest fires during the fire season. Investment in firefighting equipment and protection, such as the network of infrared (thermographic) cameras or surveillance towers, would reduce response times and make the fire-extinguishing strategy by ground or air forces easier to plan. The decrease in the number of fires was certainly, to some extent, influenced by the preventive actions of more frequent monitoring by ground fire patrols, drone surveys, and regular monitoring by firefighting aircraft, and by an increased number of surveillance cameras that can now be technically coordinated and monitored from the newly equipped Situational Operations Centre and the newly established and staffed Operational Firefighter Command of the Republic of Croatia.

Future tasks will be the development of decision support systems and tools, forest-fire risk assessment models, risk reduction, and adaptation. There is a need to provide continuous research regarding the influence of the forest fuel structure and moisture content on the probability of fire occurrence and development [94,95]. Post-fire management and measures are important for preserving forest soil quality from erosion and cost calculation.

5. Conclusions

In this paper, we analysed the current forest policy framework in order to determine, provide, valorise, and protect forests and their services. In line with the research aims, we determined the following:

(a) According to the definition of fire protection measures and activities, 29 fire-related laws and their bylaws were identified and reviewed. The roles and obligations of different stakeholders were analysed, with an emphasis on Croatian Forests Ltd. and the

Croatian Meteorological and Hydrological Service (DHMZ). The laws were grouped according to the main regulation areas: sustainable development, environmental protection, nature protection, biodiversity conservation, protected areas, fire protection, waste management, forestry, agriculture, and other related laws.

(b) The analysis shows the characteristics of different stakeholders and their obligations for the implementation of special fire protection measures, such as the Croatian Fire Association, the Ministry of Defence, the County Fire Associations, regional civil protection offices, public institutions of national parks, public institutions of nature parks, public institutions for the management of protected areas, regional civil protection offices, the Croatian Tourist Board, and the Ministry of Tourism.

(c) The economic impact of forest fires on state forest company management was calculated. In the period 2013 to 2022, there were a total of 1271 fires on forest and other land owned by the Republic of Croatia. A total of 100,168 ha was burned. The calculated damage for the first-age class was EUR 8,724,581.9, for growing stock it was EUR 20,979,957.53, and for non-wood forest functions, it was EUR 297,106,185.29 for the state forest company. The company is investing significant resources in the fight against adverse weather conditions, which requires additional operating costs. In the year 2020, the forest-fire damage exceeded EUR 176 million.

(d) Weather conditions were analysed in more detail for fire seasons with the highest and lowest Seasonal Severity Rating (SSR) values in the considered period. In almost all of Croatia, the SSR values for 2017 were above 1.5–2 times the multi-year average for the period 1981–2010. This is because the air temperatures were above average by 3–4 °C, and the precipitation amount was below average (2–31% of the average in the mid-Adriatic) in the summer months. Conversely, the SSR values in 2014 were not higher than 5 and were below average throughout the country due to a very rainy summer. The analyses of correlation among the forest fire damages spent sources for fire protection show that there is no significant benefit if only the state forest company invests in fire protection.

Unfortunately, there has not been any recent study about the negative impact of forest fires on tourism and other sectors. Because of the different databases among different institutions, the research limitations are the unavailability of relevant, timely, comprehensive, and unified forest fire damage costs. Our research recommendations are to include all of the stakeholders at regional and local levels from the very beginning for the implementation of fire protection measures. The state forest company should implement fire risk assessment through forest management activities, reduce the fuel mass in forest stands, and in relation to that, decrease the forest fire potential. The government should adopt climate-smart forestry through the relevant ministries, which will embrace and put in place a set of new EU strategies and action plans in order to ensure sustainable forest ecosystems.

Author Contributions: Conceptualisation, S.P. and D.B.; methodology, S.P., D.B., Š.P.M. and V.V.; validation, S.P., D.B., Š.P.M., D.V., I.Č.T. and V.V.; investigation, S.P., D.B. and V.V.; data curation, S.P., D.B., I.Č.T. and V.V.; writing—original draft preparation, S.P., D.B., Š.P.M., D.V., I.Č.T. and V.V.; writing—review and editing, S.P., D.B., Š.P.M., D.V., I.Č.T. and V.V.; visualisation, S.P. and I.Č.T. All authors have read and agreed to the published version of the manuscript.

Funding: This research received no external funding.

Institutional Review Board Statement: Not applicable.

Informed Consent Statement: Not applicable.

Data Availability Statement: The data presented in this study are available on request from the corresponding author.

Conflicts of Interest: The authors declare no conflict of interest.

References

1. Moreira, F.; Ascoli, D.; Safford, H.; Adams, M.A.; Moreno, J.M.; Pereira, J.M.; Catry, F.X.; Armesto, J.; Bond, W.; González, M.E.; et al. Wildfire management in Mediterranean-type regions: Paradigm change needed. *Environ. Res. Lett.* **2020**, *15*, 011001. [CrossRef]
2. Reid, W.V.; Mooney, H.A.; Cropper, A.; Capistrano, D.; Carpenter, S.R.; Chopra, K.; Dasgupta, P.; Dietz, T.; Duraiappah, A.K.; Kasperson, R.; et al. Millennium Ecosystem Assessment Synthesis Report. 2005. Available online: http://pdf.wri.org/mea_synthesis_030105.pdf (accessed on 14 April 2023).
3. Nedeljković, J.; Stanišić, M.; Avdibegović, M.; Curman, M.; Pezdevšek Malovrh, Š. Climate Change Governance in Forestry and Nature Conservation: Institutional Framework in Selected SEE Countries. *Šumarski List* **2019**, *9–10*, 445–459. [CrossRef]
4. European Commission. The European Green Deal, Communication from the Commission, COM (2019) 640 Final, Brussels, 11.12.2019. Available online: https://eur-lex.europa.eu/legal-content/EN/TXT/?qid=1588580774040&uri=CELEX%3A52019DC0640 (accessed on 15 March 2023).
5. Griggs, D.; Stafford-Smith, M.; Gaffney, O.; Rockström, J.; Öhman, M.C.; Shyamsundar, P.; Steffen, W.; Glaser, G.; Kanie, N.; Noble, I. Sustainable development goals for people and planet. *Nature* **2013**, *495*, 305–307. [CrossRef]
6. European Commission. New EU Forest Strategy for 2030, Communication from the Commission to the European Parliament, the Council, the European Economic and Social Committee and the Committee of the Region, Brussels, 16.7.2021. COM (2021) 572 Final. Available online: https://eur-lex.europa.eu/legal-content/EN/TXT/?uri=CELEX%3A52021DC0572 (accessed on 15 March 2023).
7. European Commission. EU Biodiversity Strategy for 2030. Available online: https://environment.ec.europa.eu/strategy/biodiversity-strategy-2030_en (accessed on 15 March 2023).
8. Seidl, R.; Schelhaas, M.J.; Rammer, W.; Verkerk, P.J. Increasing forest disturbances in Europe and their impact on carbon storage. *Nat. Clim. Chang.* **2014**, *4*, 806–810. [CrossRef]
9. Ganteaume, A.; Barbero, R.; Jappiot, M.; Maillé, E. Understanding future changes to fires in southern Europe and their impacts on the wildland-urban interface. *J. Saf. Sci. Resil.* **2021**, *2*, 20–29. [CrossRef]
10. Schelhaas, M.J.; Nabuurs, G.J.; Schuck, A. Natural disturbances in the European forests in the 19th and 20th centuries. *Glob. Chang. Biol.* **2003**, *9*, 1620–1633. [CrossRef]
11. Mikac, S.; Žmegač, A.; Trlin, D. *Utjecaj Klimatskih Promjena i Prirodnih Nepogoda na Šumske Ekosustave, VII*; Konferencija Hrvatske Platforme za Smanjenje rizika od Katastrofa/Holcinger, Nataša (ur.); Državna Uprava za Zaštitu i Spašavanje: Zagreb, Croatia, 2018; pp. 162–171.
12. European Environment Agency. *Nature-Based Solutions in Europe: Policy, Knowledge and Practice for Climate Change Adaptation and Disaster Risk Reduction*; EEA Report No 01/2021; Publications Office of the European Union: Luxembourg, 2021.
13. Fernández-García, V.; Fulé, P.Z.; Marcos, E.; Calvo, L. The role of fire frequency and severity on the regeneration of Mediterranean serotinous pines under different environmental conditions. *For. Ecol. Manag.* **2019**, *444*, 59–68. [CrossRef]
14. Molina, J.R.; González-Cabán, A.; Rodríguez y Silva, F. Potential Effects of Climate Change on Fire Behavior, Economic Susceptibility and Suppression Costs in Mediterranean Ecosystems: Córdoba Province, Spain. *Forests* **2019**, *10*, 679. [CrossRef]
15. Frost, S.M.; Alexander, M.E.; DeRose, R.J.; Jenkins, M.J. Fire-Environment Analysis: An Example of Army Garrison Camp Williams, Utah. *Fire* **2020**, *3*, 6. [CrossRef]
16. Rosavec, R.; Barčić, D.; Španjol, Ž.; Oršanić, M.; Dubravac, T.; Antonović, A. Flammability and Combustibility of Two Mediterranean Species in Relation to Forest Fires in Croatia. *Forests* **2022**, *13*, 1266. [CrossRef]
17. The Emergency Response Coordination Centre, European Commission, Communication from the Commission Strengthening EU Disaster Management: RescEU Solidarity with Responsibility Solidarity with Responsibility, Brussels, 23.11.2017, COM (2017) 773 Final. Available online: https://civil-protection-humanitarian-aid.ec.europa.eu/what/civil-protection/emergency-response-coordination-centre-ercc_en (accessed on 16 March 2023).
18. European Parliament, Council of the European Union. Regulation (EU) No 377/2014 of the European Parliament and of the Council of 3 April 2014 establishing the Copernicus Programme and repealing Regulation (EU) No 911/2010 Text with EEA relevance. *Off. J. Eur. Union* **2014**, *57*, 44–66.
19. European Commission. Commission Report on Forest Fires: Climate Change Is More Noticeable Every Year. Available online: https://ec.europa.eu/commission/presscorner/detail/en/ip_21_5627 (accessed on 16 March 2023).
20. European Commission. *Communication from the Commission to the European Parliament, the Council and the Committee of the Regions—Strengthening EU Disaster Management: RescEU Solidarity with Responsibility Solidarity with Responsibility*; No. COM (2017) 773 Final in Communication from the Commission to the European Parliament, the Council and the Committee of the Regions; Publications Office of the European Union: Luxembourg, 2017.
21. San-Miguel-Ayanz, J.; Costa, H.; de Rigo, D.; Libertà, G.; Artés Vivancos, T.; Durrant, T.; Nuijten, D.; Löffler, P.; Moore, P. *Basic Criteria to Assess Wildfire Risk at the Pan-European Level*; EUR 29500 EN; Publications Office of the European Union: Luxembourg, 2018; ISBN 978-92-79-98200-2. [CrossRef]
22. European Commission. Communication from the Commission to the European Parliament, the Council, the European Economic and Social Committee and the Committee of the Regions—Forging a Climate-Resilient Europe—The New EU Strategy on Adaptation to Climate Change, COM (2021) 82 Final. 2021. Available online: https://eur-lex.europa.eu/legal-content/EN/TXT/?uri=COM%3A2021%3A82%3AFIN (accessed on 18 March 2023).

23. Republic of Croatia, Croatian Firefighting Association, 2020, (Hrvatska vatrogasna zajednica), Konačno Izvješće o Realizaciji Programa Aktivnosti u Provedbi Posebnih Mjera Zaštite od Požara od Interesa za Republiku Hrvatsku u 2020. Godini. Available online: https://hvz.gov.hr/vijesti/usvojeno-konacno-izvjesce-o-realizaciji-programa-aktivnosti/2194 (accessed on 18 March 2023).
24. Barčić, D.; Dubravac, T.; Vučetić, M. Potential Hazard of Open Space Fire in Black Pine Stands (*Pinus nigra* J.F. Arnold) in Regard to Fire Severity. *SEEFOR-South-East Eur. For.* **2020**, *2*, 161–168. [CrossRef]
25. Čavlina Tomašević, I.; Cheung, K.; Vučetić, V.; Fox-Hughes, P.; Horvath, K.; Telišman Prtenjak, M.; Beggs, P.J.; Malečić, B.; Milić, V. The 2017 Split wildfire in Croatia: Evolution and the role of meteorological conditions. *Nat. Hazards Earth Syst. Sci.* **2022**, *22*, 3143–3165. [CrossRef]
26. United Nations Framework Convention on Climate Change (UNFCCC). UN Climate Change Annual Report, 2018, p. 62. Available online: https://unfccc.int/sites/default/files/resource/UN-Climate-Change-Annual-Report-2018.pdf (accessed on 16 March 2023).
27. Giorgi, F.; Lionello, P. Climate change projections for the Mediterranean region. *Glob. Planet. Chang.* **2008**, *63*, 90–104. [CrossRef]
28. Pausas, J.G.; Millian, M.M. Greening and browning in a climate change hotspot: The Mediterranean basin. *BioScience* **2019**, *69*, 143–151. [CrossRef]
29. EFFIS—Statistics Portal EU JRC. Available online: https://effis.jrc.ec.europa.eu/apps/effis.statistics/ (accessed on 16 March 2023).
30. Merlo, M.; Croitoru, L. *Valuing Mediterranean Forests: Towards Total Economic Value*; Cabi Publishing: Wallingford, UK, 2005; p. 406. [CrossRef]
31. Krznar, A.; Lindić, V.; Vuletić, D. Metodologija Vrednovanja Korisnosti Turističkorekreacijskih Usluga Šuma, Methodology for Evaluating the Usfulness of Tourist-Recreational Benefits of Forest. *Radovi* **2000**, *35*, 65–81.
32. Vuletić, D.; Vrbek, B.; Novotny, V. *The Evaluation Results of the Benefits of the Health and Landscape Forests Functions, Rezultati Vrednovanja Koristnosti Zdravstvene i Krajobrazne Funkcije Šume, XXI IUFRO World Congress, Forest and Ocietiy: The Role of Research, 7–12 August 2000*; Poster Abstracts: Kuala Lumpur, Malaysia, 2000; Volume 3, p. 325.
33. Posavec, S. Methods of Evaluation Forests—A Renewable Resource in Croatia. In *The Multifunctional Role of Forests—Policies, Methods and Case Studies, EFI Proceedings*; European Forest Institute: Joensuu, Finland, 2008; pp. 313–318.
34. Camia, A.; Amatulli, G.; San-Miguel-Ayanz, J. *Past and Future Trends of Forest Fire Danger in Europe, JRC Scientific and Technical Reports*; European Commision, Joint Research Centre, Institute for Environment and Sustainability: Luxemburg, 2008; p. 6.
35. Anon. *General Forest Management Plan 2016–2025*; Ministry of Agriculture, Republic of Croatia: Zagreb, Croatia, 2016.
36. Hrvatske Šume, d.o.o. *Business Report*; Croatian Forests Ltd.: Zagreb, Croatia, 2021.
37. Čavlović, J. *Prva Nacionalna Inventura Šuma Republike Hrvatske*; Ministarstvo Regionalnog Razvoja, Šumarstva i Vodnoga Gospodarstva i Šumarski Fakultet Sveučilišta u Zagrebu: Zagreb, Hrvatska, 2010; p. 300.
38. *OG 68/18, 115/18, 98/19*; Law on Forests. Official Gazette of the Republic of Croatia, Croatian Parliament: Zagreb, Croatia, 2019.
39. *OG 92/10*; Law on Fire Protection. Official Gazette of the Republic of Croatia, Croatian Parliament: Zagreb, Croatia, 2010.
40. *OG 80/10, 125/19*; Law on Fire Fighters. Official Gazette of the Republic of Croatia, Croatian Parliament: Zagreb, Croatia, 2010.
41. *OG 20/18 115/18, 98/19*; Law on Agricultural Land. Official Gazette of the Republic of Croatia, Croatian Parliament: Zagreb, Croatia, 2019.
42. Nunez-Mir, G.; Desprez, J.; Iannone, B., III; Clark, T.; Fei, S. An Automated Content Analysis of Forestry Research: Are Socioecological Challenges Being Addressed? *J. Forestry* **2017**, *115*, 1–9. [CrossRef]
43. *Art.5, OG 121/97, 131/06, 74/07, 18/08, 72/16*; Ordinance on the Determination of Compensation for Transferred and Restricted Rights to Forests and Forest Lands. Official Gazette of the Republic of Croatia, Croatian Parliament: Zagreb, Croatia, 2016.
44. Prpić, B. O vrijednosti općekorisnih funkcija šuma. *Šumarski List* **1992**, *6–8*, 301–312.
45. Van Wagner, C.E.; Pickett, T.L. *Equations and Fortran Program for the Canadian Forest Fire Weather Index System*; Forestry Technical Report 33; Canadian Forestry Service, Government of Canada: Ottawa, ON, Canada, 1985; p. 18.
46. Dimitrov, T. Šumski požari i sistem procjene opasnosti od požara. In *Osnove Zaštite Šuma od Požara*; Centar za Informacije i Publicitet: Zagreb, Croatia, 1982; pp. 181–256.
47. Cindrić, K.; Pasarić, Z. Modelling Dry Spells by Extreme Value Distribution with Bayesian Inference. In *Meteorology and Climatology of the Mediterranean and Black Seas, Pageoph Topical Volumes*; Vilibić, I., Horvath, K., Palau, J.L., Eds.; Springer International Publishing: Cham, Switzerland, 2019; pp. 171–188. [CrossRef]
48. Croatian Government. National Fire Protection Strategy for the Period from 2013 to 2022 (OG 68/13), Croatian Government. 2013. Available online: http://www.vlada.hr (accessed on 16 April 2023).
49. Croatian Government. National Fire Protection Action Plan. 2013. Available online: https://vlada.gov.hr/UserDocsImages/2016/Sjednice/Arhiva/81.%20-%2012.pdf (accessed on 15 June 2023).
50. Croatian Government. Report on the State of Fire Protection in the Republic of Croatia. 2022. Available online: https://vlada.gov.hr/sjednice/205-sjednica-vlade-republike-hrvatske-38059/38059 (accessed on 15 June 2023).
51. Croatian Government. Program of Activities in the Implementation of Special Fire Protection Measures of Interest to the Republic of Croatia in Year 2023. Available online: https://hvz.gov.hr/program-aktivnosti/1788 (accessed on 16 April 2023).
52. *OG 51/12*; Ordinance on the Fire Protection Plan. Official Gazette of the Republic of Croatia, Croatian Government: Zagreb, Croatia, 2012.

53. *OG 33/14;* Ordinance on Forest Fire Protection. Official Gazette of the Republic of Croatia, Ministry of Interior: Zagreb, Croatia, 2014. Available online: https://narodne-novine.nn.hr/clanci/sluzbeni/2014_03_33_599.html (accessed on 15 June 2023).
54. *OG 30/09;* Sustainable Development Strategy for Republic of Croatia. Official Gazette of the Republic of Croatia, Croatian Government: Zagreb, Croatia, 2009.
55. *OG 46/02;* National Strategy for Environmental Protection. Official Gazette of the Republic of Croatia, Croatian Government: Zagreb, Croatia, 2002.
56. *OG 46/20;* Climate Change Adaptation Strategy. Official Gazette of the Republic of Croatia, Croatian Government: Zagreb, Croatia, 2020.
57. *OG 118/18;* Law on Environmental Protection. Official Gazette of the Republic of Croatia, Croatian Parliament: Zagreb, Croatia, 2018.
58. *OG 16/19;* Law on Mitigation and Elimination of Consequences of Natural Disasters. Official Gazette of the Republic of Croatia, Croatian Parliament: Zagreb, Croatia, 2019.
59. *OG 72/17;* National Strategy and Action Plan for Nature Protection 2017–2025. Official Gazette of the Republic of Croatia, Croatian Government: Zagreb, Croatia, 2017.
60. *OG 80/13, 15/18, 14/19;* Law on Nature Protection. Official Gazette of the Republic of Croatia, Croatian Parliament: Zagreb, Croatia, 2019.
61. *OG 143/08;* Biodiversity and Landscape Conservation Strategy and Action Plan. Official Gazette of the Republic of Croatia, Croatian Government: Zagreb, Croatia, 2008.
62. *OG 6/96;* Convention on Biological Diversity. Official Gazette of the Republic of Croatia, Croatian Government: Zagreb, Croatia, 1996.
63. Croatian Government. Regulation on Protected Areas. Available online: https://mingor.gov.hr/o-ministarstvu-1065/djelokrug/uprava-za-zastitu-prirode-1180/zakoni-i-propisi-1224/zasticena-podrucja-1240/1240, (accessed on 15 June 2023).
64. *OG 51/12;* Rulebook on Fire Protection Plan. Official Gazette of the Republic of Croatia, Ministry of Interior: Zagreb, Croatia, 2012.
65. *OG 82/15,118/18;* Law on Civil Protection System. Croatian Parliament: Zagreb, Croatia, 2018.
66. *OG 14/19, 98/19;* Law on Sustainable Waste Management. Croatian Parliament: Zagreb, Croatia, 2019.
67. *OG 120/2003;* National Forest Policy and Strategy 2003. Croatian Parliament: Zagreb, Croatia, 2003.
68. *OG 20/18 115/18, 98/19;* Law on Agriculture Land. Croatian Parliament: Zagreb, Croatia, 2019.
69. *OG 108/95, 56/10;* Law on Flammable Liquids and Gases. Croatian Parliament: Zagreb, Croatia, 2010.
70. *OG 92/14;* Law on Roads. Croatian Parliament: Zagreb, Croatia, 2014.
71. *OG 70/17;* Law on Explosive Materials and Production and Trade of Weapons. Croatian Parliament: Zagreb, Croatia, 2017.
72. *OG 79/07;* Law on Dangerous Materials Transport. Croatian Parliament: Zagreb, Croatia, 2007.
73. *OG 118/18;* Law on Air Protection. Croatian Parliament: Zagreb, Croatia, 2018.
74. *OG 70/17;* Law on Safety and Interoperability of Railroad System. Croatian Parliament: Zagreb, Croatia, 2017.
75. *OG 47/19;* Ordinance on Methodology for Monitoring the Condition of Agricultural Land. Ministry of Agriculture, Republic of Croatia: Zagreb, Croatia, 2019.
76. Vučetić, V.; Anić, M. *Agrometeorološki Atlas Hrvatske u Razdoblju 1991–2020. Povodom 70 Godina Agrometeorološke Službe u DHMZ-U;* Državni Hidrometeorološki Zavod: Zagreb, Croatia, 2021; p. 241.
77. Tomašević, I.Č.; Cheung, K.K.W.; Vučetić, V.; Fox-Hughes, P. Comparison of Wildfire Meteorology and Climate at the Adriatic Coast and Southeast Australia. *Atmosphere* **2022**, *13*, 755. [CrossRef]
78. Feist, O. *Analiza Toplinskog Stresa za Potrebe Poljodjelstva u Hrvatskoj u Prošlim, Sadašnjim i Budućim Klimatskim Uvjetima, Diplomski Rad, Geofizički Odsjek Prirodoslovno-Matematičkog Fakulteta;* Sveučilište u Zagrebu: Zagreb, Croatia, 2011; p. 53.
79. Vučetić, V.; Ivatek-Šahdan, S.; Tudor, M.; Kraljević, L.; Ivančan-Picek, B.; Strelec Mahović, N. Analiza vremenske situacije tijekom kornatskog požara 30. kolovoza 2007. *Hrvat. Meteorološki Časopis* **2007**, *42*, 41–66.
80. Ferina, J.; Vučetić, V. *Ocjena Požarne Sezone u Hrvatskoj u 2017. Godini;* Državni Hidrometeorološki Zavod: Zagreb, Croatia, 2018; p. 87.
81. Tomašević, I.; Sviličić, P.; Vučetić, V. *Ocjena Požarne Sezone u Hrvatskoj u 2014. Godini;* Državni Hidrometeorološki Zavod: Zagreb, Croatia, 2015; p. 72.
82. Fernandes, P.M. Fire-smart management of forest landscapes in the Mediterranean basin under global change. *Landsc. Urban Plan.* **2013**, *110*, 175–182. [CrossRef]
83. de Rigo, D.; Libertà, G.; Houston Durrant, T.; Artés Vivancos, T.; San-Miguel-Ayanz, J. *Forest Fire Danger Extremes in Europe under Climate Change: Variability and Uncertainty;* EUR 28926 EN; Publications Office of the European Union: Luxembourg, 2017; ISBN 978-92-79-77046-3. [CrossRef]
84. EURONEWS. There Three Times More Wildfires in the EU so far This Year. 2019. Available online: https://www.euronews.com/my-europe/2019/08/15/there-have-been-three-times-more-wildfires-in-the-eu-so-far-this-year (accessed on 6 October 2023).
85. San-Miguel-Ayanz, J.; Durrant, T.; Boca, R.; Maianti, P.; LibertÀ, G.; Artes Vivancos, T.; Jacome Felix Oom, D.; Branco, A.; De Rigo, D.; Ferrari, D.; et al. *and Nuijten, D. Advance Report on Wildfires in Europe, Middle East and North Africa 2021;* EUR 31028 EN; Publications Office of the European Union: Luxembourg, 2022; ISBN 978-92-76-49633-5. [CrossRef]
86. EURONEWS; Lenarčič, J. The European Commission for Crisis Management. 2023. Available online: https://www.euronews.com/my-europe/2023/01/11/wildfire-damages-cost-2-billion-last-year-says-eu-commissioner (accessed on 6 October 2023).

87. Janota, J.J.; Broussard, S.R. Examining private forest policy preferences. *For. Policy Econ.* **2008**, *10*, 89–97. [CrossRef]
88. Anić, M.; Bašić, J.; Vučetić, V. *Meteorološka Analiza Opasnosti od Požara Raslinja u Hrvatskoj u 2021. Godini*; Državni Hidrometeorološki Zavod: Zagreb, Croatia, 2021; p. 86.
89. Barešić, D. *Utjecaj Klimatskih Promjena na Potencijalnu Opasnost od Požara Raslinja u Hrvatskoj, Diplomski Rad, Geofizički Odsjek Prirodoslovno-Matematičkog Fakulteta*; Sveučilište u Zagrebu: Zagreb, Croatia, 2011; p. 56.
90. Viegas, D.X. Weather, fuel status and fire occurrence. In *Predicting a Large Forest Fires*; Moreno, J.M., Ed.; Backhuys Publishers: Leiden, The Netherlands, 1998; pp. 31–48.
91. Mifka, B.; Vučetić, V. Vremenska analiza katastrofalnog šumskog požara na otoku Braču od 14. do 17. srpnja 2011. *Vatrog. I Upravlj. Požarima* **2012**, *3*, 13–25.
92. Richardson, D.M.; Rundel, P.W. Pine ecology and biogeography-An introduction. In *Ecology and Biogeography of Pinus*; Richardson, D.M., Ed.; Cambridge University Press: Cambridge, UK, 1999; pp. 3–49.
93. Leone, V.; Lovreglio, R.; Martinez Fernandez, J. Forest fires and anthropic influences: A study case (Gargano National Park, Italy). In *IV International Conference on Forest Fire Research and Wildland Fire Safety, Coimbra, Portugal*; electronic, edition; Viegas, D.X., Ed.; Millpress: Rotterdam, The Netherlands, 2002.
94. Bilandžija, J.; Lindić, V. Utjecaj strukture šumskog goriva na vjerojatnost pojave i razvoj požara u sastojinama alepskog bora. *Radovi* **1993**, *28*, 215–224.
95. Stipaničev, D.; Bugarić, M.; Bakšić, N.; Bakšić, D. *Fuel Moisture Content in Croatian Wildfire Spread Simulator AdriaFirePropagator*; Advances in Forest Fire Research 2022, Domingos Xavier Viegas, Luís Mário Ribeiro (ur.); Imprensa da Universidade de Coimbra: Coimbra, Portugal, 2022; pp. 216–221. [CrossRef]

sustainability

Review

Exploring the Role of ICTs and Communication Flows in the Forest Sector

Alex Vinicio Gavilanes Montoya [1,2], Danny Daniel Castillo Vizuete [1,2,*] and Marina Viorela Marcu [1]

1 Department of Forest Engineering, Forest Management Planning and Terrestrial Measurements, Faculty of Silviculture and Forest Engineering, Transilvania University of Brasov, Şirul Beethoven 1, 500123 Brasov, Romania; alex.gavilanes@unitbv.ro (A.V.G.M.)

2 Faculty of Natural Resources, Escuela Superior Politécnica de Chimborazo, Panamericana Sur, km 1 1/2, Riobamba EC-060155, Ecuador

* Correspondence: danny.castillo@unitbv.ro; Tel.: +593-987712497

Abstract: The forestry sector has used technology to improve productivity and increase service quality, reducing labor in many processes. In this sense, Information and Communication Technologies (ICTs) are having broad impacts on the forestry sector, from forestry to the marketing of forest products and the recreational use of forests. There is a wide range of technologies that can be implemented in forestry depending on the needs of each user. The objective of this study was to conduct a literature review in order to analyze the opportunities for improving ICT and communication flows in the forestry sector and to evaluate their applicability. This literature review was analyzed using the Scopus, Web of Science, and ScienceDirect databases. An overview of the importance of ICT and communication flows in the forestry sector, ICT tools, and their applications is provided. One-way and two-way communication flows coexist in forestry, integrating different communication channels, time, target audience, and message. It is clear that technologies have produced significant changes in all sectors of the forestry industry. We conclude that ICTs and communication flows contribute to forest conservation and management in the establishment of standards or policies that ensure conservation through monitoring and analysis of landscapes at different temporal and spatial scales.

Keywords: ICT tools; technology; forest management; forestry sector; forest operations

check for updates

Citation: Gavilanes Montoya, A.V.; Castillo Vizuete, D.D.; Marcu, M.V. Exploring the Role of ICTs and Communication Flows in the Forest Sector. *Sustainability* **2023**, *15*, 10973. https://doi.org/10.3390/su151410973

Academic Editors: Ivica Papa and Andreja Đuka

Received: 17 June 2023
Revised: 5 July 2023
Accepted: 10 July 2023
Published: 13 July 2023

1. Introduction

ICTs have become a fundamental part of daily life [1]. These are used to obtain, transmit, manipulate, and store data efficiently and safely [2]. For instance, these technologies include hardware and software tools, telecommunications networks, the Internet, and information management systems [3], which have enabled process automation, real-time data analysis, and more informed and accurate decision-making [4]. In this context, the rapid development of ICTs has been driven by the growing need for information in different areas [5]. Particularly, in the forestry sector, ICTs have allowed the improvement of forest management [6] and the optimization of production and marketing processes [7], as well as the monitoring and follow-up of forests and their biodiversity [8]. With this, ICTs have allowed the creation of an information society in which communication flows are fundamental [9,10].

Communication flows refer to the transfer of information between the different components of the system, including hardware, software, networks, and users [11]. These flows allow the transfer of information to occur efficiently and effectively [12], which is essential for the operation of information and communication technology systems and applications [13]. ICT information exchanges can be classified into different types, depending on the direction, purpose, and content of the information transferred [14]. For instance, the main common forms of communication are data, voice, video, text, and file communications [15].

These technologies drive major transformations and have significant potential in forest management around the world [16,17]; likewise, they can also transform the forestry industry and reduce the labor intensity of many processes [18]. Whereas the timber supply chain is highly dependent on the environment and the sustainable management of natural resources is key to its long-term viability [19], the integration of ICTs in the wood supply chain can improve efficiency, reduce costs, and improve the environmental sustainability of this industry [20,21]. In this sense, countries using ICTs in forestry are increasing the competitiveness of forest products and increasing the efficiency of forestry conservation and management functions [22]. Therefore, ICTs can have a significant impact on the development of the forestry industry as an engine of the economy [23,24].

Digital technologies have significant potential in forestry [25] and compose a major asset and transform the forest industry by providing unprecedented solutions that make forests smarter [26]. The need to implement ICTs must become a tool that serves forest users, wood buyers, operators of logging machines, or forest planners [27,28]. In addition, the implementation of ICTs and their innovations have facilitated the tasks of data collection, as well as its processing (greater accuracy and efficiency) [29]. ICTs allow large amounts of information to be processed quickly and efficiently, which can be very useful in forest management and informed decision-making [30]. In addition, they allow the monitoring of forestry operations in real time [31], the design of forest management strategies [25,32], and forest data analysis [33].

In general, data can be classified as input, output, or circumstantial data, and its analysis involves phases such as generation, collection, processing, modeling, and output information [29]. For this purpose, the forestry industry is constantly improving systems that make the most of data and make better decisions [34]. This is accomplished by allowing the information to be analyzed comprehensively in order to obtain illustrative results (geovisualization) [35]. For instance, electronic systems, and in particular Geographic Information Systems (GIS), could be used to improve forest harvesting with the prior planning of the skid trail network in order to minimize utilization impacts and risks for operators while ensuring a high level of productivity at work [34]. Some operators in the forestry sector use ICT to improve forest management results [36,37]; for instance, ICTs can be used in the planning and monitoring of forest management [38], wood production optimization [39], forest health monitoring [40], and improving traceability and transparency of the timber supply chain [41]. Considering these ICTs, efficiency and sustainability in forest management can be improved [42].

Forests provide a wide variety of services and benefits to society [43,44]. Some of the main ecosystem services that forests provide are timber and non-timber forest products, biodiversity, protection against natural disasters, climate regulation, soil and water protection, recreation, pollination, and pest and disease control [45,46]. If there is knowledge about the benefits that forests can offer for public health, actions can be implemented to manage these natural resources properly [47]. This will ensure that forests can continue providing significant long-term benefits [48]. In this manner, to guarantee the flow of forest services, it is necessary to articulate strategies that link human values with sustainability objectives [49], an aspect that is favored for the development of new technologies. ICT has enhanced productivity and reduced production costs in the forest industry and in forestry itself [50]. Data on forest growth and productivity are essential for the planning and implementation of sustainable forest management practices, which can be used within environmental applications [51]. Therefore, it is essential to have accurate and up-to-date data on the growth and productivity of forests in order to make informed decisions about their management and conservation [52].

Currently, there are various tools and technologies that allow the collection and exchange of information to occur more efficiently [53]. The use of the internet, monitoring applications, and sensors and receivers installed in equipment or machines, as well as communication between machines and between humans and machines, are some examples of these tools [54]. Remote sensing can cover a greater geographic context and time scale than other observing techniques [55,56], which can provide a more complete and detailed view of changes that occur on the Earth's surface over time [57]. For instance, are various applications for the use of ICTs to improve communication flows in the forestry sector, including the use of the following applications: (i) GIS to collect and analyze data on forests and their use [58]; (ii) mobile applications and tracking software to monitor the flow of forest products from the forest to the final consumer [59]; (iii) online platforms to share information on sustainable forest management [60]; (iv) remote monitoring systems, such as satellites and drones, to collect information on the state of forests and their evolution over time [61]; and (v) online communication tools, such as videoconferencing and social networks, to facilitate collaboration and coordination among forest sector actors in different regions and countries [62].

At this moment, there is constant growth and improvement in the implementation of ICTs and communication flows in the forestry sector [34]. There is a growing number of digital tools being developed to improve the management of forest resources and promote communication among the various actors involved [63], such as forest owners, forestry companies, government agencies, and researchers [64]. Therefore, the use of these technologies provides important improvements in forest management such as (i) the improved control of operations, (ii) automation of operations throughout the chain, (iii) improved decision-making based on data and information, (iv) combination of data on tree growth, (v) identification of timber potential, and (vi) identification of environmental conditions to plan future growth models [36,65].

The diffusion of ICTs contributes to different purposes, such as the management and conservation of forests and forest resources [66]. In addition, it can be used for the prevention of illegal logging and forest fires, raising awareness about the importance of sustainable forestry practices, and improving forest governance [67]. In general, ICTs have proven to be a valuable tool for sustainable forest management [68], and their use is expected to continue to grow in the future as new technologies are developed and access to them expands globally [69]. The use of ICTs in forest management is a topic of increasing interest [70], especially in the timber industry [9]. These technologies can enhance the efficiency and sustainability of forest production [71] while simultaneously reducing the environmental impact of this activity [72]. Therefore, it is convenient for an industry to access new technologies for exploration and transformation. Likewise, forestry as a sector should take advantage of communication technologies to be more effective in disseminating information on the benefits of forests and generating awareness.

The scope of this paper was to review information about ICTs and communication flows in the forest sector. The objectives of this study were (i) to synthesize the findings on ICTs in forestry and their applicability and (ii) to develop a systematic review of the state-of-art approaches concerning communication flows.

The document consists of five sections. Section 1 includes a review of existing literature on the topic, the importance of the research, and the relevance of the research. Section 2 details the proposed methodology. The findings are presented in Section 3. In Section 4, the discussion presents and compares the data obtained by other researchers. Finally, our work concludes in Section 5.

2. Materials and Methods

In this research, a systematic review of the literature was carried out in which information in relevant studies was identified, selected, and critically evaluated in reference to the analysis of opportunities for ICT improvement and communication flows in the forestry sector and their applicability. This review was carried out through the application of the PRISMA protocol (Preferred Reporting Items for Systematic Reviews and Meta-analyses) (Figure 1) [73]. This is a methodology applied to carry out literature reviews [74]. Information searches were carried out using key terms such as ICT and forestry, technological innovation and forestry sector, technology and forest, communication flow and forestry ICT, forestry sector and ICT and communication, information flow and forestry and communication, and ICTs in the forestry industry. Initially, a primary search for information was carried out through Google Scholar https://scholar.google.com/ (accessed on 2 January 2023). This search tool is comprehensive and accessible to identify academic studies and reports [75]. The search for scientific articles was carried out in databases such as Scopus https://www.scopus.com (accessed on 5 January 2023), Web of Science https://www.webofscience.com/ (accessed on 7 January 2023), and ScienceDirect https://www.sciencedirect.com/ (accessed on 10 January 2023). These databases offer advanced search tools that allow for easier and faster discovery of relevant and specific articles, based on the search criteria and keywords used [76]. Parameters such as author (expert in the field), affiliation (institution to which they belong), keywords (facilitating search and classification of information), year of publication (indicating the article's timeliness and relevance), and citation index were taken into account in the article selection process (reflecting the number of times the article has been cited by other authors).

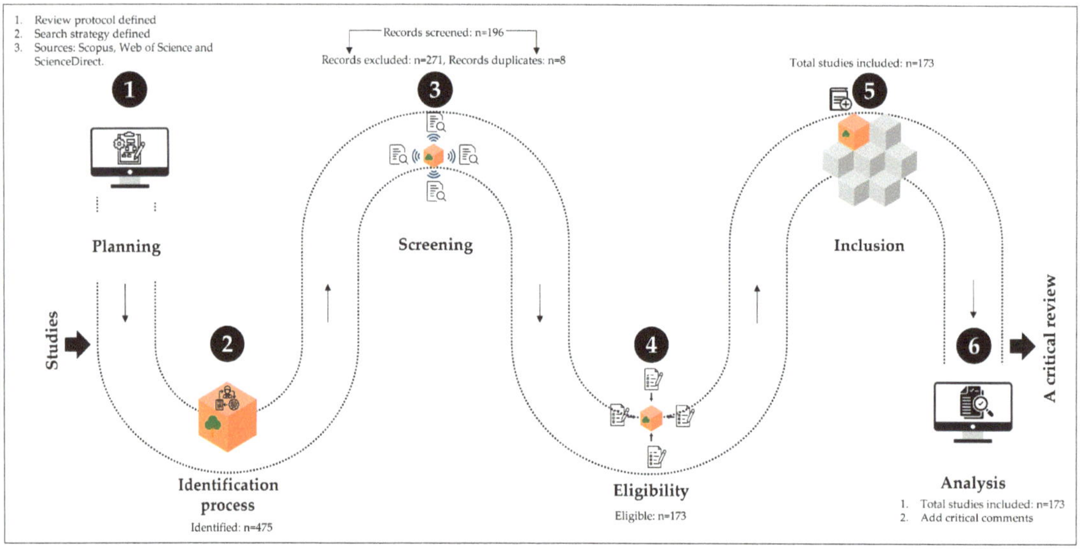

Figure 1. Methodological process.

Nevertheless, while it is recognized that studies published in other languages may contain relevant information, this study considered only scientific papers published in English and Spanish. Given the wide scope of the literature and the large number of results obtained in the search of the database, the inclusion and exclusion criteria were established based on the research topic, type of study, date of publication, language, refereed articles, and citation index. The inclusion and exclusion criteria must be clear and coherent to increase the precision and reliability of the obtained results [77]. After a preliminary review, a total of 173 papers were selected and analyzed for this research.

3. Results

It is important for us to point out some advantages and disadvantages of ICTs in the forestry sector. Several authors, including Adams and Frost [78]; Wallace et al. [79]; Fardusi et al. [80]; Gómez et al. [81]; and Tan et al. [82], indicate that among the advantages of using ICT in the forestry sector are improvements in (i) harvest planning and a reduction in costs of production; (ii) efficiency and precision in forest management; (iii) identification of forest areas at risk of fires and monitoring of forest health through the use of drones and remote sensors; (iv) the efficiency of data collection and analysis, which allows for more informed and faster decision-making; (v) innovation and technology in processes, products, and services. However, there are also disadvantages in the use of these technologies; a study carried out by Kovácsová and Antalová [83] points out, for example, the cost of its implementation and maintenance, the need for trained personnel, and the dependence on electric power and internet connectivity. In this sense, we believe that the implementation of these technologies may require a significant investment in equipment and training for the personnel in charge of their use, but at present, we consider that they are important tools to improve forest management efficiency and productivity.

3.1. ICT and Communication Flows: Definition and Importance in Forest Sector

ICTs cover a wide range of technologies that allow for obtaining, processing, analyzing, and storing information [29]. ICTs function as knowledge networks and also intervene as dissemination mechanisms because they provide effective communication channels [84]. N'dri et al. [85] suggested that the impact of ICTs is more substantial in developing countries than in developed countries; therefore, it is recommended that governments invest in infrastructure and implement ICTs progressively. Consequently, ICTs are an important tool nowadays, as they allow for the efficient acquisition, processing, analysis, and storage of information. By utilizing remote sensing technologies, such as satellite imagery and drones, ICTs enable real-time monitoring of forest cover, biodiversity, and human activities, aiding in the early detection of deforestation and illegal logging. Furthermore, ICTs empower stakeholders to make informed decisions and adopt sustainable practices for the long-term preservation of forests [86–90]. Additionally, they play a significant role in the dissemination of knowledge and communication. This makes them a key tool for economic and social development. Therefore, it is necessary to continue promoting their use and application in various productive sectors.

ICTs can play a key role in the forestry sector and in the sustainable management of forest resources. Figure 2 lists some of the ways in which ICTs can be important in the forestry sector.

Communication involves the transmission of information or ideas from a sender (person or group) to a receiver [62]. To have effective communication in the forestry area it is necessary to (a) understand the human–environment relationship, and link this concept with socioeconomic, cultural, and social aspects [91] and (b) integrate all stakeholders and maintain continuous communication between them [91,92]. Communication has become a key tool for forest management. Therefore, effective communication achieves the sustainable management of forest resources.

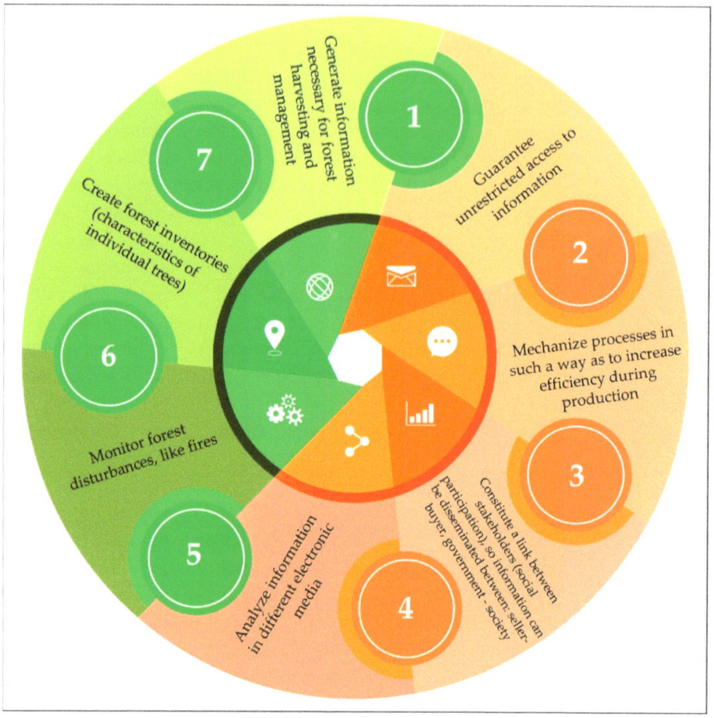

Figure 2. Importance of ICTs in the forestry area.

Stakeholders are defined as groups of people who have an impact on an organization and/or are influenced by it [91]. They can be classified into the following categories: primary stakeholders (active participants who are influenced directly by the results, such as suppliers, government, and customers) and secondary (do not have direct participation and receive a marginal effect from the situation, such as organizations and civil society in general) [93]. Therefore, primary and secondary stakeholders are important for organizations as their collaboration contributes to sustainability.

If the communication flow is linear, the elements of this process are sender, message, medium, receiver or audience, and effect [94]. Communication can be one-way when the message is delivered directly from the sender to the receiver and is considered two-way when there is interaction between the parties [91]. In the forestry area, both communication flows coexist (Figure 3); each of them integrates different mechanisms (communication channels and form of persuasion), time (short- or long-term), target audience (forest owners, general public, etc.), and message [95]. It is important to mention that if there is effective communication, it is possible to articulate and propose strategies for forest management and establish public policies focused on environmental protection and sustainable management and integrate all stakeholders' points of view into decision-making [62]. Therefore, communication flows contribute to the sustainable management of forest resources with the participation of stakeholders in forest management.

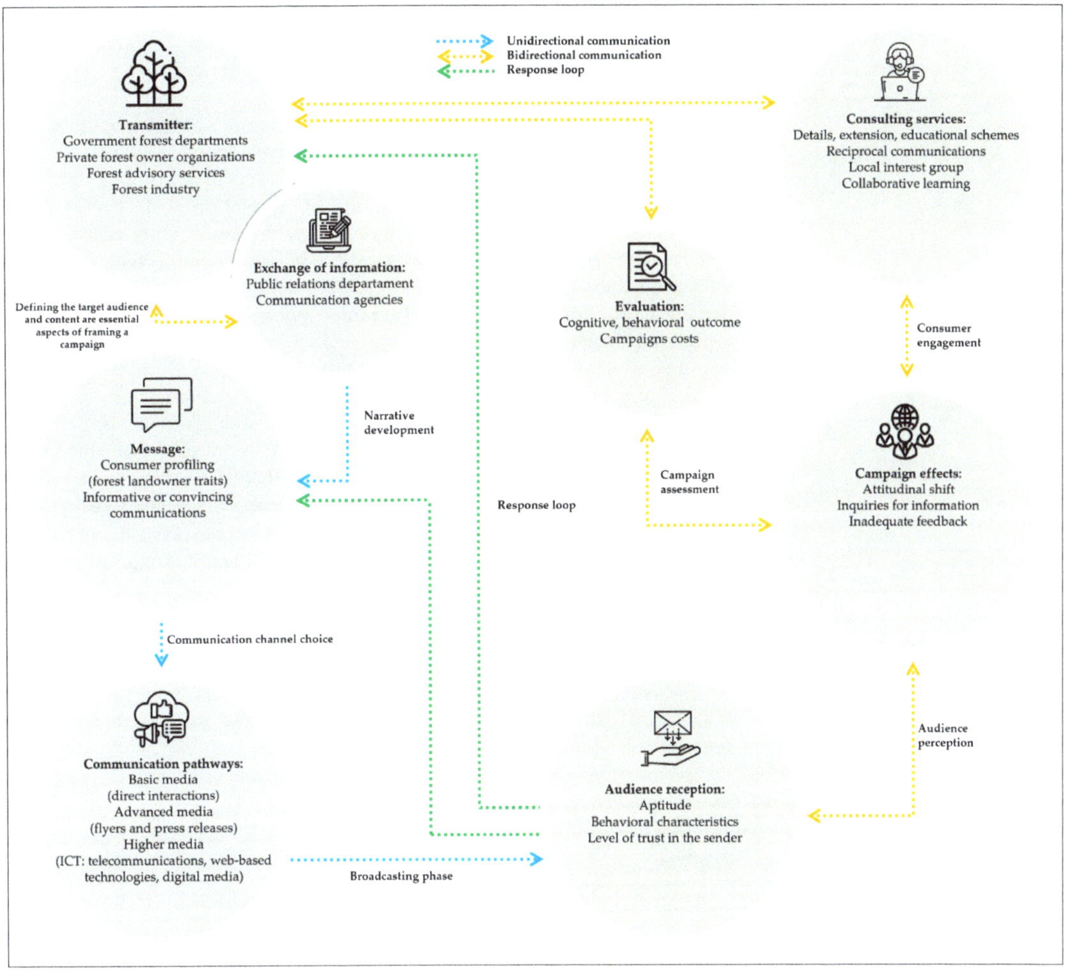

Figure 3. Communication flow in forestry.

3.2. ICT Tools and Their Applications

The adaptability of ICTs to all aspects of human life has presented the opportunity to develop diverse tools and applications focused on the forest and environment. Currently, mobile devices have furthered the common use of the internet and Services, making the monitoring of forest resources easier and more comfortable [96]. However, there remains a gap between technology and the timber industry that needs to be bridged by experiments aimed at connecting decision-makers with technology [97]. In this context, we believe that technology has great potential to improve the management and conservation of forests and the environment. However, it is necessary to bridge the gap between technology and the timber industry. Therefore, we consider it necessary to continue developing technological tools and applications for forest management, as well as integrating technology into decision-making in the forest sector.

Thus, the most common ICT tools and their respective applications in the forestry field are detailed below, such as (i) remote sensing; (ii) X-ray scanners; (iii) mobile device sensors; (iv) geographical information systems; (v) big data; (vi) radio frequency identification;

(vii) photo-trapping; (viii) techniques related to forest genetics; (ix) DNA metabarcoding; and (x) citizen science.

3.2.1. Remote Sensing

Remote sensing allows for the determination of many parameters related to the forest (productivity and state/biophysical conditions); it is a unique tool that allows a researcher to obtain repeatable observations at different temporal and spatial resolutions [98]. Remote sensing involves the acquisition, processing, and interpretation of data related to the composition of landscapes through the use of radiometric sensors, which can be active or passive [99]. Remote sensing has specific applications in forest management. For example, it enables the acquisition and processing of information about forests and other ecosystems on a large scale. Consequently, we believe that the application of this tool is important in addressing environmental and social challenges related to forest ecosystems.

This tool shows the explicit interrelation between species and habitat, allowing a researcher to characterize a specific environment, indicating changes across time and helping to predict their variations in the future. The remoting sensing application in forest management is useful in almost all cases; however, it is not precise in tele-detection in vertically and horizontally complex forest systems [100]. Therefore, the application of remote sensing could be tailored to the specific characteristics of the forest and used in conjunction with other sampling and monitoring techniques to obtain more accurate and comprehensive information for informed decision-making regarding the sustainable management of forest resources.

The functioning of active sensors is based on the emission of radiation and the subsequent measurement of energy amount and its return time; by comparison, passive sensors measure the amount of energy reflected or emitted from the matter [99]. Passive systems, unlike active systems, are affected by weather conditions [101] and cannot get details below the forest canopy [102]. Table 1 presents a list of sensors and their applications in the forestry area.

Table 1. Sensors used in forestry areas (applications).

Type of Sensor	Sensing Method	Data Type	Sensor	Measured Variables	Spatial Scale
PASSIVE	Aerial photography	2D	Photo cameras	Landscape characterization	Local
	Satellite imagery	Multispectral	Landsat TM ETM + SPOT ASTER MODIS	Landscape characterization, meteorological observations, plant productivity and chemistry.	Regional to local
		Hyperspectral	CHRIS HYPERION	Landscape characterization, meteorological observations, plant productivity and chemistry, species composition	Regional to local
		High spatial resolution	RAPID EYE (5 m) IKONOS (<1 m) WORLDVIEW (<1 m)	Identifying individuals of a landscape	Local
		High temporal resolution	SPOT (4–5 DAYS) MODIS (DAILY)	Changes in landscape over time	Global to local
ACTIVE	Airborne LiDAR and radar	Multilevel, high spatial resolution	SLICER LVIS Vertical-looking radar Side-looking radar	Landscape characterization (identifying individuals), crop production	Regional to local
	Harmonic radar	-	-	Tracking individuals, vegetation structure—3D	-
	Terrestrial lasers	High spatial, temporal resolution	Mobile Static	Characteristics of vegetation (Physical and biophysical), identifying individuals	-

Remote sensing comprises Airborne Laser Scanning (ALS), while proximal sensing is related to Terrestrial Laser scanning (TLS); both methods differ in terms of spatial resolution and coverage. ALS has a higher spatial coverage and lower resolution; in contrast, TLS has a lower spatial coverage and a higher resolution [56]. Within this framework, remote sensing includes two main methods: ALS and TLS, which differ in terms of spatial resolution and coverage. Based on this, both methods are useful according to the specific needs of the analysis and the conditions of the land surface being evaluated.

According to Coops [98], Light Detection and Ranging (LiDAR) and RADAR have a greater potential for mapping forests in terms of volume and biomass. LiDAR allows the detection of 3D forest canopy [98], so the aspects that can be determined by this tool are canopy cover, height, volume, biomass [98,102], basal area and stem density [102], forest stratification and distribution, and mean diameter [103], as well as ecological applications such as wildlife monitoring [104]. The study carried out by Borz and Proto [105] indicates that in the last decade, LiDAR-based methods have been successfully tested in several forestry-related applications, in particular in forest inventory applications, focusing mainly on data accuracy. Their usefulness for the quantitative assessment of harvested timber has been less investigated. In particular, studies on resource accounting, including the time required for different log scanning options, are still lacking.

Radar systems use electromagnetic energy (to transmit and receive pulses); therefore, to use radar data it is necessary to consider the canopy, the wavelength of the signal, and the angle [106]. Moreover, some remote sensors have a cost, but it is possible to find radar data available for free, as is the case of the Sentinel-1 Satellite (high temporal resolution: 3 to 6 days and spatial resolution: 5 × 20 m/independent of cloud cover) [107]. In essence, radar systems are a valuable tool for remote sensing because they can provide detailed information about the land surface under conditions where other remote sensing techniques may be limited. Therefore, this tool can be useful for forest management, biodiversity conservation, and the sustainable management of natural resources. The cost of some ICTs is becoming more affordable, making them more accessible to a wider audience (Table 2).

Table 2. Cost of some ICTs [98].

Type of Sensor	Cost
MODIS (Terra modis, aqua modis) SPOT Landsat TM Sentinel-1 (Radar—Europe)	Free
LiDAR	3–5 USD/ha
RADAR	<1 USD/ha
Rapid Eye, digital globe	1–3 USD/ha
CASI, AVRIS, HYPERION	3–5 USD/ha

3.2.2. X-ray Scanners

X-ray scanning is a tool used to determine the quality of wood or estimate the amount of wood inside a stem or trunk. Its operation consists of the emission of X-rays which are transmitted towards an object, and as a result of the penetration, X-ray beams are attenuated, generating a digital image of this object [108]. Therefore, X-ray scanning is a useful tool that allows for the detection of internal defects in wood, which is important for identifying areas prone to structural failure or disease propagation.

The most popular devices that are based on X-ray technology are SilviScan, Itrax, and QTRS. These devices are characterized by having a good level of accuracy (approximately 50 μm/pixel); however, the preparation of the samples involves an arduous task [109]. Nowadays, numerous industrial prototypes of X-ray scanning equipment are inappropriate for the wood industry, especially for high moisture content logs [110,111]. However, although there are popular devices using X-ray technology, the development of industrial prototypes of X-ray scanner equipment is important to meet the needs of the timber industry. This will enable the development of more accurate and efficient devices.

X-ray scanners base their functionality on the theories of Radon, theoretically demonstrated in 1970, which indicated projections of the object depending on the number of directions considered [109]. Whatever the number of directions, X-ray beams are sent and detectors measure the X-ray radiation that is transmitted through the object [111]. These studies point to the important role of X-ray scanners for wood quality assessment, tree species identification, determination of wood density, and moisture.

3.2.3. Mobile Devices Sensors

Recently, the use of mobile devices has increased, together with mobile cloud computing, allowing for data collection from various sensors in a short period of time [112]. In mobile devices such as smartphones and watches, there are various sensors that allow for the development of models related to activities that require engines such as logging; other sensors are used to track individual trees and analyze forest productivity [56]. This highlights the important role of mobile devices in enhancing efficiency and precision in decision-making across various fields, including the management and monitoring of forestry activities.

This ICT allows for the realization of in-situ observations, regardless of terrain conditions (easy transport); at the same time, it gathers measurements with an acceptable speed. However, these technologies are still in research and development, so their accuracy is not comparable to other static systems [113]. Even so, they have been a useful tool in forest degradation monitoring, especially in developing countries [114]. In conclusion, the ongoing development and refinement of sensors in mobile devices for forest monitoring could lead to significant advancements in the field of forest management. This is due to their accessibility, efficiency, and ability to collect data in hard-to-reach areas.

3.2.4. Geographical Information Systems (GIS)

GIS, in conjunction with data obtained by remote sensing (satellite images and drones), allow for the mapping of vegetation at different scales. It is also possible to assess risks such as forest fires through a multiple-criteria decision analysis (MCDA) that integrates user approaches (variety of information requirement) [115]. In the same way, GIS have played important roles in forest resource management, wood harvest planning, and forest fire management, among others [116]. Therefore, GIS are tools for the collection, processing, and analysis of spatial data. This facilitates informed and effective decision-making for the management of forest resources and the prevention of forest risks.

ArcGIS 10.5® software differ by their ability to collect data geometries (such as points, lines, polylines, or polygons) and other attributes, as well as their compatibility with data formats such as ESRI shapefile, CSV, and KML. Currently, many mapping apps were developed for mobile devices, including ArcGIS, Mapit Spatial, Qfield, SW Maps, Global Mapper Mobile, Locus GIS, and others [117]. GIS allow the integration of different data sources to obtain a complete and detailed vision of forest ecosystems. Therefore, GIS influence the planning of sustainable forest management.

3.2.5. Big Data

Big data refers to large data sets whose size exceeds the capacity of typical database software [118,119]. For a better understanding, these characteristics are represented in Figure 4.

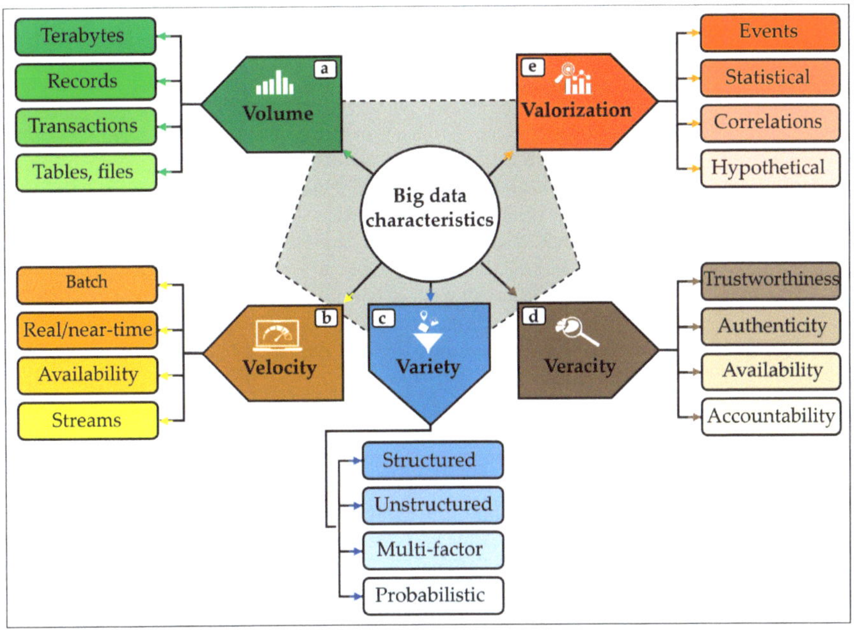

Figure 4. Big data characteristics. Volume: huge amount of storage (**a**). Velocity: the speed of data generation and processing according to demands and challenges of the analysis (**b**). Variety: different sources of data (**c**). Veracity: reliability of data (**d**). Valorization: ability to disseminate information (**e**).

In forestry, big data includes information or registers about metrics of trees, species, and volume of wood produced [119]. Additionally, it has demonstrated huge potential in forest management and ecosystem protection [120]. In agriculture, big data is largely employed in developed countries to promote the production and management of numerous products. Therefore, the use of big data in the forestry sector can have a significant impact on the economy, the environment, and society.

3.2.6. Radio Frequency Identification (RFID)

RFID identifies objects by means of radio frequency signals that correspond to the following groups: low frequency (between 30 and 300 kHz), high frequency (between 3 and 30 MHz), ultra-high frequency (between 300 and 3000 MHz), and microwaves (between 2 and 30 GHz) [121,122]. Overall, RFID technology is a versatile tool that can be used in the forestry sector to identify and track wood and other forest products along the supply chain. The choice of the appropriate frequency range will depend on the specific needs of the application.

This system includes electronic compounds, such as a microprocessor, a transponder, a reader, and a management system. The main objective of this technology is to get information about objects, animals, or plants; in this context, each microchip could be attached to a tree's base to register its localization and size, and in case of logging, it can hold the data about who cut it, so that illegal logging can be controlled [122]. Nevertheless, the application of this ICT may have some effects on ecosystems, principally on wildlife [123]. However, the use of radio frequency identification technology must be carefully considered and managed to minimize its impact on ecosystems and maximize its economic and social benefits.

3.2.7. Photo-Trapping

Photo-trapping is a technique used to monitor wildlife in their natural habitat, particularly in forested areas. [124]. This technique involves the installation of cameras in strategic locations within the forest, which are automatically triggered when they detect movement or heat [125]. Furthermore, it is a non-invasive and efficient technique [126]. This allows a researcher to obtain information about forest biodiversity, species distribution, species abundance, their habits and behaviors, and their interaction with the natural environment [125]. Consequently, it provides detailed information for assessing the status of wildlife present in a specific area. Therefore, its use can significantly contribute to the development of strategies for the sustainable management of biodiversity in forest ecosystems.

To carry out photo-trapping, special digital cameras are used that can be configured to automatically take images at different times of the day or night [127]. The cameras can be equipped with motion sensors, heat sensors, or both, to detect the presence of animals in the area [128]. These cameras are typically resistant to field conditions such as rain, sun, and cold temperatures [129]. The images captured by the cameras can be analyzed manually or through the use of specialized software to identify the animal species that appear in the images [130]. In this context, we believe it is important to place the cameras in locations where wildlife activity exists, such as trails, feeding areas, or areas with access to water. Therefore, photo-trapping does not disturb the natural behavior of animals and adheres to ethical protocols for wildlife research and monitoring.

3.2.8. Techniques Related to Forest Genetics

Forest genetics is a branch of forest biology that focuses on the study of genetics and molecular biology of trees and other forest species [131]. Some techniques related to forest genetics include DNA analysis, genetic improvement, cloning, molecular markers, and next-generation sequencing techniques [132]. Forest genetics techniques are important tools in forest management, enabling scientists to study the genetic diversity of forest species, improve wood quality, enhance disease resistance and other important traits, and conserve rare or endangered species [133]. Forest genetics employs various techniques to study the genetic variability of forest populations. Therefore, this tool contributes to sustainable forest management.

In this context, techniques related to forest genetics also have a connection with ICTs, as many of them require specialized equipment and software for their application and data analysis [134]. Some of the ways in which information and communication technology is used in forest genetics include databases, data analysis, simulation models, communication, and dissemination of results [135]. ICTs have allowed forest genetics scientists to gather and analyze large amounts of genetic information, as well as communicate the results in a clearer and more accessible manner to society [132,136–138]. In conclusion, forest genetics and ICTs are closely related, allowing for the collection and analysis of large amounts of genetic information, as well as the communication of results in a clearer and more accessible manner to society, which contributes to more efficient management and conservation of forests.

3.2.9. DNA Metabarcoding

DNA metabarcoding is a molecular technique used to identify species and communities of organisms from environmental samples such as soil, water, or air [139]. This technique is based on the sequencing of a specific region of the DNA from the organisms present in the sample, known as barcoding [140]. The obtained DNA sequence is compared to reference sequence databases to identify the corresponding species or taxon [141]. This molecular technique has applications in biodiversity research and monitoring, as well as in the management of forest ecosystems.

DNA metabarcoding has been successfully used in biodiversity studies, ecology, and conservation in various communities of organisms, ranging from plants and animals to microorganisms [142]. Some of the advantages of this technique include the ability to rapidly and accurately identify a wide range of species and taxa, as well as the capability to analyze complex environmental samples containing multiple species [143]. DNA metabarcoding is a technique that offers advantages for species identification and biodiversity analysis, benefiting ecosystem regeneration and species conservation efforts.

DNA metabarcoding is a complex technique that involves several stages: (i) Sample collection; (ii) DNA extraction, which can be done using commercial kits or standardized laboratory protocols; (iii) Amplification of DNA from organisms present in the sample; (iv) Purification of amplified DNA; (v) DNA sequencing; (vi) Data processing using specialized software, which may include filtering, assembly, and taxonomic assignment tools; and (vii) Data analysis to obtain information about the diversity and structure of the organism communities present in the sample, including species richness, relative abundance of each species, and taxonomic composition of the community [144–146]. In conclusion, this technique involves multiple stages, from sample collection to data analysis. Furthermore, it has been used in biodiversity, ecology, and conservation studies in various populations. Therefore, it becomes a promising technique to provide valuable information on biodiversity and communities of organisms present in environmental samples.

3.2.10. Citizen Science

Citizen science can be considered an ICT to the extent that it utilizes digital tools to engage people in scientific projects [147]. Technology has allowed citizen science projects to reach higher levels of participation and collaboration and has transformed the way data is collected, analyzed, and shared [148]. The collected data can be stored in online databases and analyzed by participants and scientists using specialized software and data extraction techniques [149]. ICTs have allowed citizen science projects to reach a broader and more diverse audience through online platforms that enable remote participation and online collaboration [150]. Furthermore, ICTs have enabled the communication and dissemination of results from citizen science projects [151]. For instance, social networks and mobile applications have been used to engage people in citizen science projects, enabling real-time communication and the creation of online communities to share information and data [152]. With the continuous advancement of ICTs, it is expected that citizen science will continue to grow and play an important role in scientific research and biodiversity management. In conclusion, ICTs have significantly enhanced the capacity of citizen science projects to store, analyze, and communicate collected data. This has resulted in increased efficiency in scientific research and improved accessibility and understanding of the results by society, which is of great importance in sustainable forest management.

Finally, it is important to mention that regarding the parameters for article selection in this study, the publication period ranged from 2004 to 2023. Similarly, when searching in the title, abstract, or keywords of published scientific articles, the most relevant countries were identified in terms of the origin of authors of works related to ICTs in the forestry sector: the United States, China, Canada, Spain, Finland, and Brazil. The most frequently used keywords in related scientific articles were "ICT and forestry", "technological innovation and the forestry sector", "technology and forests", "communication flow and forestry ICT", and "ICT in the forestry industry". It is also important to highlight that among the works with the highest number of citations are research papers related to the keywords "ICT and forestry" and "technology and forest", with over 3500 citations.

4. Discussion

This study focused on conducting an analysis of the opportunities for improving ICTs and communication flows in the forestry sector and their applicability based on the results reported by various studies and data repositories. Given the above information, Andreopoulou et al. [153] and Sharma et al. [154] mention that the forestry sector is one of the most important in terms of natural resources and sustainability and can significantly result from the implementation of ICTs. In this sense, for instance, CEPAL [155] indicates as one of its priorities support for the implementation of the 2030 Agenda for Sustainable

Development in Latin America and the Caribbean, which includes the use of ICTs. Therefore, we believe that ICTs can facilitate the achievement of sustainable development goals related to the management and conservation of forest resources. This fact is supported by a study carried out by Fardusi et al. [80], which mentions that the forestry sector is an area in which various ICT tools and applications have begun to be used to improve forest management and monitoring. A second study by Molinaro and Orzes [71] indicates that these technologies have been used successfully in the management of forestry companies to improve the decision-making, planning, and monitoring of forestry operations. In summary, we believe that the applications and technological tools used in the forestry sector improve the management and sustainability of forest resources by automating business management processes in the forestry sector, reducing costs and typing errors.

In this context, a study carried out by Zhang et al. [156] mentions that ICTs play a key role in the management and monitoring of forest resources, as well as in the optimization of production and marketing processes. A second study by Chen et al. [157] indicates that the use of ICTs in the forestry sector allows for more efficient management of natural resources, improving decision-making and increasing the profitability of forestry companies. For these reasons, we believe that they can be used to improve energy efficiency and digital infrastructure in the forestry sector. This fact is corroborated by a study carried out by Anastasiadou et al. [158] which points out that ICTs can help address the challenges of governance and population participation. In addition, Badiane et al. [159] state that ICTs can also provide opportunities to strengthen the capacities and skills of the forestry sector both in the public and private spheres, which can be especially important in the context of trade opening. In this sense, we believe that ICTs are essential for the forestry sector due to their ability to improve efficiency and productivity in the management of natural resources and are essential for the success and sustainability of the forestry sector in the digital age.

In view of the foregoing, we additionally believe that to further improve the use of ICTs in the forestry sector, new technologies can be implemented. In this sense, for instance, a study carried out by Palander [7] indicates that LIDAR allows for the production of digital terrain models and obtaining detailed information on topography and vegetation. A second study carried out by Galaz et al. [160] indicates that mobile applications can be developed for plantation management and decision-making in the field. A third study by Dainelli et al. [161] mentions that another option is the use of drones for the inspection of forests and early detection of pests or diseases. However, it is also important to consider the communication flows in the forestry sector. For example, a study carried out by Näyhä [162] and Castillo et al. [163] indicates that it is necessary to develop deep and wide-ranging communication strategies in relation to the needs of different stakeholders at different hierarchical levels of sustainability. Therefore, we consider that in the forestry sector, the use of different communication models is not enough to improve the acceptability of operations and competitiveness in the markets. However, they are tools that are necessary for its application.

Indeed, there are multiple ICTs used in the forestry sector to improve the collection, management, and analysis of forest data, as well as to improve supply chain monitoring, forest management, biodiversity conservation, and communication among the various stakeholders involved in the sector. In this context, Rao et al. [164] point out that ICTs can improve the efficiency of forest management and contribute to the protection and conservation of forest resources and biodiversity. Similarly, Sraku [165]; Belden et al. [166]; and Dastres and Soori [40] mention that access to ICTs in the forestry sector depends on telecommunications infrastructure, the level of economic development, public policies and regulations, the level of education and training, and the availability of financing. It is important to address these factors to improve access to ICTs in the forestry sector and harness their full potential for forest management and biodiversity conservation.

In this context, Liu et al. [167] mention that new technologies are being developed in the field of artificial intelligence. These technologies could have a significant influence on forest management [168]. For instance, a study carried out by Grabska et al. [169] mentions that machine learning algorithms can analyze large volumes of forest data and provide valuable information for decision-making. A second study by Singh et al. [170] indicates that artificial intelligence technologies are being created to improve the early detection of forest diseases and pests. However, in communication flows, artificial intelligence can also enhance communication among the various actors involved in forest management [53]. For instance, Kożuch et al. [171] reveal in their study that chatbots and virtual assistant systems can answer common questions about forest management, enabling more effective communication and efficient resource management. In summary, artificial intelligence technologies have a high potential to enhance forest management.

Regarding access to ICTs in the forestry sector, the digital divide is a reality that affects many regions of the world. A study conducted by Lowery et al. [172] states that the application of ICTs in the forestry sector faces specific challenges depending on the region and context in which they are used. These challenges include limitations in satellite coverage, varying levels of information access, diverse climatic conditions, different languages and cultures, varied approaches and priorities in forest management, lack of training and skills, costs, technical challenges, and maintenance issues. A second study by Hossain [173] indicates that in some countries, particularly the poorest and least developed ones, access to ICTs is limited or absent, which can have a negative impact on forest management and biodiversity conservation. Therefore, we believe that the challenges related to ICT access should be addressed, and their use should be integrated into a broader forest management strategy. With this, we believe that it is important to work in collaboration with the actors involved in forest management to ensure that ICTs are accessible, effective, and sustainable.

Although the PRISMA model is a rigorous and systematic methodology for conducting literature reviews, it also has certain limitations that must be considered. Some of the potential limitations of the PRISMA model in this study could be the following: (i) the limited availability of relevant studies in the literature, which could restrict the number of studies included in the review; (ii) the variable quality of the included studies, which could affect the reliability and validity of the review results; (iii) the possibility that some relevant studies have been omitted due to the selection of specific databases or the exclusion of languages other than English or Spanish; and (iv) the possibility that the inclusion and exclusion criteria are not entirely appropriate for the specific topic of the review.

The potential future impact of ICTs and communication flows in the forestry sector is promising, with significant implications for the management of forest resources. Initially, the advancement of ICTs is expected to enable broader automation of forest management processes, resulting in improved efficiency and cost reduction. Additionally, the integration of various technologies utilized in forest management is anticipated to be optimized through the application of ICTs, enhancing the accuracy and efficiency of information analysis. In terms of communication flows, ICTs are poised to facilitate increased collaboration and coordination among diverse stakeholders engaged in forest management, including forest owners, forestry companies, government agencies, and civil society members. Consequently, the utilization of ICTs has the potential to enhance the overall management of forest resources.

5. Conclusions

ICTs and communication flows generate a positive impact in the forest sector because they allow for the monitoring and analyzing of landscapes on different temporal and spatial scales, thus contributing to forest management and the establishment of norms or policies that guarantee preservation. In addition, it allows the stakeholders to be linked and establishes effective communication flows between them. Overall, there is a wide range of technologies that can be implemented in the forestry area according to the needs of each user, such as productivity improvements, monitoring, and markets, among others. Furthermore, sustainability is an important aspect of forest management, and ICTs can play a role in promoting sustainability in several ways. For instance, ICTs can be used to monitor forest health and identify areas that are at risk of degradation. This information can then be used to develop and implement management plans that will help to protect forests and their resources. Additionally, ICTs can be used to educate the public about the importance of forests and the need for sustainable forest management. This can help to raise awareness of the issue and encourage people to make choices that support sustainable forest practices.

Finally, the forestry sector can significantly benefit from the implementation of ICTs and communication flows. These technologies can improve the management and sustainability of forest resources by automating business processes, reducing costs, and minimizing errors. They can also facilitate the achievement of sustainable development goals related to the management and conservation of forest resources. However, the implementation of these technologies may require a significant investment in equipment and training for personnel. Furthermore, the use of new technologies such as LIDAR, mobile applications, and drones can further improve the efficiency and productivity of the forestry sector. It is also essential to consider communication flows and develop comprehensive communication strategies to ensure the acceptability of operations and competitiveness in the markets. Despite some disadvantages such as the cost of implementation and maintenance, the use of ICTs and communication flows are essential tools for the success and sustainability of the forestry sector in the digital age.

Author Contributions: Conceptualization, A.V.G.M. and D.D.C.V.; methodology, A.V.G.M. and D.D.C.V.; data gathering, A.V.G.M. and D.D.C.V.; writing—original draft preparation, A.V.G.M., D.D.C.V. and M.V.M.; writing—review and editing, A.V.G.M., D.D.C.V. and M.V.M.; supervision, A.V.G.M., D.D.C.V. and M.V.M.; project administration, A.V.G.M. All authors have read and agreed to the published version of the manuscript.

Funding: Alex Vinicio Gavilanes Montoya and Danny Daniel Castillo Vizuete researchers at Transilvania University of Brasov, Romania, was supported by the program "Transilvania Fellowship for Postdoctoral Research/Young Researchers".

Institutional Review Board Statement: Not applicable.

Informed Consent Statement: Not applicable.

Data Availability Statement: Not applicable.

Acknowledgments: The authors would like to thank to Transilvania University of Brasov, specially to Stelian Alexandru Borz who supervises this study and to Escuela Superior Politécnica de Chimborazo, due to this research is part of the IDIPI-266 Project from ESPOCH.

Conflicts of Interest: The authors declare no conflict of interest.

References

1. Volkodaeva, A.; Balanovskaya, A.; Rustenova, E. Trends in information and communication technologies development in context of economy digitalization. In *Digital Technologies in the New Socio-Economic Reality*; Springer International Publishing: Berlin/Heidelberg, Germany, 2022; pp. 583–592. [CrossRef]
2. Sookhak, M.; Jabbarpour, M.; Safa, N.; Yu, F. Blockchain and smart contract for access control in healthcare: A survey, issues and challenges, and open issues. *J. Netw. Comput. Appl.* **2021**, *178*, 102950. [CrossRef]
3. Øverby, H.; Audestad, J.; Øverby, H.; Audestad, J. Information and Communication Technologies. In *Introduction to Digital Economics: Foundations, Business Models and Case Studies*; Springer: Berlin/Heidelberg, Germany, 2021; pp. 17–31. [CrossRef]
4. Munirathinam, S. Industry 4.0: Industrial internet of things (IIOT). In *Advances in Computers*; Elsevier: Amsterdam, The Netherlands, 2020; Volume 117, pp. 129–164. [CrossRef]
5. Yaakob, M.; Awang, H.; Ismail, M.; Zain, F.; Kasim, M.; Adnan, A. Backward and forward reviews on technical and vocational education and training (TVET) in Malaysia: The evolution and ICT-driven future prospect. *Univers. J. Educ. Res.* **2020**, *8*, 2197–2203. [CrossRef]
6. Shivaprakash, K.; Swami, N.; Mysorekar, S.; Arora, R.; Gangadharan, A.; Vohra, K.; Kiesecker, J. Potential for Artificial Intelligence (AI) and Machine Learning (ML) applications in biodiversity conservation, managing forests, and related services in India. *Sustainability* **2022**, *14*, 7154. [CrossRef]
7. Palander, T. Outsourcing Issues of Wood Supply Chain Management in the Forest Industry. *For. Sci.* **2022**, *68*, 521–532. [CrossRef]

8. Budiman, E.; Putra, G.; Rahman, R.; Prafanto, A.; Wardhana, R.; Hairah, U. Mobile Network Experience in Forest Research and Conservation Areas. In Proceedings of the 2021 3rd East Indonesia Conference on Computer and Information Technology (EIConCIT), Surabaya, Indonesia, 9–11 April 2021; IEEE: New York, NY, USA, 2021; pp. 227–231. [CrossRef]
9. Heiskanen, A.; Hurmekoski, E.; Toppinen, A.; Näyhä, A. Exploring the unknowns–State of the art in qualitative forest-based sector foresight research. *For. Policy Econ.* **2022**, *135*, 102643. [CrossRef]
10. Adnan, S. ICTs for Development: Building the Information Society by Understanding the Consumer Market. In *Handbook of Communication for Development and Social Change*; Springer: Berlin/Heidelberg, Germany, 2020; pp. 989–1012. [CrossRef]
11. Singh, J.; Singh, S.; Kumari, M. Role of ICT in supply chain management. *J. Interdiscip. Cycle Res.* **2020**, *12*, 992.
12. Martinez, J.; Serna, M.; Sanchez, M.; Aguilar, J. The impact of ICT and knowledge management on open innovation: Empirical evidence from Mexico. *Int. J. Innov. Res. Sci. Stud.* **2023**, *6*, 193–204. [CrossRef]
13. Schilling, L.; Seuring, S. The Role of Information and Communication Technology in Managing Supply Chains in Base-of-the-Pyramid Markets. *IEEE Trans. Eng. Manag.* **2022**, *70*, 1186–1198. [CrossRef]
14. Kulkarni, V.; Sahoo, S.; Thanikanti, S.; Velpula, S.; Rathod, D. Power systems automation, communication, and information technologies for smart grid: A technical aspects review. *TELKOMNIKA (Telecommun. Comput. Electron. Control)* **2021**, *19*, 1017–1029. [CrossRef]
15. Gloria, E.; Ifeoma, E. Entrepreneurship opportunities in information and communication technology (ICT). *WATARI Multi-Discip. J. Sci. Technol. Math. Educ.* **2023**, *7*, 34–40.
16. Wu, J.; Guo, S.; Huang, H.; Liu, W.; Xiang, Y. Information and communications technologies for sustainable development goals: State-of-the-art, needs and perspectives. *IEEE Commun. Surv. Tutor.* **2018**, *20*, 2389–2406. [CrossRef]
17. Gouvea, R.; Kapelianis, D.; Kassicieh, S. Assessing the nexus of sustainability and information & communications technology. *Technol. Forecast. Soc. Chang.* **2018**, *130*, 39–44. [CrossRef]
18. Müller, F.; Jaeger, D.; Hanewinkel, M. Digitization in wood supply–A review on how Industry 4.0 will change the forest value chain. *Comput. Electron. Agric.* **2019**, *162*, 206–218. [CrossRef]
19. Luo, L.; O'Hehir, J.; Regan, C.; Meng, L.; Connor, J.; Chow, C. An integrated strategic and tactical optimization model for forest supply chain planning. *For. Policy Econ.* **2021**, *131*, 102571. [CrossRef]
20. Marchi, E.; Chung, W.; Visser, R.; Abbas, D.; Nordfjell, T.; Mederski, P.; Laschi, A. Sustainable Forest Operations (SFO): A new paradigm in a changing world and climate. *Sci. Total Environ.* **2018**, *634*, 1385–1397. [CrossRef]
21. Dalalah, D.; Khan, S.; Al-Ashram, Y.; Albeetar, S.; Abou, Y.; Alkhouli, E. An integrated framework for the assessment of environmental sustainability in wood supply chains. *Environ. Technol. Innov.* **2022**, *27*, 102429. [CrossRef]
22. Šulyová, D.; Koman, G. The significance of IoT technology in improving logistical processes and enhancing competitiveness: A case study on the World's and Slovakia's wood-processing enterprises. *Sustainability* **2020**, *12*, 7804. [CrossRef]
23. Gedik, T.; Durusoy, İ. Adoption of information and communications technology in the Turkish forest products industry: A case study. *Kastamonu Univ. J. For. Fac.* **2020**, *20*, 38–48. [CrossRef]
24. Huang, Y.; Khan, J. Has the information and communication technology sector become the engine of China's economic growth? *Rev. Dev. Econ.* **2022**, *26*, 510–533. [CrossRef]
25. Morkovina, S.; Nasriddinov, S.; Shanin, I. Forestry digital platform of Russia. *IOP Conf. Ser. Earth Environ. Sci.* **2020**, *595*, 012042. [CrossRef]
26. Maksimovic, M. Greening the future: Green Internet of Things (G-IoT) as a key technological enabler of sustainable development. In *Internet of Things and Big Data Analytics toward Next-Generation Intelligence*; Springer: Cham, Switzerland, 2018; Volume 30, pp. 283–313. [CrossRef]
27. Červený, L.; Sloup, R.; Červená, T.; Riedl, M.; Palátová, P. Industry 4.0 as an Opportunity and Challenge for the Furniture Industry—A Case Study. *Sustainability* **2022**, *14*, 13325. [CrossRef]
28. Shabaev, A. Manufacturing Information Systems for Forestry Companies Based on Integrated Modular Approach «Many Companies—One Tool». In *ICT with Intelligence Applications*; Springer: Singapore, 2022; Volume 248, pp. 633–641. [CrossRef]
29. Zhang, X.; Wang, J.; Vance, J.; Wang, Y.; Wu, J.; Hartley, D. Data Analytics for Enhancement of Forest and Biomass Supply Chain Management. *Curr. For. Rep.* **2020**, *6*, 129–142. [CrossRef]
30. Gutiérrez, E.; Lozano, S. Cross-country comparison of the efficiency of the European forest sector and second stage DEA approach. *Ann. Oper. Res.* **2022**, *314*, 471–496. [CrossRef]
31. Jäntti, M.; Aho, M. Improving the Quality of ICT and Forestry Service Processes with Digital Service Management Approach: A Case Study on Forestry Liquids. In *Quality of Information and Communications Technology: 15th International Conference, QUATIC 2022, Talavera de la Reina, Spain, 12–14 September 2022, Proceedings*; Springer International Publishing: Cham, Switzerland, 2022; pp. 175–189. [CrossRef]
32. Hu, J.; Chen, Y.; Yang, Y. The development and issues of energy-ICT: A review of literature with economic and managerial viewpoints. *Energies* **2022**, *15*, 594. [CrossRef]
33. Kalambet, M.; Yagotinceva, N.; Kolbina, O.; Yagotinceva, T.; Aksenova, A. Opportunities and prospects of IoT application in landscape architecture, design and information and communication technology. *IOP Conf. Ser. Earth Environ. Sci.* **2020**, *507*, 012007. [CrossRef]
34. Picchio, R.; Proto, A.; Civitarese, V.; Di Marzio, N.; Latterini, F. Recent contributions of some fields of the electronics in development of forest operations technologies. *Electronics* **2019**, *8*, 1465. [CrossRef]

35. Goel, R.; Vishnoi, S. Urbanization and sustainable development for inclusiveness using ICTs. *Telecommun. Policy* **2022**, *46*, 102311. [CrossRef]
36. Choudhry, H.; O'Kelly, G. Precision Forestry: A Revolution in the Woods. Basic Materials, Paper & Forest Products. 2018. Available online: https://www.mckinsey.de/~/media/McKinsey/Industries/Paper%20and%20Forest%20Products/Our%20Insights/Precision%20forestry%20A%20revolution%20in%20the%20woods/Precision-forestry-A-revolution-in-the-woods-final.pdf (accessed on 18 January 2023).
37. Mederski, P.; Borz, S.; Đuka, A.; Lazdiņš, A. Challenges in forestry and forest engineering–Case studies from four countries in East Europe. *Croat. J. For. Eng. J. Theory Appl. For. Eng.* **2021**, *42*, 117–134. [CrossRef]
38. Bhunia, G.; Shit, P. Forest Management with Advance Geoscience: Future Prospects. *Spat. Model. For. Resour. Manag. Rural Livelihood Sustain. Dev.* **2021**, 3–31. [CrossRef]
39. Panyavina, E.; Morkovina, S.; Kulakova, E.; Alyabyeva, M. Study of the readiness of forestry enterprises to use digital technologies. *IOP Conf. Ser. Earth Environ. Sci.* **2021**, *875*, 012034. [CrossRef]
40. Dastres, R.; Soori, M. The role of information and communication technology (ICT) in environmental protection. *Int. J. Tomogr. Simul.* **2021**, *35*, 24–37. Available online: https://hal.science/hal-03359776/ (accessed on 18 January 2023).
41. Rusyn, B.; Obukh, Y.; Kosarevych, R.; Lutsyk, O.; Korniy, V. Information System for Analysis of Forest Plantations and Monitoring of Ecological Condition. In Proceedings of the 2021 IEEE 4th International Conference on Advanced Information and Communication Technologies (AICT), Lviv, Ukraine, 21–25 September 2021; pp. 75–78. [CrossRef]
42. Sofia, S.; Maetzke, F.; Crescimanno, M.; Coticchio, A.; La Mela, D.; Galati, A. The efficiency of LiDAR HMLS scanning in monitoring forest structure parameters: Implications for sustainable forest management. *EuroMed J. Bus.* **2022**, *17*, 350–373. [CrossRef]
43. Dodev, Y.; Zhiyanski, M.; Glushkova, M.; Shin, W. Forest welfare services-the missing link between forest policy and management in the EU. *For. Policy Econ.* **2020**, *118*, 102249. [CrossRef]
44. O'Briena, L.; Urbaneka, R.; Gregory, J. Ecological functions and human benefits of urban forests. *Urban For. Urban Green.* **2022**, *75*, 127707. [CrossRef]
45. Wani, A.; Sahoo, G. Forest ecosystem services and biodiversity. *Spat. Model. For. Resour. Manag. Rural Livelihood Sustain. Dev.* **2021**, *1*, 529–552. [CrossRef]
46. Mori, A.; Lertzman, K.; Gustafsson, L. Biodiversity and ecosystem services in forest ecosystems: A research agenda for applied forest ecology. *J. Appl. Ecol.* **2017**, *54*, 12–27. [CrossRef]
47. Savari, M.; Khaleghi, B. The role of social capital in forest conservation: An approach to deal with deforestation. *Sci. Total Environ.* **2023**, *896*, 165216. [CrossRef] [PubMed]
48. Verkerk, P.; Costanza, R.; Hetemäki, L.; Kubiszewski, I.; Leskinen, P.; Nabuurs, G.; Palahí, M. Climate-smart forestry: The missing link. *For. Policy Econ.* **2020**, *115*, 102164. [CrossRef]
49. Djagoun, C.; Zanvo, S.; Padonou, E.; Sogbohossou, E.; Sinsin, B. Perceptions of ecosystem services: A comparison between sacred and non-sacred forests in central Benin (West Africa). *For. Ecol. Manag.* **2022**, *503*, 119791. [CrossRef]
50. Kankaanhuhta, V.; Packalen, T.; Väätäinen, K. Digital Transformation of Forest Services in Finland—A Case Study for Improving Business Processes. *Forests* **2021**, *12*, 781. [CrossRef]
51. Goodbody, T.; Coops, N.; Luther, J.; Tompalski, P.; Mulverhill, C.; Frizzle, C.; Herniman, S. Airborne laser scanning for quantifying criteria and indicators of sustainable forest management in Canada. *Can. J. For. Res.* **2021**, *51*, 972–985. [CrossRef]
52. da Silva, E.; da Silva, G.; Figueiredo, E.; de Mendonça, A.; de Oliveira, C.; Fiedler, N.; Santos, J. Optimized forest planning: Allocation of log storage yards in the Amazonian sustainable forest management area. *For. Ecol. Manag.* **2020**, *472*, 118231. [CrossRef]
53. Nitoslawski, S.; Wong, K.; Steenberg, J.; Witherspoon, K.; Nesbitt, L.; Konijnendijk, C. The digital forest: Mapping a decade of knowledge on technological applications for forest ecosystems. *Earth's Future* **2021**, *9*, e2021EF002123. [CrossRef]
54. Wu, L.; Zhang, Z. Impact and threshold effect of Internet technology upgrade on forestry green total factor productivity: Evidence from China. *J. Clean. Prod.* **2020**, *271*, 122657. [CrossRef]
55. Wang, J.; Zheng, Z.; Ma, A.; Lu, X.; Zhong, Y. LoveDA: A remote sensing land-cover dataset for domain adaptive semantic segmentation. *arXiv* **2021**, arXiv:2110.08733. [CrossRef]
56. Keefe, R.; Zimbelman, E.; Picchi, G. Use of Individual Tree and Product Level Data to Improve Operational Forestry. *Curr. For. Rep.* **2022**, *8*, 148–165. [CrossRef]
57. De Gouw, S.; Morgenroth, J.; Xu, C. An updated survey on the use of geospatial technologies in New Zealand's plantation forestry sector. *N. Z. J. For. Sci.* **2020**, *50*, 1–17. [CrossRef]
58. Negassa, M.; Mallie, D.; Gemeda, D. Forest cover change detection using Geographic Information Systems and remote sensing techniques: A spatio-temporal study on Komto Protected forest priority area, East Wollega Zone, Ethiopia. *Environ. Syst. Res.* **2020**, *9*, 1–14. [CrossRef]
59. He, Z.; Turner, P. A Systematic Review on Technologies and Industry 4.0 in the Forest Supply Chain: A Framework Identifying Challenges and Opportunities. *Logistics* **2021**, *5*, 88. [CrossRef]
60. Giannetti, F.; Laschi, A.; Zorzi, I.; Foderi, C.; Cenni, E.; Guadagnino, C.; Giambastiani, Y. Forest Sharing®as an Innovative Facility for Sustainable Forest Management of Fragmented Forest Properties: First Results of Its Implementation. *Land* **2023**, *12*, 521. [CrossRef]

61. Kinaneva, D.; Hristov, G.; Raychev, J.; Zahariev, P. Early forest fire detection using drones and artificial intelligence. In Proceedings of the 2019 42nd International Convention on Information and Communication Technology, Electronics and Microelectronics (MIPRO), Opatija, Croatia, 20–24 May 2019; IEEE: New York, NY, USA, 2019; pp. 1060–1065. [CrossRef]
62. Wilkes, J.; Deuffic, P.; Jandl, R.; Westin, K.; Lieberherr, E.; Foldal, C.; Jarský, V. Communication campaigns to engage (non-traditional) forest owners: A European perspective. *For. Policy Econ.* **2021**, *133*, 102621. [CrossRef]
63. McEwan, A.; Marchi, E.; Spinelli, R.; Brink, M. Past, present and future of industrial plantation forestry and implication on future timber harvesting technology. *J. For. Res.* **2020**, *31*, 339–351. [CrossRef]
64. Nzau, J.; Gosling, E.; Rieckmann, M.; Shauri, H.; Habel, J. The illusion of participatory forest management success in nature conservation. *Biodivers. Conserv.* **2020**, *29*, 1923–1936. [CrossRef]
65. Yachmeneva, V.; Antonova, A.; Pozharitskaya, I. Precise technologies in forestry: Problems and prospects. *IOP Conf. Ser. Earth Environ. Sci.* **2020**, *574*, 012086. [CrossRef]
66. Yılmaz, R.; Koyuncu, C. The impact of ICT penetration on deforestation: A panel data evidence. *Rev. Econ. Perspect.* **2019**, *19*, 345–354. [CrossRef]
67. Empig, E.; Sivacioğlu, A.; Pacaldo, R.; Suson, P.; Lavilles, R.; Teves, M.; Amparado, R. Climate Change, Sustainable Forest Management, ICT Nexus, and the SDG 2030: A Systems Thinking Approach. *Sustainability* **2023**, *15*, 6712. [CrossRef]
68. Kanowski, P. Multilateral forestry research and tertiary forestry education for development: Reflections on progress since the 1970s. *Int. For. Rev.* **2020**, *22*, 113–128. [CrossRef]
69. Wong, S. Gendering Information and Communication Technologies in Climate Change. In *Research Anthology on Environmental and Societal Impacts of Climate Change*; IGI Global: Hershey, PA, USA, 2022; pp. 764–779. [CrossRef]
70. Chi, N. Transforming travel motivation into intention to pay for nature conservation in national parks: The role of Chatbot e-services. *J. Nat. Conserv.* **2022**, *68*, 126226. [CrossRef]
71. Molinaro, M.; Orzes, G. From forest to finished products: The contribution of Industry 4.0 technologies to the wood sector. *Comput. Ind.* **2022**, *138*, 103637. [CrossRef]
72. Paletto, A.; Bernardi, S.; Pieratti, E.; Teston, F.; Romagnoli, M. Assessment of environmental impact of biomass power plants to increase the social acceptance of renewable energy technologies. *Heliyon* **2019**, *5*, e02070. [CrossRef]
73. Shamseer, L.; Moher, D.; Clarke, M.; Ghersi, D.; Liberati, A.; Petticrew, M.; Stewart, L. Preferred reporting items for systematic review and meta-analysis protocols (PRISMA-P) 2015: Elaboration and explanation. *BMJ* **2015**, *349*, 1–25. [CrossRef] [PubMed]
74. Rethlefsen, M.; Kirtley, S.; Waffenschmidt, S.; Ayala, A.; Moher, D.; Page, M.; Koffel, J. PRISMA-S: An extension to the PRISMA statement for reporting literature searches in systematic reviews. *Syst. Rev.* **2021**, *10*, 1–19. [CrossRef] [PubMed]
75. Delgado, E.; Orduña, E.; Martín, A. Google Scholar as a data source for research assessment. In *Springer Handbook of Science and Technology Indicators*; Springer: Berlin/Heidelberg, Germany, 2019; pp. 95–127. [CrossRef]
76. Gusenbauer, M.; Haddaway, N. Which academic search systems are suitable for systematic reviews or meta-analyses? Evaluating retrieval qualities of Google Scholar, PubMed, and 26 other resources. *Res. Synth. Methods* **2020**, *11*, 181–217. [CrossRef] [PubMed]
77. Pati, D.; Lorusso, L. How to write a systematic review of the literature. *HERD Health Environ. Res. Des. J.* **2018**, *11*, 15–30. [CrossRef] [PubMed]
78. Adams, C.; Frost, G. Integrating sustainability reporting into management practices. *Acc. Forum* **2008**, *32*, 288–302. [CrossRef]
79. Wallace, L.; Lucieer, A.; Watson, C.; Turner, D. Development of a UAV-LiDAR system with application to forest inventory. *Remote Sens.* **2012**, *4*, 1519–1543. [CrossRef]
80. Fardusi, M.; Chianucci, F.; Barbati, A. Concept to practice of geospatial-information tools to assist forest management and planning under precision forestry framework: A review. *Ann. Silvic. Res.* **2017**, *41*, 3–14. [CrossRef]
81. Gómez, C.; Alejandro, P.; Hermosilla, T.; Montes, F.; Pascual, C.; Ruiz, L.; Valbuena, R. Remote sensing for the Spanish forests in the 21st century: A review of advances, needs, and opportunities. *For. Syst.* **2019**, *28*, 1–33. [CrossRef]
82. Tan, J.; Su, X.; Wang, R. Exploring the Measurement of Regional Forestry Eco-Efficiency and Influencing Factors in China Based on the Super-Efficient DEA-Tobit Two Stage Model. *Forests* **2023**, *14*, 300. [CrossRef]
83. Kovácsová, P.; Antalová, M. Precision forestry–definition and technologies. *Šumarski List* **2010**, *134*, 603–610. Available online: https://hrcak.srce.hr/62792. (accessed on 18 January 2023).
84. Nguyen, C.; Doytch, N. The impact of ICT patents on economic growth: An international evidence. *Telecommun. Policy* **2022**, *46*, 102291. [CrossRef]
85. N'dri, L.; Islam, M.; Kakinaka, M. ICT and environmental sustainability: Any differences in developing countries? *J. Clean. Prod.* **2021**, *297*, 126642. [CrossRef]
86. Baskent, E.; Borges, J.; Kašpar, J.; Tahri, M. Marchi A Design for Addressing Multiple Ecosystem Services in Forest Management Planning. *Forests* **2020**, *11*, 1108. [CrossRef]
87. Hristov, G.; Kinaneva, D.; Raychev, J.; Zahariev, P. Emerging methods for early detection of forest fires using unmanned aerial vehicles and LoRaWAN sensor networks. In Proceedings of the 2018 28th EAEEIE Annual Conference (EAEEIE), Hafnarfjordur, Iceland, 26–28 September 2018; European Union: Brussels, Belgium, 2018. [CrossRef]
88. Grant, A.; Pawson, S.; Marzano, M. Emerging Stakeholder Relations in Participatory ICT Design: Renegotiating the Boundaries of Sociotechnical Innovation in Forest Biosecurity Surveillance. *Forests* **2019**, *10*, 836. [CrossRef]

89. Borges, J.; Nordström, E.; Garcia, J.; Hujala, T.; Trasobares, A. *Computer-Based Tools for Supporting Forest Management. The Experience and the Expertise World-Wide*; Department of Forest Resource Management, Swedish University of Agricultural Sciences: Uppsala, Sweden, 2014; ISBN 978-91-576-9236-8.
90. Castro, J.; Chirino, E. Teachers' opinion survey on the use of ICT tools to support attendance-based teaching. *Comput. Educ.* **2011**, *56*, 911–915. [CrossRef]
91. Lähtinen, K.; Toppinen, A.; Suojanen, H.; Stern, T.; Ranacher, L.; Burnard, M.; Kitek, M. Forest Sector Sustainability Communication in Europe: A Systematic Literature Review on the Contents and Gaps. *Curr. For. Rep.* **2017**, *3*, 173–187. [CrossRef]
92. Taylor, C.; Lindenmayer, D. Stakeholder engagement in a Forest Stewardship Council Controlled Wood assessment. *Environ. Sci. Policy* **2021**, *120*, 204–212. [CrossRef]
93. Pelyukh, O.; Lavnyy, V.; Paletto, A.; Troxler, D. Stakeholder analysis in sustainable forest management: An application in the Yavoriv region (Ukraine). *For. Policy Econ.* **2021**, *131*, 102561. [CrossRef]
94. Akilandeswari, V.; Pavithra, M.; Thawlath, A.; Nasreen, J. Elements of Effective Communication. *Adalya J.* **2018**, *7*, 31–32.
95. Atkin, C.; Rice, R. Advances in Public Communication Campaigns. In *The International Encyclopedia of Media Studies*, 1st ed.; Scharrer, E., Ed.; The international encyclopedia of media studies: Vol. 5: Media effects/Media psychology; Wiley-Blackwell: London, UK, 2013; pp. 526–551.
96. Zacharoula, S. Green Informatics: ICT for Green and Sustainability. *Agric. Inform.* **2012**, *3*, 1–8.
97. Niţă, M.; Borz, S. Accuracy of a Smartphone-based freeware solution and two shape reconstruction algorithms in log volume measurements. *Comput. Electron. Agric.* **2023**, *205*, 107653. [CrossRef]
98. Coops, N. Characterizing Forest Growth and Productivity Using Remotely Sensed Data. *Curr. For. Rep.* **2015**, *1*, 195–205. [CrossRef]
99. Galbraith, S.; Vierling, L.; Bosque, N. Remote Sensing and Ecosystem Services: Current Status and Future Opportunities for the Study of Bees and Pollination-Related Services. *Curr. For. Rep.* **2015**, *1*, 261–274. [CrossRef]
100. Vogeler, J.; Cohen, W. A review of the role of active remote sensing and data fusion for characterizing forest in wildlife habitat models. *Rev. De Teledetect.* **2016**, *45*, 1–14. [CrossRef]
101. Stone, C.; Mohammed, C. Application of Remote Sensing Technologies for Assessing Planted Forests Damaged by Insect Pests and Fungal Pathogens: A Review. *Curr. For. Rep.* **2017**, *3*, 75–92. [CrossRef]
102. Xu, C.; Morgenroth, J.; Manley, B. Integrating Data from Discrete Return Airborne LiDAR and Optical Sensors to Enhance the Accuracy of Forest Description: A Review. *Curr. For. Rep.* **2015**, *1*, 206–219. [CrossRef]
103. Carson, W.; Andersen, H.; Reutebuch, S.; McGaughey, R. Lidar Applications in Forestry—An Overview. *Annu. Conf. Proc.* **2004**, 1–9. Available online: https://www.firescience.gov/projects/04-1-2-02/project/04-1-2-02_04_1_2_02_deliverable_06.pdf (accessed on 18 January 2023).
104. Pittman, S.; Costa, B.; Wedding, L. LiDAR Applications. In *Coral Reef Remote Sensing*; Springer: Dordrecht, The Netherlands, 2013; pp. 145–174. [CrossRef]
105. Borz, S.; Proto, A. Application and accuracy of smart technologies for measurements of roundwood: Evaluation of time consumption and efficiency. *Comput. Electron. Agric.* **2022**, *197*, 106990. [CrossRef]
106. Hollaus, M.; Vreugdenhil, M. Radar Satellite Imagery for Detecting Bark Beetle Outbreaks in Forests. *Curr. For. Rep.* **2019**, *5*, 240–250. [CrossRef]
107. Torres, R.; Snoeij, P.; Geudtner, D.; Bibby, D.; Davidson, M.; Attema, E.; Potin, P.; Rommen, B.; Floury, N.; Brown, M.; et al. GMES Sentinel-1 mission. *Remote Sens. Environ.* **2012**, *120*, 9–24. [CrossRef]
108. Pyörälä, J.; Kankare, V.; Vastaranta, M.; Rikala, J.; Holopainen, M.; Sipi, M.; Hyyppä, J.; Uusitalo, J. Comparison of terrestrial laser scanning and X-ray scanning in measuring Scots pine (*Pinus sylvestris* L.) branch structure. *Scand. J. For. Res.* **2018**, *33*, 291–298. [CrossRef]
109. Jacquin, P.; Longuetaud, F.; Leban, J.; Mothe, F. X-ray microdensitometry of wood: A review of existing principles and devices. *Dendrochronologia* **2017**, *42*, 42–50. [CrossRef]
110. Allison, R.; Wang, X.; Senalik, C. Methods for Nondestructive Testing of Urban Trees. *Forests* **2020**, *11*, 1341. [CrossRef]
111. Wei, Q.; Leblon, B.; La Rocque, A. On the use of X-ray computed tomography for determining wood properties: A review. *Can. J. For. Res.* **2011**, *41*, 2120–2140. [CrossRef]
112. Hur, J.; Ihm, S.; Park, Y. A Variable Impacts Measurement in Random Forest for Mobile Cloud Computing. *Wirel. Commun. Mob. Comput.* **2017**, *2017*, 6817627. [CrossRef]
113. Liang, X.; Kukko, A.; Hyyppä, J.; Lehtomäki, M.; Pyörälä, J.; Yu, X.; Kaartinen, H.; Jaakkola, A.; Wang, Y. In-situ measurements from mobile platforms: An emerging approach to address the old challenges associated with forest inventories. *ISPRS J. Photogramm. Remote Sens.* **2018**, *143*, 97–107. [CrossRef]
114. Pratihast, A.; Herold, M.; Avitabile, V.; de Bruin, S.; Bartholomeus, H.; Souza, C.M., Jr.; Ribbe, L. Mobile Devices for Community-Based REDD+ Monitoring: A Case Study for Central Vietnam. *Sensors* **2012**, *13*, 21–38. [CrossRef]
115. Sivrikaya, F.; Küçük, Ö. Modeling forest fire risk based on GIS-based analytical hierarchy process and statistical analysis in Mediterranean region. *Ecol. Inform.* **2022**, *68*, 101537. [CrossRef]
116. Sonti, S. Application of Geographic Information System (GIS) in Forest Management. *J. Geogr. Nat. Disasters* **2015**, *5*, 1–5. [CrossRef]

117. Nowak, M.; Dziób, K.; Ludwisiak, Ł.; Chmiel, J. Mobile GIS applications for environmental field surveys: A state of the art. *Glob. Ecol. Conserv.* **2020**, *23*, e01089. [CrossRef]
118. Li, L.; Hao, T.; Chi, T. Evaluation on China's forestry resources efficiency based on big data. *J. Clean. Prod.* **2017**, *142*, 513–523. [CrossRef]
119. Rossit, D.; Olivera, A.; Viana, V.; Broz, D. A Big Data approach to forestry harvesting productivity. *Comput. Electron. Agric.* **2019**, *161*, 29–52. [CrossRef]
120. Hasan, S.; Zhang, Y.; Chu, X.; Teng, Y. The role of big data in China's sustainable forest management. *For. Econ. Rev.* **2019**, *1*, 96–105. [CrossRef]
121. Björk, A.; Erlandsson, M.; Häkli, J.; Jaakkola, K.; Nilsson, Å.; Nummila, K.; Puntanen, V.; Sirkka, A. Monitoring environmental performance of the forestry supply chain using RFID. *Comput. Ind.* **2011**, *62*, 830–841. [CrossRef]
122. Luvisi, A.; Lorenzini, G. RFID-plants in the smart city: Applications and outlook for urban green management. *Urban For. Urban Green.* **2014**, *13*, 630–637. [CrossRef]
123. Green, K.; Zelbst, P.; Sower, V.; Bellah, J. Impact of Radio Frequency Identification Technology on Environmental Sustainability. *J. Comput. Inf. Syst.* **2017**, *57*, 269–277. [CrossRef]
124. Chianucci, F.; Bajocco, S.; Ferrara, C. Continuous observations of forest canopy structure using low-cost digital camera traps. *Agric. For. Meteorol.* **2021**, *307*, 108516. [CrossRef]
125. Plhal, R.; Kamler, J.; Homolka, M.; Adamec, Z. An assessment of the applicability of photo trapping to estimate wild boar population density in a forest environment. *Folia Zool.* **2011**, *60*, 237–246. [CrossRef]
126. Can, Ö.; Togan, İ. Camera trapping of large mammals in Yenice Forest, Turkey: Local information versus camera traps. *Oryx* **2009**, *43*, 427–430. [CrossRef]
127. Meek, P.; Fleming, P.; Ballard, G. *An Introduction to Camera Trapping for Wildlife Surveys in Australia*; Invasive Animals Cooperative Research Centre: Canberra, Australia, 2012.
128. Meek, P.; Ballard, G.; Claridge, A.; Kays, R.; Moseby, K.; O'brien, T.; Townsend, S. Recommended guiding principles for reporting on camera trapping research. *Biodivers. Conserv.* **2014**, *23*, 2321–2343. [CrossRef]
129. Noss, A.; Polisar, J.; Maffei, L.; Garcia, R.; Silver, S. Evaluating jaguar densities with camera traps. *Wildl. Conserv. Soc.* **2013**, 1–78. Available online: https://www.researchgate.net/profile/Rony-Garcia-Anleu/publication/269709284_Evaluating_jaguar_densities_with_camera_traps/links/565604fe08ae1ef92979c1b4/Evaluating-jaguar-densities-with-camera-traps.pdf (accessed on 18 January 2023).
130. Carreira, S. Photography and biodiversity awareness: The use of images in conservation. *Metode Sci. Stud. J.* **2023**, *14*, 1–7. [CrossRef]
131. Kang, X. Research progress of forest genetics and tree breeding. *J. Nanjing For. Univ.* **2020**, *44*, 1. [CrossRef]
132. Isabel, N.; Holliday, J.; Aitken, S. Forest genomics: Advancing climate adaptation, forest health, productivity, and conservation. *Evol. Appl.* **2020**, *13*, 3–10. [CrossRef]
133. Cortés, A.; Restrepo, M.; Bedoya, L. Modern strategies to assess and breed forest tree adaptation to changing climate. *Front. Plant Sci.* **2020**, *11*, 583323. [CrossRef] [PubMed]
134. Lebedev, V.; Lebedeva, T.; Chernodubov, A.; Shestibratov, K. Genomic selection for forest tree improvement: Methods, achievements and perspectives. *Forests* **2020**, *11*, 1190. [CrossRef]
135. Wheeler, N.; Steiner, K.; Schlarbaum, S.; Neale, D. The evolution of forest genetics and tree improvement research in the United States. *J. For.* **2015**, *113*, 500–510. [CrossRef]
136. White, T.; Adams, W.; Neale, D. (Eds.) *Forest Genetics*; Cabi; Embryo Publications: Athens, Greece, 2007; ISBN 9789608002524.
137. Ruane, J.; Sonnino, A. Marker-assisted selection as a tool for genetic improvement of crops, livestock, forestry and fish in developing countries: An overview of the issues. In *Marker-Assisted Selection—Current Status and Future Perspectives in Crops, Livestock, Forestry and Fish*; Food and Agriculture Organization of the United Nations (FAO): Rome, Italy, 2007; pp. 3–13.
138. Amaral, J.; Ribeyre, Z.; Vigneaud, J.; Sow, M.; Fichot, R.; Messier, C.; Maury, S. Advances and promises of epigenetics for forest trees. *Forests* **2020**, *11*, 976. [CrossRef]
139. Evans, D.; Kitson, J.; Lunt, D.; Straw, N.; Pocock, M. Merging DNA metabarcoding and ecological network analysis to understand and build resilient terrestrial ecosystems. *Funct. Ecol.* **2016**, *30*, 1904–1916. [CrossRef]
140. Kang, Y.; Deng, Z.; Zang, R.; Long, W. DNA barcoding analysis and phylogenetic relationships of tree species in tropical cloud forests. *Sci. Rep.* **2017**, *7*, 12564. [CrossRef]
141. Liu, J.; Yan, H.; Newmaster, S.; Pei, N.; Ragupathy, S.; Ge, X. The use of DNA barcoding as a tool for the conservation biogeography of subtropical forests in China. *Divers. Distrib.* **2015**, *21*, 188–199. [CrossRef]
142. Thompson, K.; Newmaster, S. Molecular taxonomic tools provide more accurate estimates of species richness at less cost than traditional morphology-based taxonomic practices in a vegetation survey. *Biodivers. Conserv.* **2014**, *23*, 1411–1424. [CrossRef]
143. Câmara, P.; Bones, F.; Lopes, F.; Oliveira, F.; Barreto, C.; Knop, D.; Rosa, L. DNA metabarcoding reveals cryptic diversity in forest soils on the isolated Brazilian Trindade Island, South Atlantic. *Microb. Ecol.* **2023**, *85*, 1056–1071. [CrossRef] [PubMed]
144. Janzen, D.; Hajibabaei, M.; Burns, J.; Hallwachs, W.; Remigio, E.; Hebert, P. Wedding biodiversity inventory of a large and complex Lepidoptera fauna with DNA barcoding. *Philos. Trans. R. Soc. B Biol. Sci.* **2005**, *360*, 1835–1845. [CrossRef]
145. Wangensteen, O.; Turon, X. Metabarcoding techniques for assessing biodiversity of marine animal forests. *Mar. Anim. For. Ecol. Benthic Biodivers. Hotspots* **2017**, *1*, 445–503. [CrossRef]

146. Watts, C.; Dopheide, A.; Holdaway, R.; Davis, C.; Wood, J.; Thornburrow, D.; Dickie, I. DNA metabarcoding as a tool for invertebrate community monitoring: A case study comparison with conventional techniques. *Austral Entomol.* **2019**, *58*, 675–686. [CrossRef]
147. Wiggins, A.; Crowston, K. Goals and tasks: Two typologies of citizen science projects. In Proceedings of the 2012 45th Hawaii International Conference on System Sciences, Maui, HI, USA, 4–7 January 2012; IEEE: New York, NY, USA, 2012; pp. 3426–3435. [CrossRef]
148. Kasperowski, D.; Kullenberg, C. The many modes of citizen science. *Sci. Technol. Stud.* **2019**, *32*, 2–7. [CrossRef]
149. Uhlmann, E.; Ebersole, C.; Chartier, C.; Errington, T.; Kidwell, M.; Lai, C.; Nosek, B. Scientific utopia III: Crowdsourcing science. *Perspect. Psychol. Sci.* **2019**, *14*, 711–733. [CrossRef]
150. Puhan, M.; Steinemann, N.; Kamm, C.; Müller, S.; Kuhle, J.; Kurmann, R.; Von, V. A digitally facilitated citizen-science driven approach accelerates participant recruitment and increases study population diversity. *Swiss Med. Wkly.* **2018**, *148*, w14623. [CrossRef]
151. Fraisl, D.; Hager, G.; Bedessem, B.; Gold, M.; Hsing, P.; Danielsen, F.; Haklay, M. Citizen science in environmental and ecological sciences. *Nat. Rev. Methods Primers* **2022**, *2*, 64. [CrossRef]
152. Johnston, A.; Matechou, E.; Dennis, E. Outstanding challenges and future directions for biodiversity monitoring using citizen science data. *Methods Ecol. Evol.* **2023**, *14*, 103–116. [CrossRef]
153. Andreopoulou, Z.; Stiakakis, E.; Vlachopoulou, M. Green ICT applications towards the achievement of sustainable development. In *E-innovation for Sustainable Development of Rural Resources during Global Economic Crisis*; IGI Global: Hershey, PA, USA, 2014; pp. 11–21. [CrossRef]
154. Sharma, T.; Singh, A.; Kumar, N.; Chauhan, G. Digital Technologies and Tools for the Ensuing Digital Era. In *Applying Drone Technologies and Robotics for Agricultural Sustainability*; IGI Global: Hershey, PA, USA, 2023; pp. 1–11. [CrossRef]
155. CEPAL. Report on the Seminar New Challenges and New Perspectives on International Cooperation: Development Approaches from Latin America and the Caribbean. 2023. Available online: https://repositorio.cepal.org/bitstream/handle/11362/48768/S2200982_en.pdf?sequence=1 (accessed on 18 January 2023).
156. Zhang, G.; Yang, Y.; Yang, G. Smart supply chain management in Industry 4.0: The review, research agenda and strategies in North America. *Ann. Oper. Res.* **2023**, *322*, 1075–1117. [CrossRef] [PubMed]
157. Chen, C.; Ye, F.; Xiao, H.; Xie, W.; Liu, B.; Wang, L. The digital economy, spatial spillovers and forestry green total factor productivity. *J. Clean. Prod.* **2023**, *405*, 136890. [CrossRef]
158. Anastasiadou, M.; Santos, V.; Montargil, F. Which technology to which challenge in democratic governance? An approach using design science research. *Transform. Gov. People Process Policy* **2021**, *15*, 512–531. [CrossRef]
159. Badiane, O.; Hendriks, S.; Glatzel, K.; Abdelradi, F.; Admassie, A.; Adjaye, J.; Sibanda, S. Policy Options for Food System Transformation in Africa and the Role of Science, Technology and Innovation. *Sci. Innov. Food Syst. Transform* **2023**, *713*, 1–931. [CrossRef]
160. Galaz, V.; Centeno, M.; Callahan, P.; Causevic, A.; Patterson, T.; Brass, I.; Levy, K. Artificial intelligence, systemic risks, and sustainability. *Technol. Soc.* **2021**, *67*, 101741. [CrossRef]
161. Dainelli, R.; Toscano, P.; Di Gennaro, S.; Matese, A. Recent advances in Unmanned Aerial Vehicles forest remote sensing—A systematic review. Part II: Research applications. *Forests* **2021**, *12*, 397. [CrossRef]
162. Näyhä, A. Finnish forest-based companies in transition to the circular bioeconomy-drivers, organizational resources and innovations. *For. Policy Econ.* **2020**, *110*, 101936. [CrossRef]
163. Castillo, D.; Gavilanes, A.; Chávez, C.; Borz, S. Critical Review on the Perspectives of the Forestry Sector in Ecuador. *Land* **2023**, *12*, 258. [CrossRef]
164. Rao, A.; Talan, A.; Abbas, S.; Dev, D.; Taghizadeh, F. The role of natural resources in the management of environmental sustainability: Machine learning approach. *Resources Policy.* **2023**, *82*, 103548. [CrossRef]
165. Sraku, M. Enhancing Access to Forestry Information in Africa to Ensure Sustainable Forest Management. *Issues Sci. Technol. Librariansh.* **2006**, *47*, 1–4. [CrossRef]
166. Belden, C.; Surya, P.; Goyal, A.; Pruuden, P.; Etulain, T.; Bothwell, C.; Pidatala, K. ICT for data collection and monitoring and evaluation: Opportunities and guidance on mobile applications for forest and agricultural sectors. In *Agriculture and Environmental Services Technical Assistance Paper*; The World Bank: Washington, DC, USA, 2013.
167. Liu, Z.; Peng, C.; Work, T.; Candau, J.; DesRochers, A.; Kneeshaw, D. Application of machine-learning methods in forest ecology: Recent progress and future challenges. *Environ. Rev.* **2018**, *26*, 339–350. [CrossRef]
168. Nitoslawski, S.; Galle, N.; Van Den, C.; Steenberg, J. Smarter ecosystems for smarter cities? A review of trends, technologies, and turning points for smart urban forestry. *Sustain. Cities Soc.* **2019**, *51*, 101770. [CrossRef]
169. Grabska, E.; Frantz, D.; Ostapowicz, K. Evaluation of machine learning algorithms for forest stand species mapping using Sentinel-2 imagery and environmental data in the Polish Carpathians. *Remote Sens. Environ.* **2020**, *251*, 112103. [CrossRef]
170. Singh, R.; Gehlot, A.; Akram, S.; Thakur, A.; Buddhi, D.; Das, P. Forest 4.0: Digitalization of forest using the Internet of Things (IoT). *J. King Saud Univ.-Comput. Inf. Sci.* **2022**, *34*, 5587–5601. [CrossRef]
171. Kożuch, A.; Cywicka, D.; Adamowicz, K. A Comparison of Artificial Neural Network and Time Series Models for Timber Price Forecasting. *Forests* **2023**, *14*, 177. [CrossRef]

172. Lowery, B.; Dagevos, J.; Chuenpagdee, R.; Vodden, K. Storytelling for sustainable development in rural communities: An alternative approach. *Sustain. Dev.* **2020**, *28*, 1813–1826. [CrossRef]
173. Hossain, M. Role of Information and Communication Technology (ICT) in Following Global Standard for NbS: The Bangladesh Perspective. In *Regional Perspectives of Nature-based Solutions for Water: Benefits and Challenges*; Springer International Publishing: Cham, Switzerland, 2023; pp. 225–247.

sustainability

MDPI

Article

Determining the Optimal Sample Size for Assessing Crown Damage on Color Infrared (CIR) Aerial Photographs

Jelena Kolić *, Renata Pernar, Ante Seletković, Anamarija Jazbec and Mario Ančić

Faculty of Forestry and Wood Technology, University of Zagreb, 10000 Zagreb, Croatia;
rpernar@sumfak.unizg.hr (R.P.); aseletkovic@sumfak.unizg.hr (A.S.); ajazbec@sumfak.unizg.hr (A.J.);
mancic@sumfak.unizg.hr (M.A.)
* Correspondence: jkolic@sumfak.unizg.hr

Abstract: One of the priorities in sustainable forest management is monitoring the health status of trees and stands. From the aspect of remote sensing (RS), the best way of doing this is by interpreting color infrared (CIR) aerial photographs; however, this raises the issue of sample size. For this reason, to apply this method in practice, it is indispensable to determine an appropriate sample size to ensure sufficient reliability of the health status assessment of trees in CIR aerial photographs. This research was conducted in lowland forests of pedunculate oak in Croatia. To determine damage in the photographs of the main tree species, a systematic sample with varying dot grid densities—100 × 100 m, 200 × 200 m, 300 × 300 m, 500 × 500 m, 1000 × 1000 m—was used with combinations of different numbers of interpreted trees per sample. Damage indicators were also calculated based on tree distributions obtained by interpreting four trees, two trees and one tree in different sample sizes. The results of the testing showed that there were no statistically significant differences between different sample densities and numbers of interpreted trees in relation to mean damage assessment. Regardless of the fact that there were no statistically significant differences during damage assessment, it was found that by lowering sample densities, starting with 200 × 200 m, the number of trees and the number of sample points per particular sub-compartment significantly decreased, and so did the desired accuracy. Consequently, the participation (distribution) of particular species and damage degrees in the sample were lost, which significantly affected the overall tree health assessment. In contrast, grid densities of 100 × 100 m with one interpreted tree at the raster point proved to be the optimal sample size. This confirms the fact found in earlier research, that is, that the selected sample should have several spatially well-distributed points with a smaller number of trees in the point, and samples with larger numbers of trees in a smaller number of points should be avoided.

Keywords: visual interpretation of CIR aerial photographs; damage assessment; optimal sample size; digital photogrammetric workstation

Citation: Kolić, J.; Pernar, R.; Seletković, A.; Jazbec, A.; Ančić, M. Determining the Optimal Sample Size for Assessing Crown Damage on Color Infrared (CIR) Aerial Photographs. *Sustainability* **2023**, *15*, 15918. https://doi.org/10.3390/su152215918

Academic Editor: Lihu Dong

Received: 8 September 2023
Revised: 27 October 2023
Accepted: 12 November 2023
Published: 14 November 2023

1. Introduction

Sustainable management of forests under conditions of climate change and natural disasters is necessary because forests, with their economic and social functions, are of immense importance in maintaining a healthy environment and mitigating climatic extremes. Forests in Croatia, as well as in Europe, are faced with many challenges such as the increasingly frequent dieback of trees, habitat degradation, the appearance of invasive species and climate change. Tree dieback is a complex process that involves a large number of sites, stands and biotic factors at different stages and different intensities [1–6]. We contribute to the stability and preservation of forest ecosystems through systematic monitoring of the state of ecological factors. Mapping, assessing and quantifying these effects is therefore of the most importance in order to understand disease progression and develop effective forest management plans [7].

The poor health condition of forests, caused by biotic and abiotic factors, usually leads to defoliation and/or leaf discoloration [8,9]. For this reason, defoliation of the crown is considered a key indicator of the health condition [10]. The negative consequences of tree dieback are being mitigated by monitoring the status of trees, primarily by assessing tree and crown damage [11–14]. The primary task is to assess the degree of forest damage and spatial distribution of damaged trees, which can be performed by terrestrial-monitoring or by applying RS methods [15,16], most commonly by interpreting color infrared (CIR) aerial images [15,17,18]. CIR films register the visible and near-infrared (IR) radiation of the Sun reflected from the surface of vegetation, and the amount of this radiation depends on the type of tree and its physiological condition, i.e., health status [19–21]. Damage to the crown of trees (loss of leaves, death of branches and parts of the crown, change in the color of leaves) causes significant changes in reflectance, especially in the infrared part of the spectrum (wavelengths 700–1100 nm), and because of this, damaged trees can be easily recognized on CIR aerial photographs [22–28].

To identify damage on aerial photographs, it is necessary to create a photo interpretation key for each type of tree, whereby the characteristics of the appearance of an individual tree species and the degrees of damage on aerial photographs are given descriptively and with a drawing or photograph [18].

Multispectral images have also been successfully used for mapping individual forest stands [29,30], for crown reconstruction and individual tree phenotyping [31,32], for identification of tree species [33,34] and for determining the health status [15,18,35–42].

Remote sensing makes it possible to collect data on forest conditions quickly and reliably [43–45]. This procedure lessens the scope of fieldwork and saves time and money [44]. So far, there have been six forest damage assessments in the Republic of Croatia based on the photointerpretation of CIR aerial photographs. All the research was conducted by photo interpreting analogous aerial photographs using analytical stereo instruments. Trees to be interpreted were selected according to a randomly placed systematic sample, proportional to the stratum (sub-compartments) size, and with a systematic sample using the 100×100 raster method. The development of digital cameras has opened new possibilities for using digital CIR aerial photographs to assess damage to both single trees and stands on a digital photogrammetric station (DPS). However, this raises the problem of sample size; namely, in all previous research, a sample in the form of dot grids proved to be best because it is regularly distributed in space but interprets a much larger number of trees while achieving the same reliability in comparison to field assessments [46]. Thus, in order to apply the method in practice, it is necessary to determine the size of a sample that will ensure sufficient reliability in assessing the health status or tree damage in digital CIR aerial images.

Sample size has been much more extensively studied in the course of forest management assessment, during forest measurement rather than during damage assessment. The samples are taken from the entire population [47] with the aim of obtaining a cost-effective estimate of the entire population [48]. Sampling should be made for every measurement procedure since measuring the whole population would be impractical and unfeasible because of time constraints and liability for errors. For this reason, a well-conceived and executed sampling procedure efficiently substitutes total measurement as regards the quality of the acquired data and costs [49].

In their inventory of forest damage in the Spačva Basin, Kalafadžić et al. [50] tested different sampling methods and intensities in 10 randomly chosen compartments. They employed the point method and the raster method. The point method, with spacing ranging from 400 m, 200 m and 400×200 m, was used to interpret the same number of trees, while the raster method, depending on the size of the raster overlaid on the compartment, was used to interpret a different number of trees closest to the raster point. After testing, they concluded that samples with a larger number of trees in a smaller number of points should be avoided.

In all the research conducted to date, a dot grid sample proved best because it is well distributed in space. Since it interprets a much larger number of trees, it is as reliable as field assessment.

To apply the method in practice, it is necessary to determine the form and size of a sample in order to ensure sufficient reliability of tree health assessment in digital CIR aerial photographs.

The results obtained by interpreting CIR aerial imagery are most commonly presented in thematic maps using a geographic information system (GIS).

The primary goal of this study is to investigate the most appropriate sample size and the number of interpreted trees per sample in digital color infrared (CIR) aerial photographs to assess crown damage and tree health status for practical purposes. Therefore, it was necessary to carry out an inventory of damage on systematic samples with different density points (100 × 100 m, 200 × 200 m, 300 × 300 m, 500 × 500 m, 1000 × 1000 m) and a number of trees (1, 2 and 4 trees), analyze and process the data, interpret the results obtained, and based on this, determine the shape and size of the sample that would ensure sufficient reliability of crown damage assessment on CIR aerial photographs.

2. Materials and Methods

Research comprised the lowland forest area of pedunculate oak in Croatia, of the management unit (MU) "Josip Kozarac" (45°23′30″ N, 16°46′50″ E), Lipovljani Forest Office, and the Lipovljani Forest Training and Research Centre, the management unit of "Opeke" (45°22′00″ N, 16°49′45″ E) (Figure 1). The MU "Josip Kozarac" area is 5759.13 ha, while the MU "Opeke" area is 547.27 ha. In the management units, the lowest point is 94 m above sea level and the highest altitude is 96 m above sea level (MU "Opeke"), i.e., 105 m above sea level (MU "Josip Kozarac"). Four phytocoenoses are represented in the researched area in which, in addition to other species, the main types of trees are Pedunculate oak (*Quercus robur* L.) and Narrow-leaved ash (*Fraxinus angustifolia* Vahl).

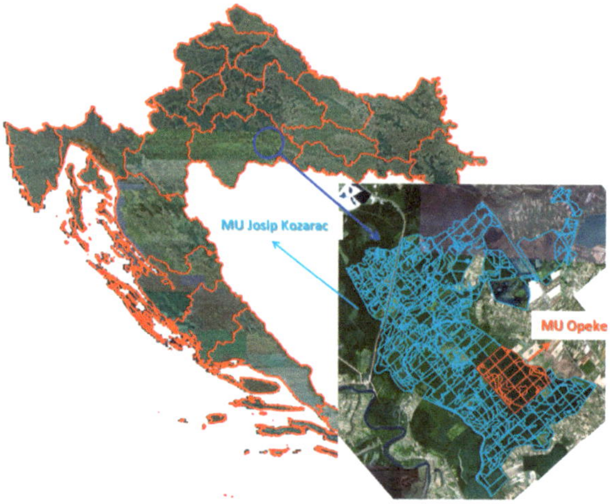

Figure 1. Research areas: MU Josip Kozarac (blue) and MU Opeke (red).

The recording was carried out in surveyed strips using a multi-spectral digital camera-UCX-Vexcel, focal length f = 100.5 mm. Digital aerial photographs (122) were taken with a spatial resolution of 0.1 m, a longitudinal overlap of 60% and a transverse overlap of 36%.

Crown damage, and thus the health status of the trees, was assessed on digital CIR aerial photographs for the main tree species—Pedunculate oak and Narrow-leaved ash.

Digital aerial photographs were interpreted on a digital photogrammetric workstation (DPS) using the PHOTOMOD Lite 4.4. software package.

Before the interpretation of the DPS, it was necessary to determine the method for mapping crown damage so that they could classify trees into individual degrees of damage to create projects and code tables.

Photointerpretation of the degree of crown damage was carried out using a carefully created photointerpretation key of the research area, and the health statuses of trees (crowns) were classified into damage degrees based on the following scale [18,35]:

Damage degree	Damage percentage
0	0–10%
1	11–25%
2.1 (2A)	26–40%
2.2 (2B)	41–60%
3.1 (3A)	61–80%
3.2 (3B)	81–100%
4	Snags

After that, the projects or sequences (strips) of two or more adjacent digital aerial photographs were created in the module PHOTOMOD Montage Desktop in four steps:

- Block forming: digital aerial photographs were uploaded into the project; the direction of the block was determined and so was the position of aerial photographs in the block.
- Aerial triangulation: camera parameters were defined, and the internal orientation of aerial photographs was conducted.
- Block adjustment: external orientation of aerial photographs.
- Block processing: the last step in the project design in which a module was selected for further use in photointerpretation.

Aerial triangulation was performed automatically on the photogrammetric workstation (in the second step) using the ray bundle method and the GPS/IMU (inertial measurement system) measurements that were recorded during aerial surveys.

A total of five projects were created in four steps—block forming, aerial triangulation, block adjustment and block processing—for photointerpretation.

After the projects were created, a code table was generated for each stereopair within the project. The table was generated in order to collect data in the PHOTOMOD Stereo Draw module. In the code table, the interpreted objects (trees) were described by a code name, number of stereopairs in which they were collected, shape (point), color, symbol and additional attributes (number of trees, position, species, damage). A total of 32 code tables were created for the needs of the research.

After the projects and code tables were created, a photointerpretation key was used to interpret digital CIR aerial photographs. In a systematic 100 × 100 m sample, the four closest crowns were interpreted in every raster point laid over the aerial photographs (Figure 2).

Data collected by interpreting digital aerial images on DPS were exported in .DXF format for further statistical processing, and in .SHP format for geospatial analyses and construction of thematic maps.

Past research shows that a randomly placed systematic sample, in which a uniform distribution of interpreted crowns is achieved with the raster method, is best for damage assessment.

In previous studies, trees to be interpreted were selected according to the randomly placed systematic sample, which was proportional to the stratum size, while the health condition was assessed using the crown closest to the raster point. In more recent research, damage was assessed using the dot grid raster method (100 × 100). In each point (sample), four crowns closest to the raster point in the top left and right corners and in the bottom left and right corners were interpreted.

Figure 2. An example of interpreting 4 crowns at each raster point using attributes entered in the code table.

Compared with earlier research, this method resulted in a much higher number of trees in the same area; therefore, a question arises as to the optimal sample size and the optimal number of interpretable trees per sample that will ensure statistically sufficient reliability of tree health assessment.

For this reason, to define the most appropriate sample sizes that will ensure statistically sufficient reliability of tree health assessment, systematic samples with different dot grid densities—200 × 200 m, 300 × 300 m, 500 × 500 m and 1000 × 1000 m—were set up (Figure 3) with combinations of different numbers of interpreted trees per sample: 1 (A), 2 (AD) and 4 (ABCD) trees (Figure 4).

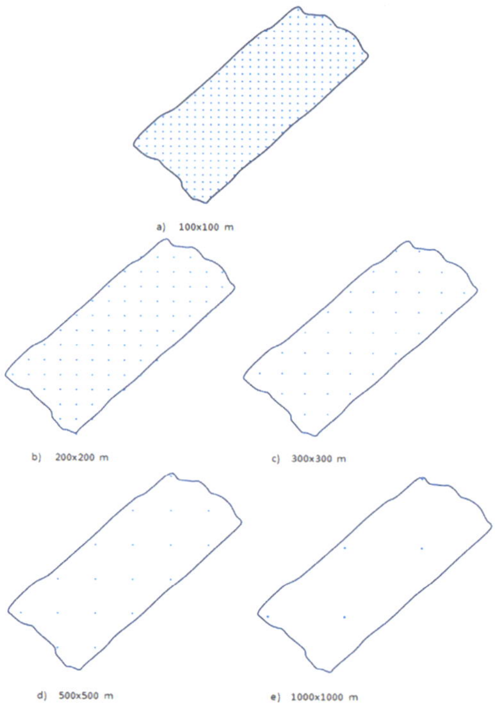

Figure 3. Systematic sampling using varying dot grid densities (segments per survey strip).

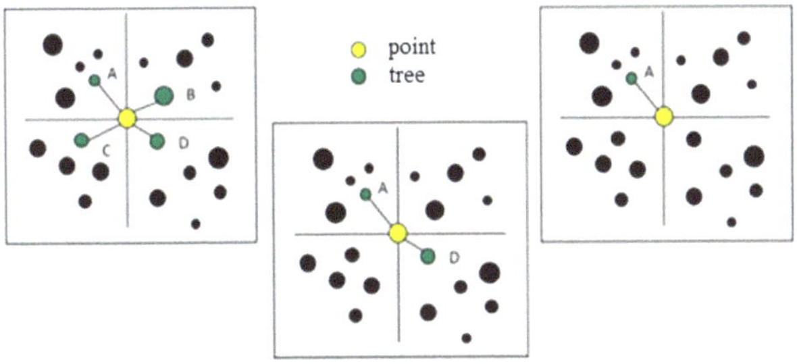

Figure 4. Positions 4 (ABCD), 2 (AD) and 1 (A) of interpreted trees for varying dot grid densities.

Based on the results of digital aerial imagery photointerpretation, damage indicators (damage, D; mean damage, MD; damage index, DI; mean damage of significantly damaged trees, MD_1) were calculated using Formulas (1)–(4) [15,35,50,51] for single tree species and all the interpreted species together, both for individual survey strips and overall for the entire surveyed area (compartments and sub-compartments covered by surveying).

1. Damage (D) is an indicator that is calculated according to the formula:

$$\mathrm{D}\ (\%) = \frac{\sum f_{(1-4)}}{\sum f_{(0-4)}} \cdot 100 \tag{1}$$

For the calculation, only the total number of damaged trees is taken into account, not include the number of damaged trees in a particular damage degree.

2. Mean Damage (MD) is calculated according to the formula:

$$\mathrm{MD}\ (\%) = \frac{\sum f_i \cdot x_i}{\sum f_i} \tag{2}$$

where f_i is the number of trees in i- damage stage

x_i–i- stage interval center in the damage stage scale for single trees

0 = 5%, 1 = 17.5%, 2.1 = 32.5%, 2.2 = 50%, 3.1 = 70%, 3.2 = 90%, 4 = 100%

To calculate the mean damage in the observed area (sample), a complex arithmetic mean is used with the number of trees in a particular damage class.

3. Damage index (DI) provides the percentage share of trees in the sample, which is classified as damage degree 2.1 (2A) or higher, i.e., severely damaged trees.

$$\mathrm{DI}\ (\%) = \frac{\sum f x_{(2-4)}}{\sum f_{(0-4)}} \tag{3}$$

4. Mean Damage1 (MD1) is the value of the mean degree of damage of trees classified into damage level 2.1 (2A) or more.

$$\mathrm{MD}_1 = \frac{\sum f_{(2-4)}}{\sum f_{(0-4)}} \cdot 100 \tag{4}$$

The calculated damage indicators and the number of interpreted trees provided input variables for statistical data processing. According to Pranjić and Lukić [52], the sample size was determined based on the variability (s_x) of the estimated mean damage and the desired accuracy ($s_{\overline{x}}$), as well as the reliability coefficient (t).

An χ^2 test was used to test different sample point densities and numbers of interpreted trees per sample with regard to tree health assessment, with 95% reliability [53].

In addition, binomial distribution was used to obtain the desired accuracies (B) for different dot grid densities and numbers of interpreted trees per sample (N).

In line with the results obtained from interpreting tree health conditions on the raster of varying dot grid densities using a combination of interpreted trees per raster point, thematic layers containing a spatial distribution of mean damage (MD) were constructed for all the species along particular survey strips and the research area (compartments/sub-compartments). Based on the acquired results, thematic maps of mean damage can also be constructed for the main tree species as well as for other damage indicators.

3. Results

The results of crown damage interpretations for 4 trees (ABCD), 2 trees (AD) and 1 tree (A) (Figure 4) for different sample sizes (100 × 100 m, 200 × 200 m, 300 × 300 m, 500 × 500 m and 1000 × 1000 m) (Figure 3) provided tree distributions per damage degree (Table 1).

Table 1. Tree distribution per specific damage degree for different sample densities and different numbers of interpreted trees.

Sample Density	Degree of Damage	Number of Trees			Sample Density	Degree of Damage	Number of Trees		
		ABCD (4)	**AD (2)**	**A (1)**			**ABCD (4)**	**AD (2)**	**A (1)**
100 × 100	0	90	46	30	500 × 500	0	6	3	1
	1	1782	931	473		1	75	48	24
	2A	1536	752	363		2A	67	27	17
	2B	973	475	225		2B	30	12	3
	3A	454	217	122		3A	11	5	3
	3B	401	199	112		3B	22	11	5
	4	22	12	5		4	2	1	0
	Σ	5258	2632	1330		Σ	213	107	53
200 × 200	0	25	12	7	1000 × 1000	0	2	1	0
	1	435	235	118		1	23	14	8
	2A	368	171	79		2A	13	3	1
	2B	261	130	65		2B	12	5	2
	3A	122	54	34		3A	3	2	1
	3B	108	57	31		3B	7	5	3
	4	5	4	2		4	0	0	0
	Σ	1324	663	336		Σ	60	30	15
300 × 300	0	11	6	2					
	1	194	92	47					
	2A	185	99	45					
	2B	119	60	26					
	3A	43	21	15					
	3B	35	16	11					
	4	3	2	2					
	Σ	590	296	148					

By interpreting a particular sample size and number of interpreted trees per sample, different tree distributions were obtained per damage degree. It was therefore necessary to determine whether there was a significant difference among the assessed damage degrees with regard to the inclusion of different numbers of trees in the sample. Testing was performed using the χ^2 test and the results are given in Table 2.

Table 2. Values of χ^2 test frequency distributions for particular sample sizes and combinations of interpreted trees in relation to the 100 × 100 m sample size and 4 interpreted trees (ABCD), with 95% reliability at a limit of 12.59.

Sample Sizes/Combinations of Numbers of Trees	Values for χ^2 Test (Total)
100 × 100 AD	2.96
100 × 100 A	8.49
200 × 200 ABCD	3.67
200 × 200 AD	5.15
200 × 200 A	6.52
300 × 300 ABCD	5.64
300 × 300 AD	6.37
300 × 300 A	3.93
500 × 500 ABCD	10.63
500 × 500 AD	11.81
500 × 500 A	7.90
1000 × 1000 ABCD	4.97
1000 × 1000 AD	9.23
1000 × 1000 A	7.90

Based on the results of the χ^2 test, there are no statistically significant differences between different sample densities and numbers of interpreted trees in relation to mean damage assessment.

In the entire research area, the interpretation of 4 trees in a 100 × 100 grid resulted in the interpretation of a total of 5258 trees per strip (Table 3). Since no statistically significant differences were found even though a large number of trees were interpreted, the next step was to determine the required sample size, i.e., the optimal number of trees required for interpretation to achieve identical assessment accuracy (Table 4).

Table 3. Determining the necessary sample size.

Survey Strip	Number of Trees	s_x	s^2	$s_x \times s^2$
8527	2144	0.40	267.98	107.05
8528	1524	0.27	225.67	61.23
8531	1590	0.33	464.09	152.80
Total	5258	1		321.07

s_x, $sx = t_{\alpha/2}{}^2/s_{\bar{x}}{}^2$ (t, distribution value for the desired confidence limit (1 − α); $s_{\bar{x},}$ standard error of the estimate of mean damage; s^2, the variance of the sample.

Table 4. Number of trees needed for interpretation to achieve identical assessment accuracy.

z	1.96	1.96	1.96	1.96
p	0.3104	0.3104	0.3104	0.3104
1 − p	0.6896	0.6896	0.6896	0.6896
B	0.01	0.012506	0.015162	0.049994
N	8223	5258	3577	329

z, table value for 95% reliability; p, assessment accuracy of mean damage; B, desired accuracy; N, number of trees.

According to Table 4, a total of 8223 trees should be interpreted in order to achieve the desired accuracy of 1% in the study area. The table also shows that an accuracy of 1.3% was achieved with 5258 interpreted trees. Therefore, along with the defined desired accuracy of 1.5%, the necessary number of trees for crown damage assessment was also determined in the study area.

In order to obtain mean damage of 31.04% in the surveyed area (sample of 100 × 100 m and 4 trees) at 1.5% accuracy and 95% reliability, 3577 trees should be interpreted. The desired accuracy of 5% was also tested, and the results obtained show that 329 trees should be interpreted. This number of trees was achieved in the 200 × 200 m sample by interpreting one tree per point (Table 1). Based on tree distribution per damage degree for different sample densities and different numbers of interpreted trees, an optimal sample size and number of interpreted trees per point was determined.

Figure 4 shows that at different sample densities and four interpreted trees per point, the same distribution trend was retained per particular damage degrees, while the total number of trees to be interpreted was significantly reduced.

During interpretation, the analysis of sample densities and numbers of interpreted trees per sample provided distribution per particular damage degree for the necessary number of trees (3577) (Figure 5).

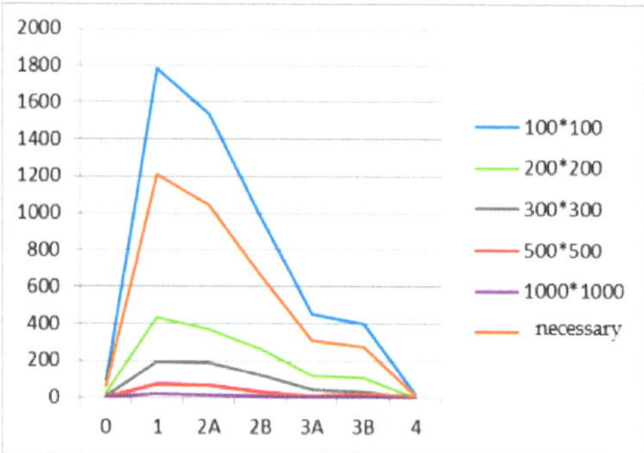

Figure 5. Distributions of tree numbers (y-axis) per damage degree (x-axis) for 4 interpreted trees at different sample sizes, and the distribution of the necessary number of trees.

In contrast, Figure 6 shows that the above trend changes as sample density reduces, starting from 200 × 200 m, if 2 or 1 tree per point is interpreted. This is why a limit threshold of 5% was taken to determine the optimal number of trees for interpretation to achieve the desired accuracy.

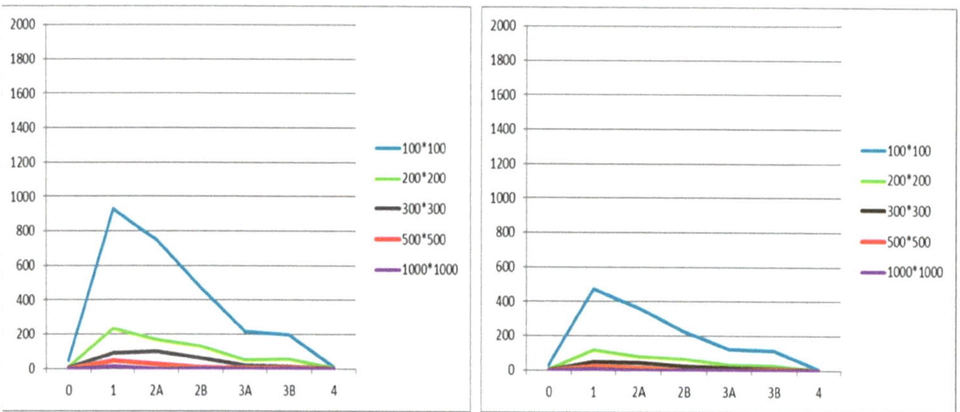

Figure 6. Distribution of tree numbers (y-axis) per damage degree (x-axis) for 2 (**left**) and 1 (**right**) interpreted trees for different sample sizes.

Damage indicators were also calculated based on tree distribution obtained by interpreting four trees (ABCD), two trees (AD) and one (A) tree for different sample sizes (Table 5).

Table 5. Calculated damage indicators (damage, D; mean damage, MD; damage index, DI; mean damage of significantly damaged trees, MD_1).

Sample	Damage Indicator (%)	Total		
		ABCD	AD	A
100 × 100 m	D	98.29	98.25	97.74
	DI	64.40	62.58	61.59
	MD	38.09	37.62	38.04
	MD1	49.81	49.84	50.99
200 × 200 m	D	98.11	98.19	97.92
	DI	64.63	61.56	60.51
	MD	38.90	38.52	39.55
	MD1	50.66	51.36	53.02
300 × 300 m	D	98.14	97.97	98.65
	DI	63.86	64.10	61.59
	MD	37.07	37.05	39.43
	MD1	47.85	47.11	50.53
500 × 500 m	D	97.18	97.20	98.11
	DI	58.52	47.15	43.48
	MD	37.42	35.26	33.73
	MD1	50.21	52.10	48.66
1000 × 1000 m	D	96.67	96.67	100.00
	DI	48.68	36.96	29.03
	MD	37.92	39.58	40.83
	MD1	53.21	55.00	67.50

Table 5 shows that the mean crown (stand) density (MD) has the same degree -2a (26–40%), regardless of the sample size and the number of interpreted trees per sample.

As a mean damage (MD) of 31.04% (sample of 100 × 100 m and 4 trees) at 1.3% accuracy and 95% reliability was achieved in the surveyed area, the desired accuracy (B) for other grid densities and numbers of interpreted trees per point (N) were also tested (Table 6).

Table 6. Numbers of interpreted trees (4 (ABCD), 2 (AD) and 1 (A) trees) and accuracies achieved for different samples.

Sample	100 × 100 m		200 × 200 m		300 × 300 m		500 × 500 m		1000 × 1000 m	
	N	B	N	B	N	B	N	B	N	B
ABCD	**5258**	**1.25**	1324	2.49	590	3.73	213	6.21	60	11.7
AD	2632	1.76	663	3.52	296	5.27	107	8.77	30	16.56
A	1330	2.49	336	4.95	148	7.45	53	12.46	15	23.41

The results obtained show that the scope for the desired accuracies ranges within limits of up to 5%. This limit value is achieved at 200 × 200 m grid density and one interpreted tree per point, as confirmed using GIS analysis (Figures 7 and 8).

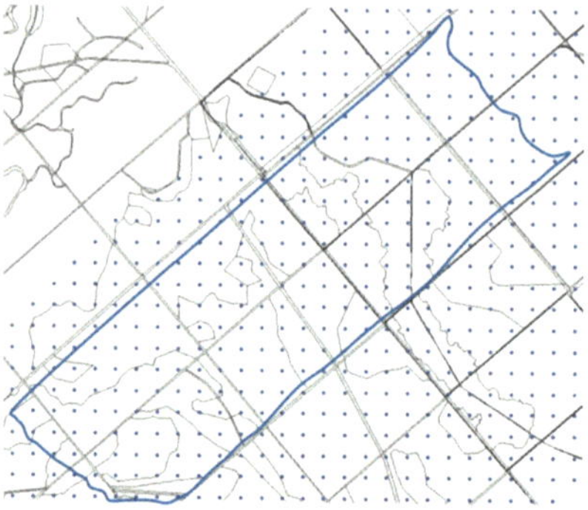

Figure 7. A systematic 100 × 100 m sample on a surveyed strip laid over the management unit plan.

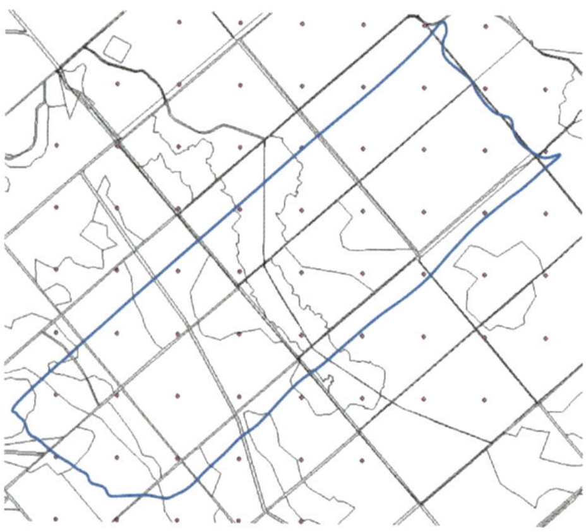

Figure 8. A systematic 200 × 200 m sample (**left**) and 300 × 300 m sample (**right**) on the surveyed strip laid over the management unit plan.

Regardless of the fact that there is no statistically significant difference in damage assessment, it is noted that the number of sample points per particular sub-compartment decreases, whereas it is within the 5% limit for the entire surveyed area and density raster of 300 × 300 m.

This way, the participation of particular species and damage degrees in the sample is lost, or in other words, particular sub-compartments are assessed based on one or two sample points, which significantly affects the health status assessment.

The results obtained confirm that it is best to use a 100 × 100 m grid with one interpreted tree per raster point since raster points are optimally spatially distributed over the sub-compartments.

4. Discussion

To define the optimal sample size that should provide statistically sufficient accuracy for crown damage assessment, different dot grid densities (100 × 100 m, 200 × 200 m, 300 × 300 m, 500 × 500 m and 1000 × 1000 m) were tested in combination with different numbers of interpreted trees per sample (1 (A), 2 (AD) and 4 (ABCD) trees).

The testing results showed that there were no statistically significant differences between different sample densities and numbers of interpreted trees in relation to mean damage assessment. Regardless of the fact that there is no statistically significant difference in damage assessment, it was found that by lowering sample densities, starting with 200 × 200 m, the number of trees and the number of sample points per particular sub-compartment significantly decreases, and so does the desired accuracy. Consequently, the participation (distribution) of particular species and damage degrees in the sample are lost, which significantly affects the overall tree health assessment. In contrast, 100 × 100 m grid densities with one interpreted tree at the raster point proved to be the optimal sample size. This confirms the fact found in earlier research, that is, that the selected sample should have several spatially well-distributed points and a smaller number of trees in the point, and samples with larger numbers of trees in a smaller number of points should be avoided [50].

It was determined that the differences in damage indicators between terrestrial data and photointerpretation data are not significant. At the same time, field damage assessments were obtained with 10.84% precision, and according to the results obtained from the recordings, this corresponds only to the 1000 × 1000 m sample. For other sample densities, the precision is much higher in favor of photointerpretation, i.e., with the same precision and number of interpreted trees per systematic sample, significantly more time and estimators need to be invested in the field.

Identifying and monitoring the spatial distribution of damaged trees and snags is one of the priorities of sustainable management. Therefore, timely location of stands with poorer health is necessary so that appropriate measures can be applied to maintain their vitality and productivity at an optimal level. CIR aerial photographs allow an insight into the field condition over a short period [36].

5. Conclusions

Crown damage assessment and determination of the optimal sample size and number of trees per sample were conducted using digital CIR aerial photographs of the MU Josip Kozarac and MU Opeke areas.

Based on the research conducted and the results obtained, the following conclusions can be drawn:

- Interpretation of different sample point densities and numbers of interpreted trees per sample showed that the optimal systematic sample size is 100 × 100 m with one interpreted tree per point, and samples with larger numbers of trees in smaller numbers of points should be avoided.
- The research contributes to the development and application of the most favorable interpretation method to be used in practice. This ensures statistically sufficient reliability of tree health status assessment on CIR aerial photographs, combined with optimal sample size and number of interpreted trees.
- The existing methods for health status assessment are thus improved and new possibilities for applying digital CIR aerial photographs to enable sustainable management of forests under conditions of climate change.

Author Contributions: Conceptualization, J.K. and R.P.; methodology, J.K. and R.P.; software, J.K. and A.S.; validation, R.P., M.A. and A.S.; formal analysis, A.J.; investigation, J.K., R.P., A.S. and M.A.; data curation, J.K.; writing—original draft preparation, J.K.; writing—review and editing, R.P. and A.J.; visualization, J.K. and M.A.; supervision, R.P. All authors have read and agreed to the published version of the manuscript.

Funding: The research was carried out at the Laboratory for Remote Sensing and GIS, Faculty of Forestry and Wood Technology. Thanks to the University of Zagreb, Faculty of Forestry and Wood Technology for supporting and funding the research.

Institutional Review Board Statement: Not applicable.

Informed Consent Statement: Not applicable.

Data Availability Statement: The data presented in this study are available on request from the corresponding author.

Conflicts of Interest: The authors declare no conflict of interest.

References

1. Harapin, M.; Androić, M. Sušenje i zaštita šuma hrasta lužnjaka. In *Hrast Lužnjak u Hrvatskoj*; Klepac, D., Ed.; HAZU Centar za Znanstveni rad Vinkovci i Hrvatske šume: Zagreb, Croatia, 1996; pp. 227–256. ISBN 953-154-079-9.
2. Prpić, B. Propadanje šuma hrasta lužnjaka. In *Hrast Lužnjak u Hrvatskoj*; Klepac, D., Ed.; HAZU Centar za Znanstveni rad Vinkovci i Hrvatske šume: Zagreb, Croatia, 1996; pp. 273–298. ISBN 953-154-079-9.
3. Wargo, P.M. Consequences of environmental stress on oak: Predisposition to pathogens. *Ann. For. Sci.* **1996**, *53*, 359–368. [CrossRef]
4. Donaubauer, E. Die Bedeutung von Krankheitserregern beim gegenwärtigen Eichensterben in Europa—Eine Literaturübersicht. *Eur. J. For. Pathol.* **1998**, *28*, 91–98. [CrossRef]
5. Führer, E. Oak decline in Central Europe: A synopsis of hypotheses. *USDA For. Serv. Gen. Tech. Rep. NE* **1998**, *247*, 7–24.
6. Thomas, F.M.; Blank, R.; Hartmann, G. Abiotic and biotic factors and their interactions as causes of oak decline in Central Europe. *For. Path.* **2002**, *32*, 277–307. [CrossRef]
7. Guerra-Hernández, J.; Díaz-Varela, R.A.; Ávarez-González, J.G.; Rodríguez-González, P.M. Assessing a novel modelling approach with high resolution UAV imagery for monitoring health status in priority riparian forests. *For. Ecosyst.* **2021**, *8*, 61. [CrossRef]
8. Auclair, A.N.D.; Worrest, R.C.; Lachance, D.; Martin, H.C. Climatic perturbation as a general mechanism of forest dieback. In *Forest Decline Concepts*; Manion, P.D., Lachance, D., Eds.; Amer Phytopathological Society: St. Paul, MN, USA, 1992; pp. 38–58.
9. De Vries, W.; Klap, J.; Erisman, J.W. Effects of environmental stress on forest crown condition in Europe. Part I: Hypotheses and approach to the study. *Water Air Soil Pollut.* **2000**, *119*, 317–333. [CrossRef]
10. Ozolinčius, R. *Lithuanian Forest Condition and Its Influencing Factors*; Lutute Publishing: Kaunas, Lithuania, 1999; p. 90.
11. Dubravac, T.; Dekanić, S.; Roth, V. Dinamika oštećenosti i struktura krošanja stabala hrasta lužnjaka u šumskim zajednicama na gredi i u nizi—Rezultati motrenja na trajnim pokusnim plohama. *Šumarski List* **2011**, *13*, 74–89.
12. Cardil, A.; Vepakomma, U.; Brotons, L. Assessing pine processionary moth defoliation using unmanned aerial systems. *Forests* **2017**, *8*, 402. [CrossRef]
13. Cardil, A.; Otsu, K.; Pla, M.; Silva, C.A.; Brotons, L. Quantifying pine processionary moth defoliation in a pine-oak mixed forest using unmanned aerial systems and multispectral imagery. *PLoS ONE* **2019**, *14*, e0213027. [CrossRef]
14. Navarro-Cerrillo, R.M.; Varo-Martínez, M.Á.; Acosta, C.; Rodriguez, G.P.; Sanchez-Cuesta, R.; Gomez, F.J.R. Integration of WorldView-2 and airborne laser scanning data to classify defoliation levels in Quercus ilex L. Dehesas affected by root rot mortality: Management implications. *For. Ecol. Manag.* **2019**, *451*, 117564. [CrossRef]
15. Kalafadžić, Z.; Kušan, V. Definiranje stupnja oštećenosti šumskog drveća i sastojina. *Šumarski List* **1990**, *114*, 517–526.
16. Pádua, L.; Hruška, J.; Bessa, J.; Adao, T.; Martins, L.M.; Goncalves, J.A.; Peres, E.; Sousa, A.M.R.; Castro, J.P.; Sousa, J.J. Multi-temporal analysis of forestry and coastal environments using UASs. *Remote Sens.* **2017**, *10*, 24. [CrossRef]
17. Kalafadžić, Z.; Kušan, V. Opadanje prirasta jele (*Abies alba* L.) kao posljedica novonastalih oštećenja šuma u Gorskom kotaru. *Šumarski List* **1989**, *113*, 415–422.
18. Pernar, R. Način i pouzdanost određivanja oštećenosti hrasta lužnjaka (*Quercus robur* L.) na infracrvenim kolornim (ICK) aerosnimkama. *Glas. Za Šumske Pokuse* **1994**, *31*, 1–34.
19. Kalafadžić, Z. Današnje mogućnosti primjene fotointerpretacije u zaštiti šuma. *Šumarski List* **1973**, *97*, 149–165 str.
20. Donassy, V.; Oluić, M.; Tomašegović, Z. *Daljinska Istraživanja u Geoznanostma*; JAZU: Zagreb, Croatia, 1983; pp. 281–329.
21. Huss, J.; Akca, A.; Hildebrandt, G.; Kenneweg, H.; Peerenboom, H.G.; Rhody, B. *Luftbildmessung und Fernerkundung in der Forstwirtschaft*; Herbert Weichmann Verlag GmbH: Karlsruhe, Germany, 1984; p. 396.
22. Harris, J.G. A new approach to surface reconstruction: The coupled depth/slope model. In Proceedings of the First International Conference on Computer Vision, London, UK, 8–11 June 1987; pp. 277–283.
23. Hildebrandt, G.; Gross, K.P. Inventarizacija i opažanje sadašnjih oštećenosti šuma u SR Njemačkoj. *Šumarski List* **1989**, *113*, 279–298.
24. Leckie, D.G.; Teillet, P.M.; Fedosejevs, G.; Ostaff, D.P. Reflectance characteristics of cumulative defoliation of balsam fir. *Can. J. For. Res.* **1988**, *18*, 1008–1016. [CrossRef]
25. Atzberger, C.; Werner, W. Needle reflectance of healthy and diseased spruce stands. In *1st EARSeL Workshop on Imaging Spectroscopy*; Schaepman, M., Schlapfer, D., Itten, K.I., Eds.; Remote Sensing Laboratories, University of Zurich (Switzerland), Impression Dumas: Saint-Etienne, France, 1998; pp. 271–283.
26. Daniulis, J. *Aerial Photography*; Enciklopedija: Vilnius, Lithuania, 1998; p. 248.
27. Entcheva Campbell, P.K.; Rock, B.N.; Martin, M.E.; Neefus, C.D.; Irons, J.R.; Middleton, E.M.; Albrechtova, J. Detection of initial damage in Norway spruce canopies using hyperspectral airborne data. *Int. J. Remote Sens.* **2004**, *25*, 5557–5583. [CrossRef]
28. Barry, K.M.; Stone, C.; Mohammed, C.L. Crown-scale evaluation of spectral indices for defoliated and discoloured eucalypts. *Int. J. Remote Sens.* **2008**, *29*, 47–69. [CrossRef]
29. Näsi, R.; Honkavaara, E.; Lyytikäinen-Saarenmaa, P.; Blomqvist, M.; Litkey, P.; Hakala, T.; Viljanen, N.; Kantola, T.; Tanhuanpaa, T.; Holopainen, M. Using UAV-based photogrammetry and hyperspectral imaging for mapping bark beetle damage at tree-level. *Remote Sens.* **2015**, *7*, 15467–15493. [CrossRef]
30. Näsi, R.; Honkavaara, E.; Blomqvist, M.; Lyytikainen-Saarenmaa, P.; Hakala, T.; Viljanen, N.; Kantola, T.; Holopainen, M. Remote sensing of bark beetle damage in urban forests at individual tree level using a novel hyperspectral camera from UAV and aircraft. *Urban For. Urban Green.* **2018**, *30*, 72–83. [CrossRef]

31. Díaz-Varela, R.A.; de la Rosa, R.; León, L.; Zarco-Tejada, P.J. High-resolution airborne UAV imagery to assess olive tree crown parameters using 3D photo reconstruction: Application in breeding trials. *Remote Sens.* **2015**, *7*, 4213–4232. [CrossRef]
32. Guerra-Hernández, J.; González-Ferreiro, E.; Monleón, V.; Faias, S.P.; Tome, M.; Diaz-Varela, R.A. Use of multi-temporal UAV-derived imagery for estimating individual tree growth in *Pinus pinea* stands. *Forests* **2017**, *8*, 300. [CrossRef]
33. Laliberte, A.S.; Herrick, J.E.; Rango, A.; Winters, C. Acquisition, orthorectification, and object-based classification of unmanned aerial vehicle (UAV) imagery for rangeland monitoring. *Photogramm Eng. Remote Sens.* **2010**, *76*, 661–672. [CrossRef]
34. Ahmed, O.S.; Shemrock, A.; Chabot, D.; Dillon, C.; Williams, G.; Wasson, R.; Franklin, S.E. Hierarchical land cover and vegetation classification using multispectral data acquired from an unmanned aerial vehicle. *Int. J. Remote Sens.* **2017**, *38*, 2037–2052. [CrossRef]
35. Kalafadžić, Z.; Kušan, V. Ustanovljavanje stanja šuma na velikim površinama primjenom ICK aerosnimki. *Glas. Za Šumske Pokuse* **1990**, *26*, 447–459.
36. Kolić, J.; Pernar, R.; Seletković, A.; Ančić, M. The Importance of CIR Aerial Imagery in Inventory, Monitoring and Predicting Forest Condition. *South-East Eur. For.* **2015**, *6*, 191–200. [CrossRef]
37. Lehmann, J.R.K.; Nieberding, F.; Prinz, T.; Knoth, C. Analysis of unmanned aerial system-based CIR images in forestry–a new perspective to monitor pest infestation levels. *Forests* **2015**, *6*, 594–612. [CrossRef]
38. Lisein, J.; Michez, A.; Claessens, H.; Lejeune, P. Discrimination of deciduous tree species from time series of unmanned aerial system imagery. *PLoS ONE* **2015**, *10*, e0141006. [CrossRef]
39. Dash, J.P.; Watt, M.S.; Pearse, G.D.; Heaphy, M.; Dungey, H.S. Assessing very high resolution UAV imagery for monitoring forest health during a simulated disease outbreak. *ISPRS J. Photogramm. Remote Sens.* **2017**, *131*, 1–14. [CrossRef]
40. Nevalainen, O.; Honkavaara, E.; Tuominen, S.; Viljanen, N.; Hakala, T.; Yu, X.W.; Hyyppa, J.; Saari, H.; Polonen, I.; Imai, N.N.; et al. Individual tree detection and classification with UAV-based photogrammetric point clouds and hyperspectral imaging. *Remote Sens.* **2017**, *9*, 185. [CrossRef]
41. Senf, C.; Seidl, R.; Hostert, P. Remote sensing of forest insect disturbances: Current state and future directions. *Int. J. Appl. Earth Obs. Geoinf.* **2017**, *60*, 49–60. [CrossRef] [PubMed]
42. Safonova, A.; Tabik, S.; Alcaraz-Segura, D.; Rubtsov, A.; Maglinets, Y.; Herrera, F. Detection of fir trees (*Abies sibirica*) damaged by the bark beetle in unmanned aerial vehicle images with deep learning. *Remote Sens.* **2019**, *11*, 643. [CrossRef]
43. Kalafadžić, Z.; Kušan, V. Visoka tehnologija u inventuri šuma. *Šumarski List* **1991**, *115*, 509–520.
44. Pernar, R.; Ančić, M.; Seletković, A.; Kolić, J. Važnost daljinskih istraživanja pri procjeni šteta na šumskim sastojinama uzrokovanih velikim prirodnim nepogodama. In Proceedings of the Gospodarenje Šumama u Uvjetima Klimatskih Promjena i Prirodnih Nepogoda, Zagreb, Croatia, 20 April 2018; Anić, Igor; HAZU: Zagreb, Croatia, 2020; pp. 143–160. [CrossRef]
45. Fraser, B.T.; Congalton, R.G. Monitoring Fine-Scale Forest Health Using Unmanned Aerial Systems (UAS) Multispectral Models. *Remote Sens.* **2021**, *13*, 4873. [CrossRef]
46. Kolić, J.; Pernar, R.; Seletković, A.; Samaržija, M.; Ančić, M. Comparison of crown damage assessment on analog and digital infrared color (CIR) aerial photographs. *Šumarski List* **2022**, *5–6*, 187–199. [CrossRef]
47. Gregorie, T.G.; Valentine, H.T. *Sampling Strategies for Natural Resorces and the Envioronment*; Chapman and Hall/CRC: Boca Raton, FL, USA, 2008; p. 474.
48. Spurr, S.H. *Forest Inventory*, Roland Press Co.: New York, NY, USA, 1952; p. 476.
49. Vedriš, M. *Utjecaj Različitih Metoda Uzorkovanja na Izmjeru i Procjenu Strukturnih Elemenata Bukovo-Jelovih Sastojina*; Faculty of forestry, University of Zagreb: Zagreb, Croatia, 2010.
50. Kalafadžić, Z.; Kušan, V.; Horvatić, Z.; Fintić, R.; Hrašovec, B. *Određivanje Stupnja Oštećenosti Šuma Bukve i Jele Zajednice Općina Rijeka*; Faculty of Forestry, Study: Zagreb, Croatia, 1989.
51. Pernar, R.; Seletković, A.; Ančić, M.; Vedriš, M.; Teslak, K. Assessing the health status of beech-fir forests using remote sensing methods. *Period. Biol.* **2008**, *110*, 157–161.
52. Pranjić, A.; Lukić, N. *Izmjera Šuma*, 1st ed.; Faculty of Forestry: Zagreb, Croatia, 1997; pp. 1–405.
53. Pranjić, A. *Šumarska Biometrika*, 1st ed.; Faculty of Forestry: Zagreb, Croatia, 1986; pp. 1–205.

sustainability

MDPI

Article

Invisible Frost Stress on Introduced *Dalbergia odorifera*: A Bioassay on Foliar Parameters in Seedlings from Six Provenances

Xiaowen Li [1], Yu Liu [1], Sheng Yang [1], Jinwang Wang [1], Haitao Xia [1], Xiaojin Liu [2] and Qiuxia Chen [1,*]

[1] Wenzhou Key Laboratory of Resource Plant Innovation and Utilization, Zhejiang Institute of Subtropical Crops, Zhejiang Academy of Agricultural Sciences, Wenzhou 325005, China; lixiaowen1979@126.com (X.L.); liuyu@zaas.ac.cn (Y.L.); yangsheng0072001@sina.com (S.Y.); kingwwang@163.com (J.W.); xht.11@163.com (H.X.)

[2] Research Institute of Tropical Forestry, Chinese Academy of Forestry, Guangzhou 510520, China; xjliu@caf.ac.cn

* Correspondence: yzscqx@163.com

Abstract: Valuable trees are frequently taken from their original habitat and introduced to a different location in the pursuit of better economic development. Global climate change imposes a higher probability of warm spells during chilly seasons; these may increase the threat posed by frost to newly introduced, valuable species. In this study, *Dalbergia odorifera* was cultured as a valuable tree species that was introduced from an original provenance in Sanya (1° N) to the northern mountains in Pingxiang (22° N), Guangzhou (23° N), Zhangpu (24° N), Xianyou (25° N), and up to the northernmost limit in Wenzhou (28° N). Seedlings of these six provenances were tested in a field study conducted in Wenzhou (control) to examine their resistance to local frost stress and to detect the driving forces related to meteorological factors in the winter–spring period of 2015–2016. The leaves sampled over seven days exhibited the typical characteristics of frost impairment. The daily maximum temperature delivered warm spells, increasing by ~7 °C. The daily minimum temperature (−4.3 to −2.0 °C) did not reach freezing point until the early spring of 2016. The controlled seedlings showed lower malondialdehyde content than those from the southern locations, and no mortality occurred. Invisible frost stress was caused by low nitrogen utilization during the earlier stages during warm spells, as well as damage to membrane integrity during the later stage when the minimum temperature suddenly declined. A warm spell was found to impose a negative driving force five days before a sudden chill, which led to frost having an impact on superoxide accumulation and electrical leakage. We conclude that the *D. odorifera* seedlings that dwell effectively in Wenzhou obtained stronger resistance to local frost stress than those from the southern locations. Low cell membrane integrity and high electrical leakage in leaf cells accounted for the frost damage.

Keywords: chilly stress; foliar variable; silviculture; subtropical forest; warm spell

Citation: Li, X.; Liu, Y.; Yang, S.; Wang, J.; Xia, H.; Liu, X.; Chen, Q. Invisible Frost Stress on Introduced *Dalbergia odorifera*: A Bioassay on Foliar Parameters in Seedlings from Six Provenances. *Sustainability* **2023**, *15*, 14097. https://doi.org/10.3390/su151914097

Academic Editors: Ivica Papa and Andreja Đuka

Received: 13 August 2023
Revised: 3 September 2023
Accepted: 11 September 2023
Published: 23 September 2023

1. Introduction

The results have been contradictory concerning the characterization of the effect of frost risk on forest phenology and sustainability. Certain findings have illustrated that trends with reduced frequency and intensity in frost events can offset some of the negative effects of a sudden chill on forest plants [1,2]. This argument has mainly been supported via modeling with past data and via predictions according to the monitored changes. However, it has also been argued that climate warming has increased frost risks for forest plants due to the advanced phenology and aggravated impairment of organs during dehardening [3–5]. This risk has been predicted to exist with continuous warming at an expected probability as high as 20% in a thoroughly projected period that runs up to 2090 [6]. The frost risk mainly threatens the buds of dormant plants by advancing bud burst, and this occurs during warm spells in the early spring or late winter [7–9]. These pieces of evidence have mostly been obtained from studies on temperate forests, but particularly rare instances can be referred to in investigations of forests subjected to subtropical and tropical climates.

Forest trees acclimating to a warm climate are not fully dormant in chilly seasons, at least not to the same degree as in the induced dormancy when hardening, as is found in trees subjected to a temperate climate [10]. In neotropical forests, it has been revealed that the incursions of Arctic cold waves may cause mild frost at subtropical latitudes [11]. Frost stresses affecting tropical forests are increasing following the processes induced by the collapsing Arctic cryosphere [12]. However, in lowland tropical forests, cold waves trigger extensive tree mortality because of heavier cold air masses moving downhill in advective frosts [11,13]. Although a general study revealed that woody plants do not always show apparent frost-damaged symptoms in tropical climates [14], in savannas, both trees and shrubs exhibit large-scale diebacks following severe cold waves [15,16]. Therefore, the threat of forest stress may be a provenance-dependent phenomenon for tropical forests, which can be supported by a mosaic forest–grassland pattern (which are dominated by plants of different provenances that follow frost along low tree lines [17]). Theoretically, trees originating from low-latitude locations can dwell in a new habitat with chillier temperatures, but they could still encounter a scenario where there is the possibility of suffering from frost attack due to exposure to a cold-wave invasion. Information is still limited regarding the provenance-specific effects of frost stress on subtropical and tropical forest trees.

Differences that are large enough to activate essential botanical variations in plants of contrasting provenances are mostly shown in two locations over a distance across latitudes [18,19]. The transfer of a planting location changes the corresponding plant's acclimation and adaption to local frost stress [11,13]. Climate warming increases the risk of exposure to frost due to an advanced phenology in a warm spell and less of a hardening dormancy immediately following a sudden chill. A microclimate near a city can further strengthen the frost risk posed to non-endemic dwellers from other locations due to the heat island effect (HIE) [20,21]. The socioeconomic dimensions of a host city account for the magnitude of the HIE [22], which may also be a promotor of frost risk for local forest plants. Therefore, cities with varied socioeconomic states should be a source of variation in the frost risks for forests that have economically valuable plants introduced from other locations. Furthermore, montane regions near cities that are located in a vast range across latitudes can be an ideal set of regions for testing the provenance-specific responses of introduced plants.

Frost damage in montane plants is mainly quantified through the phenological data collected from field observations [8,23–25], near-surface remote sensing [1,7,8,26], digital photograph monitoring [27], radial growth measurements [28–32], and field sampling and chemical analysis [3,33]. These studies provide references across a vast geographical range of the forest stratosphere, but the results are still limited by a failure to explain physiological changes in the process of impairment that is caused by frost damage. More needs to be uncovered about the provenance-dependent responses of specific physiological variables [3,33] that are used to assess plant growth and development in response to the threat of a cold spell [34]. To the best of our knowledge, the well-demonstrated parameters of forest trees subjected to late spring frosts have mainly been reported to be the non-structural carbohydrate metabolism and cell electrolyte leakage [3,33]. More physiological parameters need to be investigated to reveal their responses to cold resistance against a frost threat.

Dalbergia odorifera T. Chen is a leguminous species from the family Fabaceae that is endemic to the tropical montane areas of China. The wood of this species is of high economic value for uses in furniture and folk medicine [35]. In 2012, the average international price of *Dalbergia* timber ranged from USD 16,575 to USD 49,656 per m^3, which was evaluated to be higher in comparison with the temporally local prices of this timber's use in Indonesia (which ranged from USD 1412 to USD 2500 per m^3 [36]). The annual demand for raw

D. odorifera heartwood is over 300 tons, and the annual production value exceeds USD 700 million [37]. The high demand for wood resources has rendered its natural reserves vulnerable under criteria A1d [38]. Therefore, this species is frequently introduced northward from its original location to achieve high economic profits through timber trades. Cultural practices have been studied to improve the adaptation of *D. odorifera* stocks introduced from southern locations to the local freezing that occurs in northern mountains [39–41]. However, there are still uncertainties surrounding the mechanism of the physiological responses of trees with southern provenances to the local frost shock that occurs in northern regions. A winter chilly wave may cause severe damage to non-endemic *D. odorifera*, but the differences in physiological responses across different provenances has rarely been documented.

In this study, *D. odorifera* stocks from six locations were targeted as the study objects. One provenance was the stock's original habitat in Hainan, and the other five were introduced to the northern mountains up to the northern location at Wenzhou (which is the most developed region and can enlarge the value of its development to a higher level than that of any of the southern locations). However, Wenzhou is also a place that has experienced several freezing events in recent years, and its local *D. odorifera* has been reported to show symptoms of frostbite, although its mortality rates were not obvious. An over-year winter–spring episode was investigated in relation to foliar physiology in order to examine responses to combined temperature across provenances. The locations of the provenances tested in this study were adapted from montane fields near cities with highly varied socioeconomic states and HIEs. Our objective was to reveal the invisible mechanism of frost stress on *D. odorifera* specimens, which were introduced from southern provinces using foliar parameters in Wenzhou. Our study exhibits significant novelty in that it is the first to reveal the meteorological mechanism that accounts for the frost stress on *D. odorifera* when using foliar parameters. We hypothesized that northern provenances would show better adaptation to subtropical frost shock than southern provenances.

2. Materials and Methods

2.1. Subsection

Seedlings from six *D. odorifera* provenances were chosen as the study materials in this study. The plantation established in Jianfengling National Forest Park (18°41′ N, 108°46′ E), Sanya city, Hainan province, China, was characterized as having an original provenance. The second plantation was established in Pingxiang city (22°07′ N, 106°53′ E), Guangxi province, China, through introducing seedlings transferred from the first plantation. The third plantation was established by introducing seedlings to Guangzhou city (23°12′ N, 113°23′ E), Guangdong province, China. The fourth and fifth plantations were established by introducing seedlings to Zhangpu city (24°07′ N, 117°37′ E) and Xianyou city (25°22′ N, 118°41′ E), Fujian province, China, respectively. The sixth plantation was our objective control, and it was established by introducing stocks from all the abovementioned provenances to a northern mountain near Wenzhou city (28°00′ N, 120°37′ E), Zhejiang province, China. Seedlings introduced to Wenzhou have established themselves to the local conditions and dwell in a local edaphic environment, but certain individuals still suffer from frost stresses and show degraded symptoms following chilly seasons. The geographical distributions of the six plantations are shown in Figure 1. All plantations were located in montane areas belonging to the municipal lands of host cities, of which the socioeconomic conditions are shown in Table 1. Overall, the provenances in Hainan and Guangxi have tropical monsoon climates, and those in Guangdong, Fujian, and Zhejiang are located in a subtropical maritime monsoon climate. All regional climates were largely controlled by the monsoon, which resulted in a general common temperature fluctuation during a winter–spring cross-year episode.

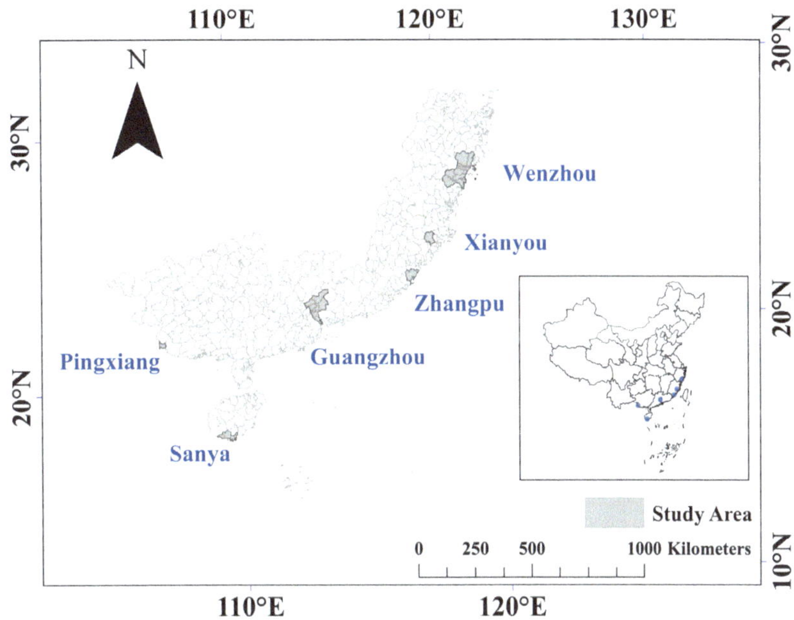

Figure 1. Spatial distribution of the municipal areas of *Dalbergia odorifera* provenances along latitudes, as well as across subtropical and tropical climates in South China.

Table 1. Socioeconomic conditions of the host prefecture regions where sampling plots of *Dalbergia odorifera* were placed in regions of South China.

Plot Order	Province	Municipal	GDP [1] (USD Billion)	Industrial Proportion (%)			Resident Population (Million)
				1st	2nd	3rd	
1	Hainan	Sanya	43.58	13.73	20.54	65.73	0.75
2	Guangxi	Pingxiang	5.69	8.77	28.84	62.39	0.12
3	Guangdong	Guangzhou	1810.04	1.25	31.64	67.11	13.50
4	Fujian	Zhangpu	276.74	13.40	48.50	38.10	5.00
5	Fujian	Xianyou	30.97	9.87	51.34	38.79	1.15
6	Zhejiang	Wenzhou	461.81	2.80	43.80	53.40	8.11

[1] GDP, gross domestic product.

In March of 2015, a total of 1200 *D. odorifera* seedlings—where every 200 seedlings collected had a different provenance, meaning that all six provenances were represented (Figure 1)—were transplanted to a forest nursery in Wenzhou. Then, 100 seedlings were chosen from one provenance to frame their initial growing states to an even extent, and a total of 600 seedlings were selected from six provenances. The criteria for screening the chosen seedlings were the growing morphologies of height in a 50–80 cm range and their root-collar diameter (RCD) in a 0.5–0.8 cm range (Figure 2). These screening criteria were established according to the general growth state for most seedlings. Seedlings of all six provenances were mixed together and planted in three plots at a spacing of 1.0 m × 1.5 m. For every provenance, a total of 100 seedlings were randomly distributed to three groups, and each group contained about 33 of them. Three groups of seedlings were planted to three different plots, which were taken as three replicated blocks for the seedlings from a specific provenance. Every individual seedling was labeled using a hanging tag to mark its provenance and to show whether it was used for sampling. Six individual seedlings were randomly selected for one provenance per site and assigned as six sampling objectives, and it was these results that were averaged for the plot.

Figure 2. A typical view of the *D. odorifera* stocks from the provenances transplanted to Wenzhou in one site, where stocks from all the six provenances were mixed together.

2.2. Sampling Dates

The experimental duration lasted from 1 September 2015 to 31 January 2016. Sampling dates were determined by synthesizing the data of real-time forecasting meteorological factor changes and summarizing the recent temperature fluctuations. Every date for sampling was chosen as a day on which frost is likely occur in one of immediately following days.

According to previous studies on forest trees that were subjected to a late spring frost in Europe [7,8,23,32], frost damage becomes more harmful when there is an earlier advanced warming spell that is followed by a sudden decline in the daily temperature. We employed the theoretical model put forth by Gu et al. [42] and Augspurger [43] to describe these factors. In their model, they determined that the initial (as early as the occurrence of no budburst during a deep hardening phase) temperature was elevated in advance by a warming spell for about 3–5 days; thereafter, the temperature suddenly declined to a freezing level and persisted for 1–2 days. All fluctuations repeated 2–3 times until frost damage became visible on the newly growing organs and tissues. It is unreasonable to expect temperature decline to be lower than freezing in subtropical and tropical climates as frequently as in a temperate climate. Therefore, we referred to the pattern of dynamic temperature fluctuations that accounted for frost events recorded to most likely happen in temperate forests [7,8,27,43]. Therefore, the leaves were sampled on days that fit with the following rules:

(1) At least 5 days remained before leaf sampling was to take place, with an expectation of a dual occurrence of advanced warming and a sudden temperature decline.
(2) The last 1–2 days prior to sampling had to be accompanied by sharp declines in both the lowest and highest daily temperatures, and this had to happen before the occurrence of a frost.
(3) A sampling day had to immediately follow a decline in the highest daily temperature, which was also the day with the lowest daily temperature in the most recent 5 days.

(4) Any day that did not have characteristics of temperature fluctuation following any of rules (1)–(3) could not be chosen for foliar sampling.

As a result, seven days were chosen for sampling because they obeyed all of the above-mentioned rules (Figure 3). These days were 15 September, 1 November, 11 November, 27 November, and 18 December in 2015, and 9 January and 25 January in 2016. The same-day meteorological conditions are also shown for the sampling days.

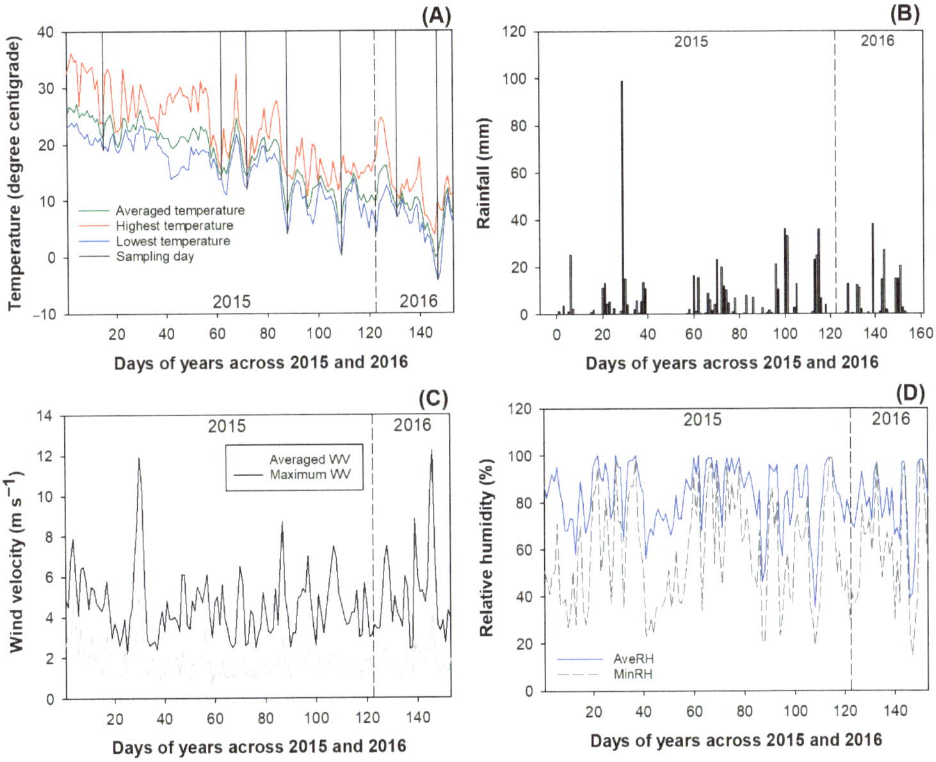

Figure 3. Dynamic changes in the meteorological factors (temperature, (**A**); rainfall, (**B**); wind velocity, (**C**); and relative humidity, (**D**)) in Wenzhou, where the *D. odorifera* stocks were transplanted from five other locations during the period from 1 September 2015 to 31 January 2016. Specific sampling days are marked in cell A: 15 September (15 d after experimental commencement); 1 November (62 d); 11 November (72 d); 27 November (88 d); and 18 December (109 d) in 2015. In addition, we include 9 January (131 d) and 25 January in 2016 (147 d).

2.3. Sampling and Chemical Analysis

In one plot, six labeled seedlings were used for leaf sampling as a bulk per provenance, and three bulked averages were taken as three replicated values from three repeated blocks. At the same time, in the first sampling, all of the seedlings had experienced a growing season of about six months, and this was recognized as a sufficient time for new leaves to grow. Fresh new leaves were sampled since they were especially vulnerable to frost shocks [4,27]. At each sampling, ten fresh leaves were collected from the twigs and branches that were fully exposed to sunlight. This eliminated the possibility that wilted leaves were fully accounted for by a frost shock, but not by a response to low-light stress. The sampled leaves all looked healthy, without any of the symptoms that appear when leaves are attacked by insects. The leaves were excised and transported to the laboratory on ice (0–2 °C).

Photosynthetic pigments were assessed by determining the chlorophyll-a and -b and carotenoid contents using a method adapted from Hiscox and Israelstam [44] with minor modifications. The leaves were approximately 0.05 g in weight, and they were cut to pieces and immersed in a tube with 2.5 mL of dimethyl sulfoxide. The tubes were heated at 65 °C for 1 h in a water bath under dark conditions for 1 h. Chlorophyll-a and -b contents were measured using a spectrophotometer (UV-Visible 8453, gilent Tech. Inc., Santa Clara, CA, USA) at 663 nm and 645 nm, respectively. The carotenoid content was estimated using the equations of Lichenthaler and Wellburn [45].

Soluble sugar content, soluble protein content, peroxidase (POD) activity, and malondialdehyde (MDA) content were measured using methods adapted from Liu et al. [46]. A leaf sample of around 0.1 g in weight was heated in a hot water bath for 30 min, and this was centrifuged at 5000 rpm and measured for soluble sugar content via a spectrophotometer at 490 nm. For the assays of soluble protein content and POD activity, another 0.1 g sample was ground in a mortar with liquid nitrogen as the freezing reagent. This was then moved to a tube, which was placed in an ice bath, and 1 mL of a mixture of ice-cold sodium phosphate buffer solution (50 mmol L^{-1}, pH = 7.0) and 2% polyvinylpyrrolidone was added. The homogenate was kept in ice for 1 min and centrifuged at 5000 rpm (4 °C) for 10 min, after which all of the supernatant was collected. The soluble protein was measured using the Coomassie Brilliant Blue staining method [47]. The POD activity was assayed using a classical method adapted from Ryu and Dordick [48]. To assay the MDA content, a 0.5 g leaf sample was mixed with 5 mL of 10% trichloroacetic acid and centrifuged at 5000 rpm for 10 min. The supernatant was mixed with 2 mL of 0.67% thiobartituric acid, heated in a boiling water bath for 30 min, cooled down to room temperature, centrifuged at 5000 rpm, and detected for absorbances at 450 (A_{450}), 532 (A_{532}), and 600 (A_{600}). Therefore, the MDA content (C_{MDA}) could be calculated using the following equation:

$$C_{MDA} = 6.452 \times (A_{532} - A_{600}) - 0.559 \times A_{450} \tag{1}$$

Freezing-impaired cell leakage was assayed using electrical conductance (EC), which was performed according to a method used by Wang et al. [3]. Fresh leaves were cut into segments approximately 5 mm in length, which were then placed in a capped tube (10 mL volume) with 5 mL of distilled water. This treatment was assigned as C_1. Another 10 mL tube was filled with 5 mL of distilled water and assigned as the control of B_1. All C_1 and B_1 tubes were placed in the dark for 24 h to enable a full electrolyte leakage at a natural rate for water; they were then measured for their levels of EC. Thereafter, C_1- and B_1-labeled tubes were heated in an oven at 90 °C for 2 h, cooled down to room temperature, and then kept in the dark for 24 h; following this, they were measured for EC and labeled as post-stress treatments C_2 and B_2, respectively. Finally, the EC gauge for freezing injury can be calculated as follows [49,50]:

$$EC = \frac{LCC_1 - LCB_1}{LCC_2 - LCB_2} \times 100\% \tag{2}$$

where LCC_1, LCB_1, LCC_2, and LCB_2 are the leakage conductance values for the labeled samples C_1, B_1, C_2, and B_2, respectively.

2.4. Statistical Analysis

All data passed tests of normality and variance homogeneity; hence, no transformation was needed. SAS software (ver. 9.4, SAS Inst. Inc., Cary, NY, USA) was used for the data analysis. Analysis of variance (ANOVA) was used with a mixed model to detect the differences for every leaf parameter among the provenances on seven repeated measuring days [51]. When a significant difference was detected ($p < 0.05$) in a specific sampling day, the results were arranged and compared between the provenances with a Tukey test. Meteorological variables were extracted for every sampling day as the daily maximum temperature five days prior to the sampling day and the daily minimum temperature three days prior to the sampling day, as well as according to the rainfall, wind velocity, and RH on the sampling day. These meteorological factors were pooled as independent variables and used to test the contributions to every leaf parameter via multivariable linear regression models.

3. Results

3.1. Photosynthetic Pigments

Repeated variations in the different provenances had significant effects on the contents of chlorophyll-a, chlorophyll-b, and carotenoids (Table 2). The chlorophyll-a content in the Wenzhou sample was no different from that in Sanya for most sampling days except for 72 (Figure 4(A3)) and 109 days after the experiment began (Figure 4(A5)). On both days, the seedlings from Wenzhou had higher chlorophyll-a content than those from the Sanya sample. The chlorophyll-a content in the seedlings from Wenzhou was not different from that in the two samples of Zhangpu and Xianyou in Fujian for most sampling days. The chlorophyll-a content in the Guangzhou sample was lower than that for the northern provenances for the initial two sampling days (Figure 4(A1,A2)), but this increased to be higher than that in the Sanya and Pingxiang samples on the last sampling day (Figure 4(A7)).

Table 2. Parameters (*F* and *p* values) from the analysis of variance (ANOVA) of the repeated effects of provenance variation on the foliar parameters in *D. odorifera* individuals that were subjected to different sampling days across 2015 and 2016.

Source of Variation	ANOVA [1]	2015					2016	
		15 Sep.	01 Nov.	11 Nov.	27 Nov.	18 Dec.	09 Jan.	25 Jan.
Chla [2]	*F* [3]	**11.89** [4]	**21.49**	**17.28**	**6.91**	**33.45**	**9.99**	**5.98**
	P	**0.0003**	**<0.0001**	**<0.0001**	**0.0030**	**<0.0001**	**0.0006**	**0.0053**
Chlb [5]	*F*	**11.33**	**19.41**	**13.20**	**6.82**	**31.15**	**9.00**	**5.73**
	P	**0.0003**	**<0.0001**	**0.0002**	**0.0031**	**<0.0001**	**0.0009**	**0.0063**
Carotenoid	*F*	**10.24**	**23.30**	**23.40**	**7.15**	**37.23**	**12.53**	**4.01**
	P	**0.0005**	**<0.0001**	**<0.0001**	**0.0026**	**<0.0001**	**0.0002**	**0.0227**
MDA [6]	*F*	**18.81**	**3.46**	**4.64**	1.66	**3.48**	1.01	**6.89**
	P	**<0.0001**	**0.0362**	**0.0138**	0.2190	**0.0355**	0.4509	**0.0030**
POD [7]	*F*	**3.96**	2.76	**6.82**	1.93	**7.33**	**8.29**	**4.82**
	P	**0.0235**	0.0694	**0.0031**	0.1620	**0.0023**	**0.0014**	**0.0119**
Sugar	*F*	**3.75**	**18.48**	2.68	0.15	**27.91**	**10.85**	**12.72**
	P	**0.0283**	**<0.0001**	0.0749	0.9762	**<0.0001**	**0.0004**	**0.0002**
Protein	*F*	**4.34**	**12.54**	**5.34**	1.26	**37.23**	**4.62**	**6.89**
	P	**0.0173**	**0.0002**	**0.0082**	0.3423	**<0.0001**	**0.0140**	**0.0030**
EC [8]	*F*	1.52	**24.46**	**50.12**	**8.34**	**27.95**	**3.56**	**3.93**
	P	0.2548	**<0.0001**	**<0.0001**	**0.0013**	**<0.0001**	**0.0331**	**0.0289**

[1] ANOVA, analysis of variance; [2] Chla, chlorophyll-a content; [3] *F* and *P* values represent ANOVA significance; [4] bold values indicate significant results; [5] Chlb, chlorophyll-b content; [6] MDA, level of malondialdehyde; [7] POD, peroxidase activity; [8] EC, electrical conductance.

The chlorophyll-b content was no different between the Wenzhou and Sanya provenances for all sampling days (Figure 4(B1–B7)). Compared to the chlorophyll-b content in the Wenzhou sample, the content in the Guangzhou sample was lower 15 d (Figure 4(B1)) and 62 d post-experiment commencement (Figure 4(B3)), and those in the Pingxiang sample were lower 62 d (Figure 4(B3)), 109 d (Figure 4(B5)), and 131 d (Figure 4(B6)) post-experiment commencement.

Carotenoid content in the Sanya sample was lower than that in the Wenzhou sample 72 d (Figure 4(C3)) and 109 d (Figure 4(C5)) after the experiment began. Carotenoid content in the Guangzhou samples was lower than that in the Wenzhou samples for the early sampling days at 15 d (Figure 4(C1)) and 62 d (Figure 4(C3)) after the experiment began.

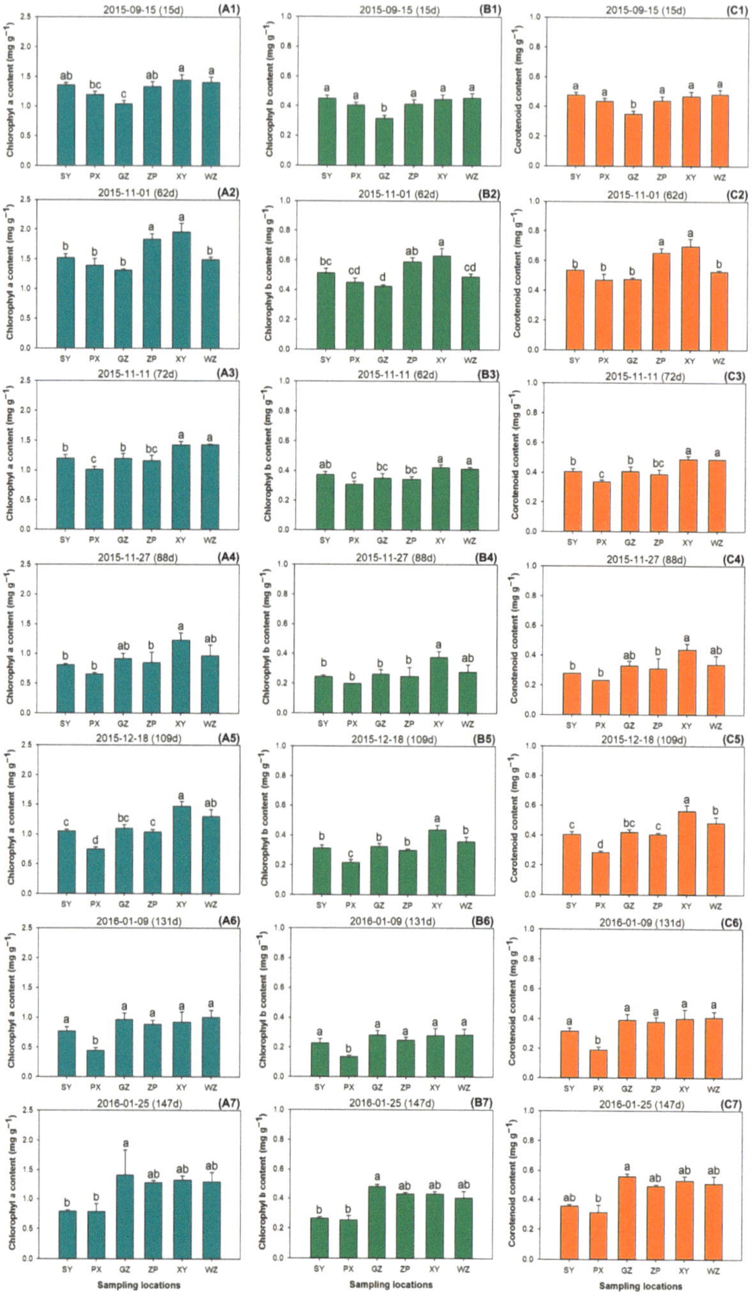

Figure 4. Foliar contents of the chlorophyll-a (**A1–A7**), chlorophyll-b (**B1–B7**), and carotenoid (**C1–C7**) contents in the *D. odorifera* provenances that were subjected to different sampling days across 2015 and 2016. Provenance names: SY, Sanya; PX, Pingxiang; GZ, Guangzhou; ZP, Zhangpu; XY, Xianyou; and WZ, Wenzhou. Different lower-case letters indicate a significant difference between the provenances according to Tukey's test at the 0.05 level.

3.2. Antioxidant Activity

The foliar MDA content was different among the provenances for most sampling days except for the two days of 27 November 2015 and 9 January 2016 (Table 2). The foliar MDA content was lower in the Sanya sample than in Wenzhou on the sampling day 72 d after the experiment began (Figure 5(A3)), but contrary results occurred 147 d (Figure 5(A7)) after experiment commencement. The foliar MDA content was higher for the Wenzhou provenance than that in the two provenances (Zhangpu and Xianyou) in Fujian on the early sampling days of 15 d (Figure 5(A1)), 62d (Figure 5(A2)) and 72 d (Figure 5(A3)) after experiment commencement. The foliar MDA content in Guangzhou was higher than that in Wenzhou on 109 d (Figure 5(A5)) and 147 d (Figure 5(A7)) after the experiment began.

The foliar POD activity did not differ between the provenances on 1 November and 27 November 2015, with all of the rest being significant on the rest of the sampling days (Table 2). The foliar POD activity did not show significant a difference between the Sanya and Wenzhou provenances (Figure 5(B1–B7)). On the 72nd day after the experiment began, the POD activity in Xianyou was higher than that in Wenzhou (Figure 5(B3)); however, 131 days after commencement, the results were reversed (Figure 5(B6)).

3.3. Soluble Sugar, Protein, and EC

The foliar soluble sugar content was different among the provenances for most sampling days except for 11 and 27 November 2015 (Table 2). On 15 September 2015, the soluble sugar content was higher for the Guangzhou and Zhangpu provenances than for the Wenzhou, Sanya, and Xianyou provenances (Figure 6(A1)). Soluble sugar was higher in all southern provenances, expect for Xianyou, than in Wenzhou on 1 November 2015 (Figure 6(A2)). The degree of soluble sugar did not show any differences among the provenances on 11 (Figure 6(A3)) and 27 November 2015 (Figure 6(A4)). On 18 December 2015, the soluble sugar was higher in the samples from Pingxiang, Zhangpu, and Xianyou than Wenzhou (Figure 6(A5)). On 19 January 2016, the soluble sugar was higher for the Sanya and Pingxiang provenances than for the Wenzhou provenance (Figure 6(A6)). On 25 January 2016, the soluble sugar was lower for Pingxiang and Guangzhou than Wenzhou.

The soluble protein content was significantly different among the provenances for most sampling days except for 27 November 2015 (Table 2). On 15 September 2015, the soluble protein in the Wenzhou provenance samples did not differ from that in the other provenances (Figure 6(B1)). Compared to the samples with a Wenzhou provenance, the Sanya provenance samples had a lower soluble protein content on the 62nd day of the experiment (Figure 6(B2)). Again, the soluble protein did not differ between Wenzhou and the other provenances, but that in the Xianyou samples was higher than in the Sanya and Guangzhou samples (Figure 6(B3)). On 27 November 2015, no difference in the degree of soluble protein was detected among the provenances (Figure 6(B4)). On 18 December 2015, the soluble protein in the Wenzhou samples was lower than that in the Xianyou samples, but it was higher than in the Sanya, Pingxiang, and Zhangpu samples (Figure 6(B5)). The soluble protein in the Wenzhou provenance samples was lower than that in the Guangzhou and Xianyou samples on 9 January 2016 (Figure 6(B6)). On 25 January 2016, the soluble protein in the Wenzhou samples was only higher that in the Guangzhou samples (Figure 6(B7)).

The EC was only different among the provenances on 15 September 2015 (Figure 6C1). The EC was lower in the Wenzhou samples than in most of the other provenances except for Sanya (Figure 6(C2)). On 11 November 2015, the EC in the Wenzhou provenance samples was lower than that in the Sanya, Guangzhou, and Zhangpu provenance samples (Figure 6(C3)). The Guangzhou provenance was unique in exhibiting higher EC than the Wenzhou provenance (Figure 6(C4)). The EC in the Wenzhou samples was lower than that in the Guangzhou, Zhangpu, and Xianyou samples (Figure 6(C5)). The samples with a provenance in Sanya were unique in having a higher EC than the Wenzhou provenance samples on 9 January 2016 (Figure 6(C6)). The EC in the Wenzhou provenance samples was no different from that for the other provenances, except for the samples with provenances in Pingxiang and Guangzhou (Figure 6(C7)).

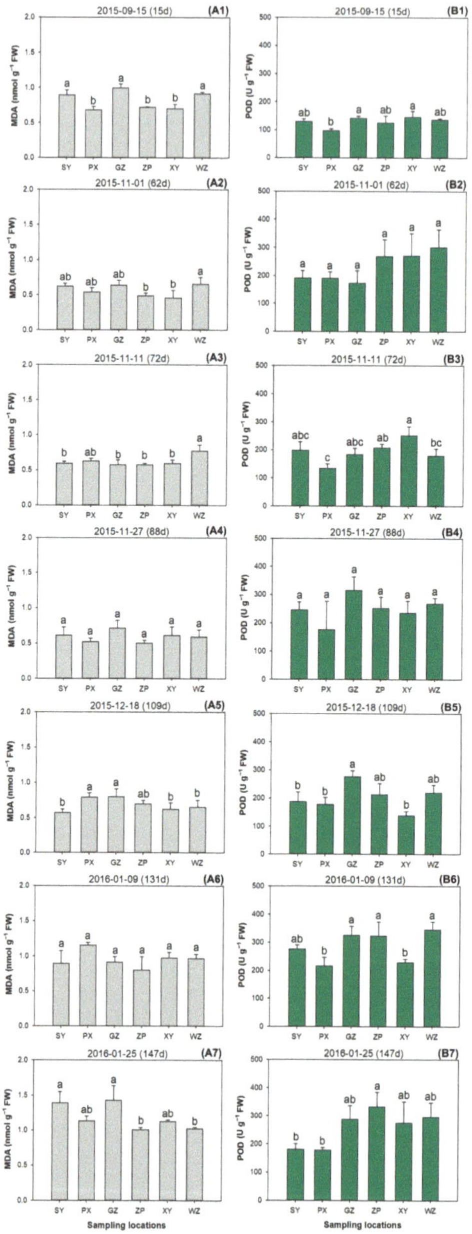

Figure 5. Foliar malondialdehyde (MDA) content (**A1** to **A7**) and peroxidase (POD) activity (**B1** to **B7**) in the *D. odorifera* provenances that were subjected to different sampling days across 2015 and 2016. Provenance name: SY, Sanya; PX, Pingxiang; GZ, Guangzhou; ZP, Zhangpu; XY, Xianyou; and WZ, Wenzhou. Different lower-case letters indicate significant differences among the provenances according to Tukey's test at the 0.05 level.

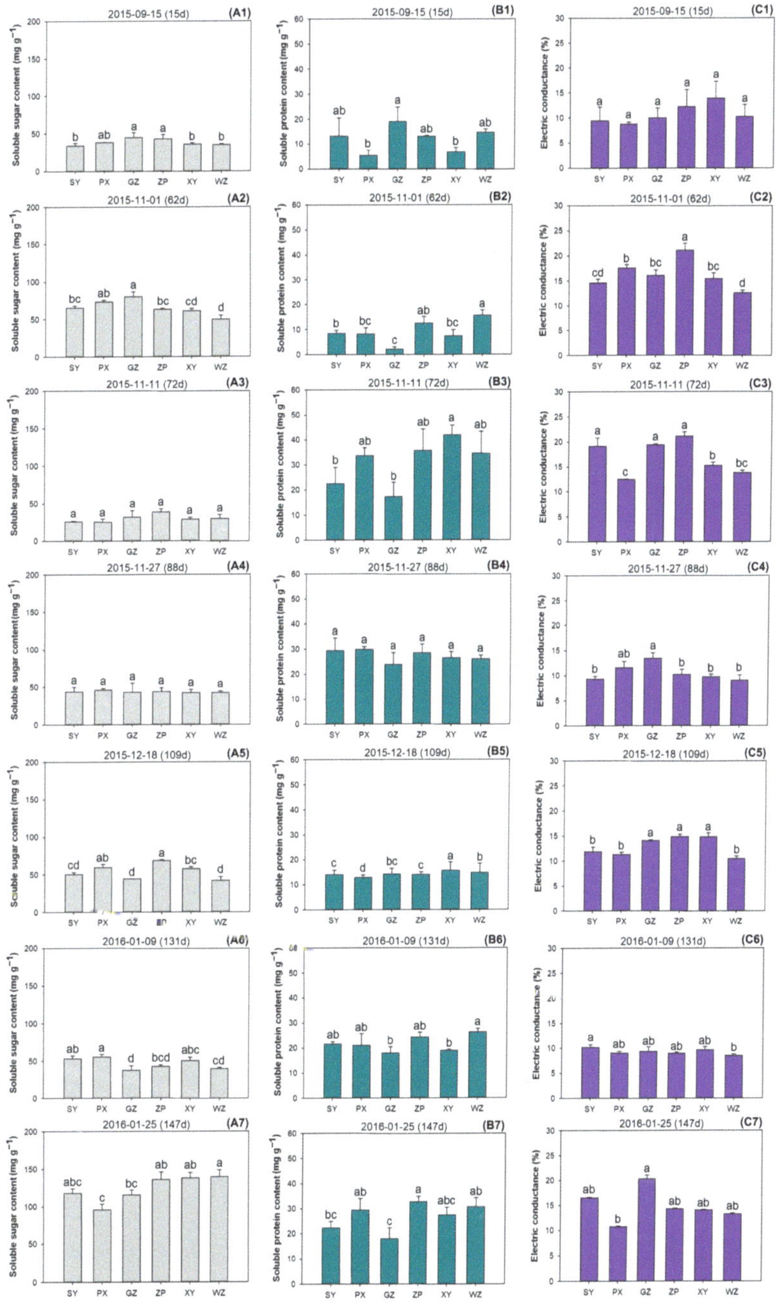

Figure 6. Foliar soluble sugar content (**A1** to **A7**), soluble protein content (**B1** to **B7**), and EC (**C1** to **C7**) in the *D. odorifera* provenances that were subjected to different sampling days across 2015 and 2016. Provenance name: SY, Sanya; PX, Pingxiang; GZ, Guangzhou; ZP, Zhangpu; XY, Xianyou; and WZ, Wenzhou. Different lower-case letters indicate significant differences among the provenances according to Tukey's test at the 0.05 level.

3.4. Regression of Foliar Parameters against Meteorological Factors

The maximum and minimum daily temperatures on the sampling days (MaxT0 and MinT0, respectively) showed contrasting contributions to the chlorophyll-a and -b contents (Figure 7(A1,A2)). MaxT0 generated positive contributions to two chlorophyll contents, while MinT0 generated a negative. MinT0 also showed a negative contribution to the carotenoid content with a positive contribution from the maximum daily temperature five days prior to sampling (MaxT5) (Figure 7(A3)).

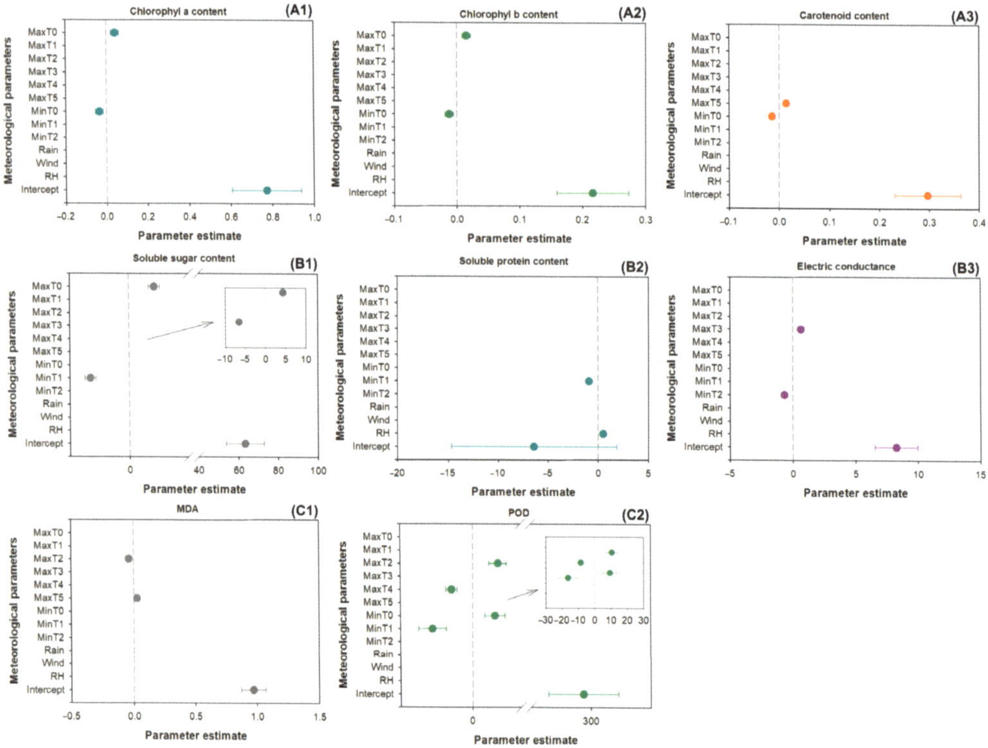

Figure 7. Multivariate linear regressions of the foliar chlorophyll-a content (**A1**), chlorophyll-b content (**A2**), carotenoid content (**A3**), soluble sugar content (**B1**), soluble protein content (**B2**), EC (**B3**), malondialdehyde (MDA) content (**C1**), and peroxidase activity (POD) (**C2**) activity in *D. odorifera* from the six provenances that were assessed on different sampling days across 2015 and 2016. Investigation days: MaxT0, maximum daily temperature on the sampling day; MaxT1, maximum daily temperature one day prior to sampling; MaxT2, maximum daily temperature two days prior to sampling; MaxT3, maximum daily temperature three days prior to sampling; MaxT4, maximum daily temperature four days prior to sampling; MaxT5, maximum daily temperature five days prior to sampling; MinT0, minimum daily temperature on the sampling day; MinT1, minimum daily temperature one day prior to sampling; MinT2, minimum daily temperature two days prior to sampling; Rain, average daily rainfall on the sampling day; Wind, average daily wind velocity on the sampling day; and RH, relative humidity on the sampling day. Colored dots indicate a parameter that was estimated by the regression model for specific dependent leaf parameters. Error bars mark the standard errors for the dependent variables.

Again, MaxT0 generated a positive contribution to soluble sugar content, while the minimum daily temperature one day prior to sampling (MinT1) generated a negative contribution (Figure 7(B1)). MinT1 also made a negative contribution to soluble protein content with a positive contribution from RH (Figure 7(B2)). The minimum daily temperature two days prior to sampling (MintT2) made a negative contribution to the EC, while the maximum daily temperature three days prior to sampling (MaxT3) made a positive contribution (Figure 7(B3)).

MaxT2 and MaxT5 made negative and positive contributions to MDA, respectively (Figure 7(C1)). In contrast, MaxT2 made a positive contribution to POD with another positive contribution from MinT0, and there were also two negative contributions from the maximum daily temperature four days prior to sampling (MaxT4) and MinT1 (Figure 7(C2)).

3.5. Analyses of the Meteorological Driving Forces Generating Frost

In a synthesis of all the abovementioned results, the highest and lowest daily temperatures were the most significant meteorological factors that affected frost resistance five days prior to sampling (Figure 8). The highest daily temperature resulted in upregulations of MDA content and electronical conductance five and three days prior to sampling, respectively, both of which were characterized to be symptoms of the negative effects of temperature fluctuation on frost resistance. However, four days prior to sampling, the highest daily temperature induced a decline in POD activity, and this was characterized as having a positive effect on frost resistance. In contrast, two days prior to sampling, POD activity increased by the daily highest temperature in synchronization with a decline in the MDA content, and these were together characterized as a positive effect on frost resistance. On the same day, the daily lowest temperature also induced a positive effect due to the decline of EC. One day before sampling, the daily highest temperature lost its effect on frost resistance, but the daily lowest temperature resulted in a strong negative effect because of a collection of decreases in soluble sugar content, protein content, and POD activity. On the sampling day, the contents of chlorophyll-a, chlorophyll-b, and soluble sugars were all increased in response to the highest daily temperature, which together suggested a positive effect on frost resistance. In contrast, the daily lowest temperature on the sampling days resulted in a negative effect due to triple decreases in chlorophyll-a, chlorophyll-b, and carotenoid contents, as well as an increase in POD activity (Figure 8).

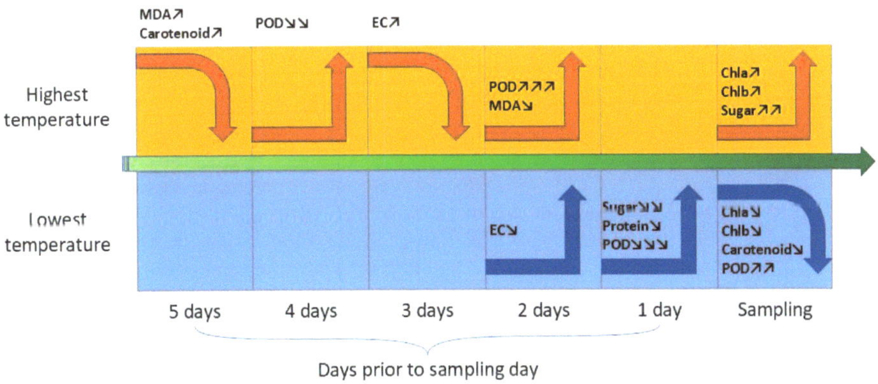

Figure 8. A summary of the driving forces from the meteorological factors and their changes in the five days prior to sampling on the responses of leaf parameters in *D. odorifera* from the six provenances that were assessed on different investigation days across 2015 and 2016. Each cell arranged in a vertical column indicates a day 1–5 days prior to the sampling day. Cells along the supernatant line are colored in claybank and contain the mean effects caused by the following highest daily temperatures; those along the line at the bottom are colored in dark blue and contain the mean effects caused by the following lowest daily temperatures. Upward arrows in thick lines indicate the positive responses that were characterized by changes in leaf parameters. The downward arrows in thick lines indicate the negative responses. Arrows in thin lines indicate the responses of changes in the

leaf parameters that were assessed by parameter estimates in the regressed models (Figure 7). An upward thin arrow indicates a positive response in frost-resistant performance; a downward thin arrow indicates a negative response to frost stress. The number of thin arrows indicates the absolute value in magnitude of the regressed parameter estimate: one arrow, the absolute value of the parameter estimate ranging 0–1; two arrows, from 1–10; and three arrows, over 10.

4. Discussion

4.1. Characterization of Frost Impairment

The plant physiological symptoms of newly growing organs that were employed in the trees of temperate forests were determined [26,42] and screened for the frost-responsive period in order to determine the seven days for sampling (as they were expected to cause frost impairments). The late spring frost used to be reported as a failure of spring due to it generating large-scale greening hiatuses [26,52]. This was expected to be linked with an episode of frequent frost events in the latter stages of the days in our study in spring 2016. This period included days with a minimum temperature that was lower than the freezing point, which can cause frost impairment, i.e., this was the last sampling day on 25 January 2016, which is when the MDA was the only parameter that responded to the provenance variation. Seedlings of southern provenances in Sanya and Guangzhou showed higher MDA levels than those in the local location in Wenzhou. When facing freezing stress, the MDA can be up-regulated as a response to the interruption of cell membrane integrity [53,54]. However, the responsive change in MDA did not synchronize with any of the symptoms of visible frost injuries that were reported in birch (*Betula pendula*) [55] and Norwegian spruce (*Picea abies*) trees [56]. Therefore, it can be surmised that, when exposed to a freezing event in a sub-tropical latitude, seedlings of southern tropical provenances may suffer frost stress that damages their cell membranes. The null responses of the EC suggested that no frost-induced cell damage was so severe as to cause leakage. The null responses for the POD activity also suggested no scavenging mechanism was activated. Overall, we can attribute the abovementioned changes of a natural decline of cell membrane integrity in the leaves of the southern seedlings to a local freezing event in Wenzhou, specifically, one that was not destructive enough to cause any physiological damage following frost stress.

4.2. Invisible Effect of Frost on the Leaves of the Southern Seedlings

Again, we do not challenge the idea that the effect of frost on trees at subtropical and tropical latitudes can be invisible. This realization partially resulted from the awareness of the side-effect of elevation on the magnitude of frost-induced mortality in neotropical forests [11,13]. We also referred to an observation made regarding landscape tree species in a temperate urban forest where frost occurred, but the damage was invisible [3]. In the period before 147 post-transplant days had passed, none of the records of the daily minimum temperature reached freezing point, but some of the leaf parameters still showed significant responses to provenance variation. The EC is a widely measured physiological parameter that changes in conjunction with the magnitudes of visible injuries of frost on tree plants [49]. Although reactive oxygen species (ROS) are also a widely used parameter for assessing stressed plants that have been subjected to frost, it has also been reported to generate rare effects on *D. odorifera* [57]. However, it was found that the chilled *D. odorifera* seedlings showed significant responses in terms of MDA content [58,59] and POD activity [59].

Compared to the EC in Wenzhou, the EC in the Pingxiang, Guangzhou, Zhangpu, and Xianyou samples were all higher on the 62nd post-transplant day. The soluble sugar content was mostly higher, but the soluble protein was mostly lower in these four provenances on the same day. This means that the starch was hydrolyzed by the frost stress to the sugars, which can fuel enzyme activity to counter chilly stress [60]. However, the amount of N used as the protein was insufficient. On that day, the higher contents of chlorophyll-a, chlorophyll-b, and carotenoids in the Zhangpu and Xianyou samples suggested that N made a greater contribution to the syntheses of the photosynthetic pigments. Accordingly, the POD activity did not differ between the provenances, and the MDA was lower or unchanged in the four southern provenances. Together, these findings support the idea that the N used for enzymes was not affected at this time because the membrane integrity was not also affected. On the 72nd post-transplant day, the EC was higher in the samples from Sanya, Guangzhou, and Zhangpu than in those from Wenzhou. The seedlings from these three southern provenances were also found to fall in a state with lower MDA, chlorophyll-a, and carotenoid contents, which together suggest a negative effect of frost on photosynthetic pigment synthesis, but better cell membrane integrity. The EC in the Guangzhou provenance samples was higher than in Wenzhou on both day 88 and day 109 after the transplant, which was in accordance with the membrane damage on the 109th day after transplant. Overall, we consider the samples from Guangzhou to have resulted in the most fragile invisible response to frost stress, and this resulted from low N utilization in the earlier stages, which was replaced by damage to membrane integrity in the latter stages.

4.3. Invisible Frost Effect on Leaves of Southern Seedlings

Although we monitored the changes not only in temperature, but also in other meteorological factors, the regression models indicated that it was only extreme daily temperature fluctuations that accounted for the major frost effect. The invisible frost effect was determined to be significant from the fifth day before sampling. Although the daily highest temperature can promote the synthesis and accumulation of carotenoids five days before sampling, it can also cause cell membrane damage throughout wintertime, which is followed by electrical leakage two days thereafter. Five of the days were also reported to be a critical time for late spring frost, before which the first instance of the highest daily temperature broke the record in the temperate forests [42,43]. Thus, trees experiencing acclimation to chilly temperatures are deeply dormant, and they need a warm spell to trigger the physiological responses that are needed for the process of dehardening. However, high temperatures on the second day before sampling can have a positive effect, which originates from the scavenging of superoxides. This effect continues to the next day, and the acclimation to chills increases. In contrast, the decline in the daily minimum temperature on the sampling day exhausts the continuous accumulation of positive effects, thus generating the frost effect. Overall, the effect of frost on subtropical and tropical trees also depends on warming in advance, which is followed by a sudden decline in the daily temperature. Local trees can acclimate to an earlier warming five to two days before the occurrence of frost injury, but this acclimation will be interrupted by the accumulation of superoxides on cell membranes if a sudden chill arrives.

4.4. Limits of This Study

Our study has three limits that have not been overcome. First, the meteorological models of the threat caused by late spring frosts in temperate forests were used in this study to investigate the possible effects of wintertime frosts on tropical and subtropical trees. Frost events were established by comparing the temperature fluctuations between the current monitoring year and those in the past 30–80 years [8,23,31,43]. These models were regressed depending on long historical meteorological records. However, in this study, we did not document and use the chronicles of the meteorological records for the past few years. A long-term series of data should be used in future works for tree species dwellings in southern habitats. Second, we focused on an introduced species, *D. odorifera*, which have obtained the ability to acclimate to wintertime frost. Certain other species that are sensitive to local temperature fluctuations, and which experience rare records of introduction, would be a better choice, at least as the reference. Finally, the provenance of Wenzhou is not a place with harsh winter temperatures for subtropical plants. Some of the more northern regions should be tested in future works to examine a chillier environment in relation to the frost effect.

5. Conclusions

Based on the findings of this study, we conclude that the *Dalbergia odorifera* introduced from southern provenances can acclimate to frost stress in northern areas such as Wenzhou. However, the seedlings from the Wenzhou area exhibit higher resistance to frost than those with tropical provenances. A period of frost imposed freezing stress a total of five times, and these frosts started with an abnormal increase in the highest daily temperature. These frost events damaged cell membrane integrity and imposed frost damage on the plants' cytoplasm and photosynthetic capacity through a sudden decline in the lowest daily temperature. We recommend that further work makes use of our design and the layout of the experiment, but with deeper explanations that use data from the gene expressions of key enzymes.

Author Contributions: Conceptualization, X.L. (Xiaowen Li) and Q.C.; methodology, Q.C.; software, X.L. (Xiaowen Li); validation, X.L. (Xiaowen Li); formal analysis, S.Y. and X.L. (Xiaowen Li); investigation, S.Y., X.L. (Xiaowen Li), and J.W. and X.L. (Xiaojin Liu); resources, X.L. (Xiaowen Li) and Y.L.; data curation, Y.L. and H.X.; writing—original draft preparation, X.L. (Xiaowen Li); writing—review and editing, Q.C.; visualization, X.L. (Xiaowen Li) and H.X.; supervision, Q.C.; project administration, X.L. (Xiaojin Liu) and Q.C.; funding acquisition, Q.C. All authors have read and agreed to the published version of the manuscript.

Funding: This study was financially supported by the Department of Science and Technology of Zhejiang (Key Scientific and Technological Grant for Breeding New Agricultural Varieties, grant number: 2021C02070-9) and the Guangdong Forestry Bureau (Innovation Project of Forestry Science and Technology, grant number: 2020KJCX007).

Institutional Review Board Statement: Not applicable.

Informed Consent Statement: Not applicable.

Data Availability Statement: Not applicable.

Acknowledgments: The authors thank the editors and reviewers for their contributions to the current form of this manuscript.

Conflicts of Interest: The authors declare no conflict of interest.

References

1. Tao, Z.X.; Xu, Y.J.; Ge, Q.S.; Dai, J.H.; Wang, H.J. Reduced frost hardiness in temperate woody species due to climate warming: A model-based analysis. *Clim. Chang.* **2021**, *165*, 35. [CrossRef]
2. He, Z.; Du, J.; Chen, L.; Zhu, X.; Lin, P.; Zhao, M.; Fang, S. Impacts of recent climate extremes on spring phenology in arid-mountain ecosystems in China. *Agric. For. Meteorol.* **2018**, *260*, 31–40. [CrossRef]
3. Wang, X.; Meng, L.; Wei, H. Biomass, carbohydrate, and leakage conductance in buds of six ornamental tree species subjected to a "false spring" in Northeast China. *Ann. For. Res.* **2022**, *65*, 15–30. [CrossRef]
4. Vitasse, Y.; Schneider, L.; Rixen, C.; Christen, D.; Rebetez, M. Increase in the risk of exposure of forest and fruit trees to spring frosts at higher elevations in Switzerland over the last four decades. *Agric. For. Meteorol.* **2018**, *248*, 60–69. [CrossRef]
5. Vitra, A.; Lenz, A.; Vitasse, Y. Frost hardening and dehardening potential in temperate trees from winter to budburst. *New Phytol.* **2017**, *216*, 113–123. [CrossRef] [PubMed]
6. Marquis, B.; Bergeron, Y.; Houle, D.; Leduc, M.; Rossi, S. Variability in frost occurrence under climate change and consequent risk of damage to trees of western Quebec, Canada. *Sci. Rep.* **2022**, *12*, 7220. [CrossRef] [PubMed]
7. Nole, A.; Rita, A.; Ferrara, A.M.S.; Borghetti, M. Effects of a large-scale late spring frost on a beech (*Fagus sylvatica* L.) dominated Mediterranean mountain forest derived from the spatio-temporal variations of NDVI. *Ann. For. Sci.* **2018**, *75*, 83. [CrossRef]
8. Hufkens, K.; Friedl, M.A.; Keenan, T.F.; Sonnentag, O.; Bailey, A.; O'Keefe, J.; Richardson, A.D. Ecological impacts of a widespread frost event following early spring leaf-out. *Glob. Chang. Biol.* **2012**, *18*, 2365–2377. [CrossRef]
9. Eysteinsson, T.; Karlman, L.; Fries, A.; Martinsson, O.; Skulason, B. Variation in spring and autumn frost tolerance among provenances of Russian larches (Larix Mill.). *Scand. J. For. Res.* **2009**, *24*, 100–110. [CrossRef]
10. Sklenar, P.; Jaramillo, R.; Wojtasiak, S.S.; Meneses, R.I.; Muriel, P.; Klimes, A. Thermal tolerance of tropical and temperate alpine plants suggests that 'mountain passes are not higher in the tropics'. *Glob. Ecol. Biogeogr.* **2023**. [CrossRef]
11. Bojórquez, A.; Álvarez-Yépiz, J.C.; Búrquez, A.; Martínez-Yrízar, A. Understanding and predicting frost-induced tropical tree mortality patterns. *Glob. Chang. Biol.* **2019**, *25*, 3817–3828. [CrossRef] [PubMed]

12. Miner, K.R.; D'Andrilli, J.; Mackelprang, R.; Edwards, A.; Malaska, M.J.; Waldrop, M.P.; Miller, C.E. Emergent biogeochemical risks from Arctic permafrost degradation. *Nat. Clim. Chang.* **2021**, *11*, 809–819. [CrossRef]
13. Bojórquez, A.; Martínez-Yrízar, A.; Álvarez-Yépiz, J.C. A landscape assessment of frost damage in the northmost Neotropical dry forest. *Agric. For. Meteorol.* **2021**, *308–309*, 108562. [CrossRef]
14. Das, A.A.; Ratnam, J. The thermal niche and phylogenetic assembly of evergreen tree metacommunities in a mid-to-upper tropical montane zone. *Proc. R. Soc. B-Biol. Sci.* **2022**, *289*, 20220038. [CrossRef] [PubMed]
15. Hoffmann, W.A.; Flake, S.W.; Abreu, R.C.R.; Pilon, N.A.L.; Rossatto, D.R.; Durigan, G. Rare frost events reinforce tropical savanna-forest boundaries. *J. Ecol.* **2019**, *107*, 468–477. [CrossRef]
16. Pilon, N.A.L.; Cava, M.G.B.; Hoffmann, W.A.; Abreu, R.C.R.; Rossatto, D.R.; Durigan, G. Effects and response of the Cerrado ground-layer to frost along the canopy cover gradient. *Oecologia* **2022**, *200*, 199–207. [CrossRef]
17. Finckh, M.; Wendefeuer, J.; Meller, P. Frost-driven lower treelines in Angola and their implications for tropical forest-grassland mosaics. *J. Veg. Sci.* **2021**, *32*, e13084. [CrossRef]
18. Duan, Y.D.; Guo, B.T.; Zhang, L.; Li, J.X.; Li, S.; Zhao, W.B.; Yang, G.; Zhou, S.; Zhou, C.W.; Song, P.H.; et al. Interactive climate-soil forces shape the spatial distribution of foliar N:P stoichiometry in Vaccinium uliginosum planted in agroforests of Northeast China. *Front. Ecol. Evol.* **2022**, *10*, 1065680. [CrossRef]
19. Mashizi, A.K.; Sharafatmandrad, M.; Naghavi, S. Ecosystems multiple-use management: An approach based on change in economic, social, and ecological values of plant communities. *Int. J. Environ. Sci. Technol.* **2022**, *19*, 3845–3858. [CrossRef]
20. Liao, S.B.; Cai, H.; Tian, P.J.; Zhang, B.B.; Li, Y.P. Combined impacts of the abnormal and urban heat island effect in Guiyang, a typical Karst Mountain City in China. *Urban Clim.* **2022**, *41*, 101014. [CrossRef]
21. Xue, J.S.; Zong, L.; Yang, Y.J.; Bi, X.Y.; Zhang, Y.H.; Zhao, M.H. Diurnal and interannual variations of canopy urban heat island (CUHI) effects over a mountain-valley city with a semi-arid climate. *Urban Clim.* **2023**, *48*, 101425. [CrossRef]
22. Li, Y.; Sun, Y.W.; Li, J.L.; Gao, C. Socioeconomic drivers of urban heat island effect: Empirical evidence from major Chinese cities. *Sustain. Cites Soc.* **2020**, *63*, 102425. [CrossRef]
23. Zohner, C.M.; Mo, L.D.; Renner, S.S.; Svenning, J.C.; Vitasse, Y.; Benito, B.M.; Ordonez, A.; Baumgarten, F.; Bastin, J.F.; Sebald, V.; et al. Late-spring frost risk between 1959 and 2017 decreased in North America but increased in Europe and Asia. *Proc. Natl. Acad. Sci. USA* **2020**, *117*, 12192–12200. [CrossRef]
24. Tessier, J.T. Early spring warming may hasten leaf emergence in *Erythronium americanum*. *Am. J. Bot.* **2019**, *106*, 1392–1396. [CrossRef] [PubMed]
25. Wang, Y.F.; Case, B.; Rossi, S.; Dawadi, B.; Liang, E.Y.; Ellison, A.M. Frost controls spring phenology of juvenile Smith fir along elevational gradients on the southeastern Tibetan Plateau. *Int. J. Biometeorol.* **2019**, *63*, 963–972. [CrossRef] [PubMed]
26. Bascietto, M.; Bajocco, S.; Mazzenga, F.; Matteucci, G. Assessing spring frost effects on beech forests in Central Apennines from remotely-sensed data. *Agric. For. Meteorol.* **2018**, *248*, 240–250. [CrossRef]
27. Menzel, A.; Helm, R.; Zang, C. Patterns of late spring frost leaf damage and recovery in a European beech (*Fagus sylvatica* L.) stand in south-eastern Germany based on repeated digital photographs. *Front. Plant Sci.* **2015**, *6*, 110. [CrossRef]
28. Meiners, S.J.; Presley, G.N. Differential damage of a late frost to Ozark tree species. *J. Torrey Bot. Soc.* **2015**, *142*, 12–17. [CrossRef]
29. Principe, A.; van der Maaten, E.; van der Maaten-Theunissen, M.; Struwe, T.; Wilmking, M.; Kreyling, J. Low resistance but high resilience in growth of a major deciduous forest tree (*Fagus sylvatica* L.) in response to late spring frost in southern Germany. *Tree Struct. Func.* **2017**, *31*, 743–751. [CrossRef]
30. Hadad, M.; Tardif, J.C.; Conciatori, F.; Waito, J.; Westwood, A. Climate and atmospheric circulation related to frost-ring formation in *Picea mariana* trees from the Boreal Plains, interior North America. *Weather Clim. Extremes* **2020**, *29*, 100264. [CrossRef]
31. Puchalka, R.; Koprowski, M.; Przybylak, J.; Przybylak, R.; Dabrowski, H.P. Did the late spring frost in 2007 and 2011 affect tree-ring width and earlywood vessel size in Pedunculate oak (*Quercus robur*) in northern Poland? *Int. J. Biometeorol.* **2016**, *60*, 1143–1150. [CrossRef] [PubMed]
32. Vitasse, Y.; Bottero, A.; Cailleret, M.; Bigler, C.; Fonti, P.; Gessler, A.; Levesque, M.; Rohner, B.; Weber, P.; Rigling, A.; et al. Contrasting resistance and resilience to extreme drought and late spring frost in five major European tree species. *Glob. Chang. Biol.* **2019**, *25*, 3781–3792. [CrossRef] [PubMed]
33. D'Andrea, E.; Scartazza, A.; Battistelli, A.; Collalti, A.; Proiett, S.; Rezaie, N.; Matteucci, G.; Moscatello, S. Unravelling resilience mechanisms in forests: Role of non-structural carbohydrates in responding to extreme weather events. *Tree Physiol.* **2021**, *41*, 1808–1818. [CrossRef] [PubMed]
34. Hauer, R.J.; Wei, H.X.; Koeser, A.K.; Dawson, J.O. Gas Exchange, Water Use Efficiency, and Biomass Partitioning among Geographic Sources of *Acer saccharum* Subsp. saccharum and Subsp. nigrum Seedlings in Response to Water Stress. *Plants* **2021**, *10*, 742. [CrossRef]
35. USDA. "Dalbergia odorifera". Germplasm Resources Information Network (GRIN). Available online: https://encyclopedia.thefreedictionary.com/Dalbergia+odorifera#cite_note-Furniture-5 (accessed on 11 April 2017).
36. Atikah, T.D.; Wardani, W.; Surya, D. Population, economic value and the trade of *Dalbergia latifolia* in Lombok, West Nusa Tenggara, Indonesia. *IOP Conf. Ser. Earth Environ. Sci.* **2021**, *762*, 012015. [CrossRef]
37. Cui, Z.; Li, X.; Xu, D.; Yang, Z. Changes in Non-Structural Carbohydrates, Wood Properties and Essential Oil During Chemically-Induced Heartwood Formation in *Dalbergia odorifera*. *Front. Plant Sci.* **2020**, *11*, 1161. [CrossRef]

38. WCMC. Dalbergia odorifera. The IUCN Red List of Threatened Species 1998: e.T32398A9698077. Available online: https://www.iucnredlist.org/species/32398/9698077 (accessed on 21 April 2017).
39. Li, X.W.; Chen, Q.X.; Lei, H.Q.; Wang, J.W.; Yang, S.; Wei, H.X. Nutrient Uptake and Utilization by Fragrant Rosewood (*Dalbergia odorifera*) Seedlings Cultured with Oligosaccharide Addition under Different Lighting Spectra. *Forests* **2018**, *9*, 29. [CrossRef]
40. Li, X.W.; Gao, Y.; Wei, H.X.; Xia, H.T.; Chen, Q.X. Growth, biomass accumulation and foliar nutrient status in fragrant rosewood (*Dalbergia odorifera* TC Chen) seedlings cultured with conventional and exponential fertilizations under different photoperiod regimes. *Soil Sci. Plant Nutr.* **2017**, *63*, 153–162. [CrossRef]
41. Li, X.W.; Xia, H.T.; Wang, J.W.; Chen, Q.X. Nutrient uptake and assimilation in fragrant rosewood (*Dalbergia odorifera* TC Chen) seedlings in growing media with un-composted spent mushroom residue. *PLoS ONE* **2021**, *16*, e0249534.
42. Gu, L.; Hanson, P.J.; Mac Post, W.; Kaiser, D.P.; Yang, B.; Nemani, R.; Pallardy, S.G.; Meyers, T. The 2007 eastern US spring freezes: Increased cold damage in a warming world? *Bioscience* **2008**, *58*, 253–262. [CrossRef]
43. Augspurger, C.K. Spring 2007 warmth and frost: Phenology, damage and refoliation in a temperate deciduous forest. *Funct. Ecol.* **2009**, *23*, 1031–1039. [CrossRef]
44. Hiscox, J.D.; Israelstam, G.F. A method for the extraction of chlorophyll from leaf tissue without maceration. *Can. J. Bot.* **1979**, *57*, 1332–1334. [CrossRef]
45. Lichenthaler, H.K.; Wellburn, A.R. Determinations of total carotenoids and chlorophylls a and b of leaf extracts in different solvents. *Biochem. Soc. Trans.* **1983**, *11*, 591–592. [CrossRef]
46. Liu, X.; Lu, X.; Yang, S.; Liu, Y.; Wang, W.Q.; Wei, X.; Ji, H.J.; Zhang, B.; Xin, W.Z.; Wen, J.X.; et al. Role of exogenous abscisic acid in freezing tolerance of mangrove *Kandelia obovata* under natural frost condition at near 32 degrees N. *BMC Plant Biol.* **2022**, *22*, 593. [CrossRef] [PubMed]
47. Bradford, M.M. Rapid and sensitive method for quantitation of microgram quantitles of protein utilizing principle of protein-dye binding. *Anal. Biochem.* **1976**, *72*, 248–254. [CrossRef] [PubMed]
48. Ryu, K.; Dordick, J.S. How do organic solvents affect peroxidase structure and function? *Biochemistry* **1992**, *31*, 2588–2598. [CrossRef]
49. L'Hirondelle, S.J.; Simpson, D.G.; Binder, W.D. Overwinter storability of conifer planting stock: Operational testing of fall frost hardiness. *New For.* **2006**, *32*, 307–321. [CrossRef]
50. Burr, K.E.; Tinus, R.W.; Wallner, S.J.; King, R.M. Comparison of three cold hardiness tests for conifer seedlings. *Tree Physiol.* **1990**, *6*, 351–369. [CrossRef]
51. Wei, H.; Ma, B.; Hauer, R.J.; Liu, C.; Chen, X.; He, X. Relationship between environmental factors and facial expressions of visitors during the urban forest experience. *Urban For. Urban Green.* **2020**, *53*, 126699. [CrossRef]
52. Liu, Y.; Meng, G.Y.; Suo, S.F.; Li, D.; Wang, A.M.; Cheng, X.H.; Yang, J. Spring Failure Analysis of Mining Vibrating Screens: Numerical and Experimental Studies. *Appl. Sci.* **2019**, *9*, 3224. [CrossRef]
53. Jalili, I.; Ebadi, A.; Askari, M.A.; KalatehJari, S.; Aazami, M.A. Foliar application of putrescine, salicylic acid, and ascorbic acid mitigates frost stress damage in *Vitis vinifera* cv.'Giziluzum'. *BMC Plant Biol.* **2023**, *23*, 135. [CrossRef] [PubMed]
54. Norozi, M.; ValizadehKaji, B.; Karimi, R.; Solgi, M. Potassium and zinc-induced frost tolerance in pistachio flowers is associated with physiological and biochemical changes. *Tree Struct. Funct.* **2020**, *34*, 1021–1032. [CrossRef]
55. Prozherina, N.; Freiwald, V.; Rousi, M.; Oksanen, E. Interactive effect of springtime frost and elevated ozone on early growth, foliar injuries and leaf structure of birch (*Betula pendula*). *New Phytol.* **2003**, *159*, 623–636. [CrossRef]
56. Polle, A.; Kroniger, W.; Rennenberg, H. Seasonal fluctuations of ascorbate-related enzymes: Acute and delayed effects of late frost in spring on antioxidative systems in needles of Norway spruce (*Picea abies* L). *Plant Cell Physiol.* **1996**, *37*, 717–725. [CrossRef]
57. Yang, Y.J.; Liu, T.; Zhang, S.B.; Huang, W. Photoinhibition of oxygen-evolving complex and photosystem II at chilling stress in the tropical tree species Dalbergia odorifera. *Photosynthetica* **2020**, *58*, 245–252. [CrossRef]
58. Li, Y. *Cold-Resistance Analysis and Suitability of Cultivation Areas for Dalbergia Odorifera Seedlings in Southern Zhejiang*; Zhejiang A&F University: Lin'an, China, 2021; (In Chinese with English Abstract and Legends).
59. Li, J. *Study on Cold Resistance of Different Geographical Provenance of Dalbergia Odorifera Seedlings*; Guangxi University: Nanning, China, 2018; (In Chinese with English Abstract and Legends).
60. Wei, H.X.; Guo, P. Carbohydrate metabolism during new root growth in transplanted *Larix olgensis* seedlings: Post-transplant response to nursery-applied inorganic fertilizer and organic amendment. *iFor. Biogeosci. For.* **2017**, *10*, 15–22. [CrossRef]

sustainability

Article

Altitudinal Distribution, Seasonal Dynamics and *Borrelia burgdorferi* Sensu Lato Infections in Hard Ticks (Acari: Ixodidae) in Different Forest Communities in Inland Croatia

Marko Vucelja [1,*], Stjepan Krčmar [2,*], Josipa Habuš [3], Vesna Mojčec Perko [3], Marko Boljfetić [4], Linda Bjedov [1] and Josip Margaletić [1]

[1] Faculty of Forestry and Wood Technology, University of Zagreb, 10000 Zagreb, Croatia
[2] Department of Biology, Josip Juraj Strossmayer University of Osijek, 31000 Osijek, Croatia
[3] Department of Microbiology and Infectious Diseases, with Clinic, Faculty of Veterinary Medicine, University of Zagreb, 10000 Zagreb, Croatia
[4] Oikon Ltd.—Institute of Applied Ecology, 10020 Zagreb, Croatia
[*] Correspondence: marko.vucelja@sumfak.unizg.hr (M.V.); stjepan@biologija.unios.hr (S.K.); Tel.: +385-01-2352-450 (M.V.); +385-31-399-922 (S.K.)

Abstract: Altitudinal distributions, population structures and seasonal dynamics of tick fauna at three localities in Continental Croatia (Medvednica and Papuk) and an alpine biogeographic region (Gorski Kotar) were studied. Sampling of questing ticks was performed twice a year (spring and autumn, from 2019 to 2021) at different altitudes (200, 400, 600, 800 and 1000 m above sea level) using the flagging method. In total, 2942 ticks (53.9% larvae, 40.1% nymphs, 6.0% adults) were sampled and 2937 (99.83%) were determined as *Ixodes ricinus*, 4 (0.14%) as *Haemaphysalis concinna* and 1 (0.03%) as *Ixodes frontalis*. *Ixodes ricinus* was the only species found at all altitudes and sampling sites. The highest tick abundance was recorded at higher altitudes (800–1000 m asl.) on Medvednica and in Gorski Kotar within mixed forests of European beech and European silver fir, while on Papuk most of the ticks were sampled at lower altitudes (200 m asl.) in Sessile oak forest. From 27 pools containing 305 ticks, 1 (3.7%) was positive for *Borrelia burgdorferi* sensu lato infection. *Borrelia burgdorferi* s.l. was detected in 20% (1/5) of the pools containing nymphs and adults collected in Gorski Kotar (600 and 800 m asl.). Sequencing of the ospA gene and phylogenetic analysis revealed the presence of the *Borrelia burgdorferi* sensu stricto genotype.

Keywords: ticks; *Ixodes ricinus*; *Ixodes frontalis*; *Haemaphysalis concinna*; *Borrelia burgdorferi* s.l.; *Borrelia burgdorferi* sensu stricto; Croatia; forests

Citation: Vucelja, M.; Krčmar, S.; Habuš, J.; Perko, V.M.; Boljfetić, M.; Bjedov, L.; Margaletić, J. Altitudinal Distribution, Seasonal Dynamics and *Borrelia burgdorferi* Sensu Lato Infections in Hard Ticks (Acari: Ixodidae) in Different Forest Communities in Inland Croatia. *Sustainability* **2023**, *15*, 4862. https://doi.org/10.3390/su15064862

Academic Editor: María Pilar González-Hernández

Received: 13 January 2023
Revised: 7 March 2023
Accepted: 7 March 2023
Published: 9 March 2023

1. Introduction

Hard ticks (Acari: Ixodidae) are obligatory hematophagous ectoparasites and one of the main arthropod-borne disease vectors worldwide. More than a century ago, they were the first arthropods proved as vectors of infectious disease [1,2]. Tick-borne diseases (e.g., Lyme borreliosis, tick-borne encephalitis, tularemia, Mediterranean spotted fever and human granulocytic anaplasmosis) are extremely diverse, while infectious agents transmitted by ticks include numerous bacteria, viruses and protozoa [3]. As ticks and tick-borne diseases (TBDs) infest a wide variety of wild vertebrate species, as well as livestock, companion animals and humans, they are also a cause of significant economic losses worldwide [4]. In recent decades, ticks have gained more public attention due to the expansion of their (altitudinal and latitudinal) ranges and the increasing incidence of tick-borne diseases [5–7]. Such increase is a result of different environmental and social factors, e.g., climate changes (prolonged growing seasons, warmer and shorter winters, etc.) [8–10], loss of predators [11], reforestation of agricultural lands, woodland expansion [12,13], urbanization, human mobility and population growth [14–16]. The higher risk of TBDs is typically associated with woodland habitats that provide ticks with the most favorable microclimatic conditions and high availability of hosts [17–19]. Even so, a rising number of medically important tick species populations are also being recorded in urban and suburban areas across Europe [10]. One tick species well-known for its high abundance and wide distribution in Europe and Croatia is *Ixodes ricinus* (Castor bean tick, Sheep tick, Deer tick, Forest tick) [20]. It is a major vector for a great number of pathogens (e.g., *Borrelia burgdorferi* sensu lato, *Borrelia miyamotoi*, tick-borne encephalitis virus, *Rickettsia* spp., *Anaplasma phagocytophilum*, *Babesia divergens*, *Babesia microti* and *Francisella tularensis*) [21,22]. *Ixodes ricinus* is the main vector of *Borrelia burgdorferi* s.l. in inland Croatia [23,24], especially in the northern and northwestern parts, where high prevalences (40–86%) of *Borrelia burgdorferi* have been recorded within tick populations [25–27]. Lyme borreliosis or Lyme disease, recognized in 1975 in the town of Old Lyme, Connecticut, in North America [28,29], is currently the most common tick-borne and vector-borne disease in Europe and in the USA [30,31]. *Borrelia burgdorferi* sensu lato (*B. burgdorferi* s.l.) is a bacterial complex transmitted by hard ticks that contains multiple genospecies. Of 21 genospecies identified, 8 are present in Europe: *B. burgdorferi* sensu stricto (s.s.)., *B. afzelii*, *B. garinii*, *B. bissettiae*, *B. bavariensis*, *B. spielmanii*, *B. lusitaniae* and *B. valaisiana* [32], and the first 6 cause disease in humans. The incidence of Lyme borreliosis has more than doubled over the last decades in Europe [33]. In Croatia, it has shifted from 255 to more than 400 reported human cases per year in the last 20 years [34]. It is recognized now that *Borrelia burgdorferi* exists at quite variable rates wherever the vector exists and that maximum prevalence is recorded in large regions of Central Europe, where the infection rates of *Ixodes ricinus* ticks range from 3% to 45% [10,35,36]. While there is an environmental niche that accounts for the prevalence of the causative agents, such as *Borrelia burgdorferi* s.l. [36], there is a necessity for a deeper understanding of tick fauna and their zoonotic potential in our rapidly changing environment. In their recent review, Kahl and Gray (2023) pointed out that over the last 20 years there were 2.4 times more publications on ticks and tick-borne diseases than in the previous 20 years [37]. The latter is not so surprising given that vector-borne diseases have been recognized as some of the most important public health problems of this century [10]. The most comprehensive ecological studies on the diversity of tick fauna in Croatia have focused mainly on the Mediterranean region (the Adriatic Coast, along with the islands known for their karst topography) and they were conducted during the second half of the twentieth century [20,38–42]. Croatian northwestern and western regions were only partly included [26,43,44]. Novel studies, carried out during the last two decades, have mainly been made at lower altitudes in central and eastern parts of inland Croatia [45–47]. A total of 23 hard tick species, belonging to five genera (*Ixodes*, *Haemaphysalis*, *Rhipicephalus*, *Dermacentor* and *Hyalomma*) have been recorded in Croatia so far [20,48]. There have been no studies on the altitudinal distributions of ticks in mountainous regions of Croatia for various reasons (i.e., terrains being unreachable by vehicles, small human populations, etc.). This research dealt with the question of the diversity of tick fauna, their spatio-temporal distribution and the zoonotic threat in the continental and alpine regions of Croatia. The aim of this study was to determine tick species, their altitudinal distributions, population structures, seasonal dynamics and the prevalence of Lyme disease causative agents (*Borrelia burgdorferi* s.l.) within tick fauna at three mountain sites in inland Croatia. We hypothesized that the most prevalent tick species would be from the genera *Ixodes* and *Dermacentor*, that they would appear in higher abundances during springtime sampling, that their numbers would decrease with higher altitudes and also that the highest prevalences of *Borrelia burgdorferi* s.l. would be found within tick populations on the Medvednica mountain in the northwestern part of the country, as it was previously known as a Lyme disease focus in Croatia.

2. Study Area

In this study, research on tick fauna was carried out from 2019 to 2021 at three locations within two biogeographic regions in inland Croatia [49]. Tick species diversity and their altitudinal distributions, population structures and seasonal dynamics were monitored at two mountain sites (Medvednica and Papuk) in the continental biogeographic region and one site in the alpine region (Gorski Kotar) (Figure 1).

Figure 1. Study areas and locations of tick sampling sites. (Gorski kotar mountain site: 1—Donje Tihovo, 2—Marija Trošt, 3—Lučice, 4—Štari Laz; Medvednica mountain site: 5 —Dotršćina, 6–9 Medvednica [a,b,c,d]; Papuk mountain site: 10—Markovac, 11, 12—Slatinski Drenovac [a,b], 13—Radovanci).

According to the geographical distribution of climatic types by Köppen, all three mountain study sites belong to Cfb—a temperate climate without a dry season and with a warm summer [50,51]. The temperature of the coldest month of the year (January) is above −3 °C, while the summers are fresh, with the average monthly temperature of the hottest month being below 22 °C [52]. Nomenclature of forest communities and soil types correspond to data specified in "Forest phytocenology and forest communities in Croatia" by Vukelić and Rauš (1998) and "Croatian forest vegetation" by Vukelić (2012) [53,54] and "Management of forest soils in Croatia" by Martinović, respectively [55]. Overall, the study included 13 sampling sites and nine different forest communities. The list of tick sampling sites, their GPS positions, altitudes and forest communities are shown in Table 1. Photography of every sampling site is given in Appendix A (Figures A1–A13).

Table 1. List of tick sampling sites with their GPS positions, altitudes and forest soil and forest community types.

Mountain Site	Tick Sampling Site	Altitude, Latitude (Degree/min/s)	Altitude (m asl.)	Forest Soil	Forest Community
GOK	1 Donje Tihovo	N45 25.927 E14 50.814;	400	Pseudogley on sloping terrain	EBDF
	2 Marija Trošt	N45 24.798 E14 49.351;	600	Dystric cambisol	EFDF
	3 Lučice	N45 22.359 E14 48.402;	800	Dystric cambisol	DBF
	4 Stari Laz	N45 20.740 E14 51.486;	1000	Dystric cambisol	MPF
MED	5 Dotrščina	N45 51.174 E16 01.172;	200	Dystric cambisol	POEH
	6 Medvednica [a]	N45 52.748 E15 58.568:	400	Dystric cambisol	EBW
	7 Medvednica [b]	N45 53.201 E15 58.330;	600	Dystric cambisol	EBW
	8 Medvednica [c]	N45 53.394 E15 57.634;	800	Dystric cambisol	EBW
	9 Medvednica [d]	N45 53.916 E15 56.753;	1000	Dolomitic cambisol	PBF
PAP	10 Markovac	N45 25.583 E17 32.227;	200	Luvisol	SOFD
	11 Slatinski Drenovac [a]	N45 31.553 E17 41.306;	400	Dystric cambisol	EBW
	12 Slatinski Drenovac [b]	N45 30.850 E17 39.990	600	Dystric cambisol	EBW
	13 Radovanci	N45 30.673 E17 39.430;	800	Dystric cambisol	MEBBLA

Legend: GOK: Gorski Kotar; MED: Medvednica; PAP: Papuk; EBDF: European beech forest with Deer fern (*Blechno- fagetum sylvaticae* Ht. 1950); EFDF: European fir forest with Deer fern (*Blechno—Abietetum* Ht. 1950); DBF: Dinaric beech–fir forest (*Omphalodo—Fagetum* Marinčak et al. 1992); MPF: Mixed private forest stand with European beech (*Fagus sylvatica* L.), European silver fir (*Abies alba* Mill.), Sycamore (*Acer pseudoplatanus* L.), European ash (*Fraxinus excelsior* L.) and Common hazel (*Coryilus avellana* L.); POEH: Pedunculate oak forest with European hornbeam (*Carpino betuli—Quercetum roboris typicum* Rauš 1969); EBW: European beech forest with Woodruff (*Galio odorati-Fagetum* Sougnez et Thill 1959); PBF: Pannonian beech–fir forest (*Festuco drymeiae—Abietetum* Vukelić et Baričević 2007); SOFD: Sessile oak forest with *Festuca drymeia* (*Festuco drymeiae—Quercetum petraeae*/Jank. 1968/Hruška 1974); MEBBLA: Mountain European beech forest with balm-leaved archangel (*Lamio orvalae—Fagetum sylvaticae* Ht. 1938).

Tick sampling was performed from 2019 to 2021, twice a year (spring and autumn), at all three mountain sites at different altitudes. Mean air temperatures and air humidity data for the spring (March–June) and autumn (September–December) sampling seasons are shown in Table 2. The data from the meteorological stations nearest to the sampling locations were provided by the Croatian Meteorological and Hydrological Service.

Table 2. Mean air temperatures and air humidity data for spring and autumn tick sampling seasons at three mountain sites.

	GOK/Air Temperature (°C) $^{(min–max)}$			GOK/Air Humidity (%) $^{(min–max)}$		
Year Season	**2019**	**2020**	**2021**	**2019**	**2020**	**2021**
S	10.7 $^{(0.2–14.9)}$	10.2 $^{(-4.5–22.2)}$	9.9 $^{(-3.9–25.3)}$	82 $^{(57–97)}$	78.5 $^{(45–97)}$	85.8 $^{(71–98)}$
A	8.5 $^{(-4.1–18.6)}$	7.4 $^{(-4.8–18.0)}$	6.7 $^{(-3.7–18.8)}$	89.3 $^{(78–98)}$	89.8 $^{(75–98)}$	90 $^{(76–98)}$
	MED/air temperature (°C) $^{(min–max)}$			MED/air humidity (%) $^{(min–max)}$		
S	9.73 $^{(-1.0–24.4)}$	9.01 $^{(-5.9–21.2)}$	8.25 $^{(-5.6–23.6)}$	74.5 $^{(36–100)}$	67.5 $^{(29–99)}$	72.5 $^{(31–100)}$
A	7.53 $^{(-6.6–21.3)}$	6.63 $^{(-6.4–18.4)}$	5.9 $^{(-5.4–19.0)}$	82.5 $^{(45–100)}$	67.1 $^{(48–100)}$	84.5 $^{(32–100)}$
	PAP/air temperature (°C) $^{(min–max)}$			PAP/air humidity (%) $^{(min–max)}$		
S	12.3 $^{(0.2–24.6)}$	11.3 $^{(-3.0–22.7)}$	10.8 $^{(-2.0–26.4)}$	76.3 $^{(67–98)}$	82.5 $^{(59–96.6)}$	81.4 $^{(63–97.3)}$
A	8.95 $^{(-4–21.1)}$	8.1 $^{(-2.6–19.4)}$	7.2 $^{(-2.7–19.6)}$	88.3 $^{(77–98)}$	88.8 $^{(70.2–96.2)}$	88.5 $^{(73.6–97.1)}$

Legend: GOK: Gorski Kotar; MED: Medvednica; PAP: Papuk; S: spring; A: autumn.

The area of Gorski Kotar covers 1275 square kilometers, and it is located in the zone of Central European climate characterized by long and snowy winters, low average annual temperature, high air humidity and abundant precipitation with quite strong winds from the northeast and southwest. The flora and fauna of Gorski Kotar are known for their richness, while 63% of the total area of Gorski Kotar is covered by forests ("Green lungs of

Croatia or Croatian Switzerland"), with a predominance of fir–beech forest communities. The highest peak is Bjelolasica (1534 m).

Medvednica (pronounced [mĕdvednitsa], lit. "Bear Mountain") is a mountain area just north of Croatia's capital, the city of Zagreb. It extends from the southwest to the northeast and is 42 km in length. In 1981, it was declared a nature park with a protected area of 228 km^2. The plant cover of Medvednica is mostly represented by natural and preserved forests (63% of the area). Due to the uneven terrain, various geological substrates and soil types, as many as 12 forest communities appear there, with distinct zonations depending on altitude and exposure. The predominant species of trees are European beech, holm oak, hornbeam, sweet chestnut, maple and ash. Medvednica is a home to various types of mammals: roe deer, wild boar, wild cat, fox, marten and weasel. The highest point is Sljeme (1033 m asl).

Papuk is a mountain in Eastern Croatia, on the northern and northwestern border of the Požega basin. In 1999, the the Papuk Nature Park was founded, and it encompasses the largest part of the Papuk mountain (33,600 ha), with a general NW-SE extension of about 45 km. Forest habitats, with 11 different forest communities, cover around 95% of the park. The dominant tree species are European beech (47%), sessile oak, fir, hornbeam, sycamore, Austrian oak, pubescent oak, wild cherry, etc. The dense forests of Papuk are the habitats of deer, roe deer, wild boars, foxes and martens. Its highest peak, also named Papuk, is 953 m asl.

3. Materials and Methods

3.1. *Tick Sampling and Identification*

Ticks were sampled by the flag dragging method [56–58], using white flannel cloth (1 m × 1 m) on two transects (2 × 100 m in length) at every sampling site. The flag was pulled over the ground surface, over leaves and low vegetation, and it was inspected every 5–10 m, depending on the catch. Ticks were collected from the flag with tweezers and stored (alive) in plastic tubes (eppendorf, 1.5 mL) with a safety cap.

During the field work, a mobile application and the online database "krpelji.info" (https://www.krpelji.info/) (accessed on 5 December 2021) were used while sampling ticks. The application was developed (and is still in "test mode") by the Faculty of Forestry and Wood Technology (FFWT), University of Zagreb. It allows the recording of GPS positions (waypoints and routes/transects), altitudes, date and time of sampling, vegetation type, sampling methods, number of collected specimens, tick species and their life stages. Additional uses of the application included photographing the sampling site, collected ticks, etc.

The sampled ticks were transported to FFWT, where a visual determination of the ticks and their life stages (larva, nymph and adult) was performed. For the determination of ticks, Leica Wild Stereo Microscope MZ8 (Leica Microsystems, Mannheim, Germany) was used (magnification 50×) equipped with an object micrometer together with the Quick Photo software package, ModellCamera 2 and a Dino-Lite digital microscope (magnification 20×–220×; 500×) (AnMo Electronics Corporation, Taiwan, China) with DinoCapture 2.0 software, version 1.5.17. B. Identification of sampled ticks based on different morphological characteristics was performed using current keys for tick identification [59,60]. After determining species and life stages, the ticks were stored in a freezer (−80 °C). The fieldwork was carried out under permits issued by the Ministry of Economy and Sustainable Development of the Republic of Croatia (UP/I-612-07/19-48/154, 517-05-1-1-19-3, 26 June 2019).

3.2. *Borrelia burgdorferi* s.l. *Detection*

Ticks were organized in pools (2–20 individuals per pool; median pool size = 11) in accordance with the locality, season (spring or autumn), altitude (200, 400, 600, 800 or 1000 m asl.) and tick life stage (nymph or adult; larvae were not included). When necessary, due to the size of the samples, pools included tick samples from two altitudes and both developmental stages. The study site of Medvednica included 18 pools with 165 nymphs and 69 adults; in Gorski Kotar, there were 54 nymphs and adults grouped together in 5 pools, while the study site of Papuk included 4 pools with 18 nymphs and adults combined. For specimens collected on Medvednica and Papuk, grouping according to altitudes was performed for ticks sampled at 200 and 400 m asl. and at 600 and 800 m asl. For ticks from Gorski Kotar, the same type of grouping was performed for specimens collected at 600 and 800 m asl.

Ticks were homogenized in a Qiagen tissuelyser II (Qiagen, Retsch, Germany) for 10 min at 30 Hz (using 1–3 mm corundum beads (Macherey-Nagel), 3 mm steel beads (Macherey-Nagel), 5 mm metal beads (IKA) and 500 µL of PBS). After homogenization, genomic DNA was extracted using a commercial kit (NucleoSpin® Tissue kit, Macherey-Nagel, Düren, Germany) following the manufacturer's instructions. *B. burgdorferi* s.l. DNA was detected by the nested PCR method targeting a specific ospA gene, as already described [61]. Nucleotide sequencing of the PCR products was performed by Macrogen Europe, Inc. (https://macrogenlab.com) (accessed on 9 January 2023). The resulting sequences were analyzed using BioEdit software v7.0 (4 https://bioedit.software.informer.com) (accessed on 9 January 2023). Phylogenetic analyses were conducted using MEGA version 11 (https://www.megasoftware.net) (accessed on 9 January 2023).

3.3. *Statystical Analysis*

The differences observed between the numbers of ticks collected at different altitudes were tested by Chi-square analysis. This was used to test the null hypothesis that the collected numbers of ticks would be similar for all altitudes. Data were analyzed using Microsoft Excel 365 and Statistica ver. 14.0.1.25., TIBCO Software, Inc. (Palo Alto, CA, USA).

4. Results

During this study (2019–2021), 2942 questing ticks were collected by the flag dragging method at three mountain sites in inland Croatia. The collected ticks belonged to three species: two were from the genera *Ixodes* and one was from *Haemaphysalis* (Table 3). *Ixodes ricinus* L., 1758, (Castor bean tick, Sheep tick, Deer tick, Forest tick) was by far the most dominant species (N = 2937; 99.83%), while *Haemaphysalis concinna* Koch, 1844 and *Ixodes frontalis* Panzer, 1798 represented less than 1% of the catch. Overall, 94.05% of all sampled specimens, including every tick species, were subadults. The largest number of collected *Ixodes ricinus* were in larval stage (53.90%), closely followed by nymphs (40.14%), while adults represented a little under 6% of the catch. Regarding the *Ixodes ricinus* sex ratio, males were slightly predominant (3.51%) over females (2.45%). *Ixodes ricinus* was the only tick species that was sampled at all life stages (larval, nymphal and adult) at every location, i.e., in the study area, while the other two species were collected only as subadults (Table 3).

Table 3. Species identifications, sex ratios and developmental stages of hard ticks collected in the study areas.

Species	Females	Males	Nymphs	Larvae	Total
Ixodes ricinus	72	103	1179	1583	2937
Ixodes frontalis	0	0	1	0	1
Haemaphysalis concinna	0	0	3	1	4
Σ	72	103	1183	1584	2942

The highest number of ticks was collected at the study site of Gorski Kotar (N = 1430; 48.6%), followed by Medvednica (N = 1129; 38.4%) and Papuk (N = 383; 13.0%). The list of tick species and their sex ratios and life stages according to study sites are shown in Table 4.

Table 4. List of collected tick species and their life stages according to sampling location, altitude and forest community type.

Mountain Site	Altitude (m asl.)	Forest Community	*Ir* ♀	*Ir* ♂	*Ir* n	*Ir* l	*If* n	*Hc* n	*Hc* l	Σ	Mean Density per 100 m² (min–max)
GOK	400	EBDF	0	2	116	250	0	0	0	368	30.7 (1–131.5)
	600	EFDF	5	4	84	42	0	0	0	135	11.3 (0.5–32.5)
	800	DBF	6	4	66	383	0	0	0	459	38.3 (0–207.5)
	1000	MPF	2	5	51	410	0	0	0	468	39 (0–177.5)
	Σ		13	15	317	1085	0	0	0	1430	29.8
MED	200	POEH	3	5	34	38	1	0	0	81	6.75 (0–21.5)
	400	EBW	25	29	94	14	0	0	0	162	13.5 (0–22.5)
	600	EBW	4	14	107	25	0	0	0	150	12.5 (0.5–20.5)
	800	EBW	8	18	237	225	0	0	0	488	40.7 (13.5–109.5)
	1000	PBF	15	16	208	9	0	0	0	248	20.7 (1.5–63)
	Σ		55	82	680	311	1	0	0	1129	18.8
PAP	200	SOFD	3	3	89	78	0	3	0	176	14.7 (0–42.5)
	400	EBW	1	2	29	10	0	0	0	42	3.5 (0–11.5)
	600	EBW	0	0	23	69	0	0	1	93	7.8 (0–38)
	800	MEBBLA	0	1	41	30	0	0	0	72	6 (0–23.5)
	Σ		4	6	182	187	0	3	1	383	8.0
	Σ Σ		72	103	1179	1583	1	3	1	2942	18.9

Legend: GOK: Gorski Kotar; MED: Medvednica; PAP: Papuk; *Ir*: *Ixodes ricinus*; *If*: *Ixodes frontalis*; *Hc*: *Haemaphysalis concinna*; ♀(female), ♂(male), n (nymph), l (larva); EBDF: European beech forest with Deer fern (*Blechno-fagetum sylvaticae* Ht. 1950); EFDF: European fir forest with Deer fern (*Blechno—Abietetum* Ht. 1950); DBF: Dinaric beech–fir forest (*Omphalodo—Fagetum* Marinčak et al. 1992); MPF: Mixed private forest stand with European beech (*Fagus sylvatica* L.), European silver fir (*Abies alba* Mill.), Sycamore (*Acer pseudoplatanus* L.), European ash (*Fraxinus excelsior* L.) and Common hazel (*Coryilus avellana* L.); POEH: Pedunculate oak forest with European hornbeam (*Carpino betuli—Quercetum roboris typicum* Rauš 1969); EBW: European beech forest with Woodruff (*Galio odorati-Fagetum* Sougnez et Thill 1959); PBF: Pannonian beech–fir forest (*Festuco drymeiae—Abietetum* Vukelić et Baričević 2007); SOFD: Sessile oak forest with *Festuca drymeia* (*Festuco drymeiae—Quercetum petraeae*/Jank. 1968/Hruška 1974); MEBBLA: Mountain European beech forest with balm-leaved archangel (*Lamio orvalae—Fagetum sylvaticae* Ht. 1938).

Ixodes ricinus was the one species found at all altitudes and sampling sites, while *Haemaphysalis concinna* was recorded only on Papuk (at 200 m asl. and 600 m asl.) and *Ixodes frontalis* only on Medvednica (200 m asl.) (Table 4).

The Chi-square analysis showed that altitude significantly influenced the number of collected ticks in all three mountain areas—Gorski Kotar ($\chi^2 = 201.73$, $p < 0.05$), Medvednica ($\chi^2 = 442.95$, $p < 0.05$) and Papuk ($\chi^2 = 103.38$, $p < 0.05$). However, sampling of the same altitudes (600 m asl and 800 m asl) on the mountains of Gorski Kotar and Medvednica resulted in similar numbers of collected ticks (Table 4). At these altitudes, there was no significant difference in the numbers of collected specimens ($\chi^2 = 0.78$, $p > 0.05$; $\chi^2 = 0.88$, $p > 0.05$).

Ticks were least numerous in their adult developmental stage in tick samples at all three study sites (Gorski Kotar: 1.96%; Medvednica: 12.13%; Papuk: 2.61%). The highest overall share of adults (33.33%) was recorded on Medvednica, at 400 m asl., in European beech forest with Woodruff (Table 4). Larvae dominated overall catches in Gorski Kotar (75.87%), while nymphs were the most dominant (60.23%) on Medvednica (Table 4). On Papuk, both larvae and nymphs had approximately equal overall shares (48%) in tick samples. The tick species, their sex ratios and life stages according to study sites and sampling seasons are shown in Table 5.

During the springtime sampling, 89.1% of all the ticks were collected. Nymphs were sampled both in spring and autumn at all mountain sites during the whole research period. Larvae were also sampled in spring and autumn at all sampling sites (Table 5).

Table 5. List of collected tick species and their life stages according to sampling location, altitude, forest community type and sampling season.

Mountain Site	Altitude (m asl.)	Forest Community	Sp.	2019		2020		2021	
				S	A	S	A	S	A
GOK	400	EBDF	*Ir*	0♀, 0♂, 26 n, 22 l	0♀, 0♂, 2 n, 0 l	0♀, 0♂, 41 n, 222 l	0♀, 1♂, 26 n, 0 l	0♀, 1♂, 4 n, 0 l	0♀, 0♂, 17 n, 6 l
	600	EFDF	*Ir*	0♀, 2♂, 16 n, 10 l	0♀, 0♂, 1 n, 0 l	4♀, 2♂, 38 n, 21 l	1♀, 0♂, 12 n, 5 l	0♀, 0♂, 4 n, 0 l	0♀, 0♂, 13 n, 6 l
	800	DBF	*Ir*	1♀, 0♂, 24 n, 0 l	0♀, 0♂, 0 n, 0 l	0♀, 2♂, 30 n, 383 l	0♀, 0♂, 0 n, 0 l	2♀, 1♂, 5 n, 0 l	2♀, 0♂, 1 n, 0 l
	1000	MPF	*Ir*	1♀, 3♂, 21 n, 330 l	0♀, 0♂, 0 n, 0 l	1♀, 1♂, 20 n, 62 l	0♀, 0♂, 6 n, 18 l	0♀, 1♂, 4 n, 0 l	2♀, 5♂, 51 n, 410 l
MED	200	POEH	*Ir*	2♀, 3♂, 15 n, 3 l	0♀, 1♂, 2 n, 7 l	1♀, 1♂, 13 n, 28 l	0♀, 0♂, 0 n, 0 l	0♀, 0♂, 4 n, 0 l	0♀, 0♂, 0 n, 0 l
			If	0♀, 0♂, 1 n, 0 l	0♀, 0♂, 0 n, 0 l	0♀, 0♂, 0 n, 0 l	0♀, 0♂, 0 n, 0 l	0♀, 0♂, 0 n, 0 l	0♀, 0♂, 0 n, 0 l
	400	EBW	*Ir*	5♀, 9♂, 18 n, 0 l	0♀, 0♂, 0 n, 0 l	4♀, 2♂, 52 n, 14 l	1♀, 1♂, 6 n, 0 l	15♀, 15♂, 15 n, 0 l	0♀, 2♂, 3 n, 0 l
	600	EBW	*Ir*	3♀, 12♂, 26 n, 0 l	0♀, 0♂, 2 n, 0 l	0♀, 0♂, 48 n, 24 l	0♀, 0♂, 1 n, 0 l	0♀, 1♂, 26 n, 1 l	1♀, 1♂, 4 n, 0 l
	800	EBW	*Ir*	4♀, 6♂, 98 n, 0 l	0♀, 0♂, 9 n, 20 l	1♀, 1♂, 22 n, 195 l	1♀, 1♂, 47 n, 0 l	1♀, 7♂, 42 n, 6 l	1♀, 3♂, 19 n, 4 l
	1000	PBF	*Ir*	3♀, 2♂, 36 n, 1 l	0♀, 0♂, 0 n, 3 l	2♀, 2♂, 117 n, 5 l	0♀, 0♂, 5 n, 0 l	10♀, 12♂, 39 n, 0 l	0♀, 0♂, 11 n, 0 l
PAP	200	SOFD	*Ir*	0♀, 0♂, 18 n, 57 l	0♀, 1♂, 5 n, 0 l	1♀, 0♂, 62 n, 21 l	0♀, 0♂, 0 n, 0 l	2♀, 1♂, 1 n, 0 l	0♀, 1♂, 3 n, 0 l
			Hc	0♀, 0♂, 2 n, 0 l	0♀, 0♂, 0 n, 0 l	0♀, 0♂, 1 n, 0 l	0♀, 0♂, 0 n, 0 l	0♀, 0♂, 0 n, 0 l	0♀, 0♂, 0 n, 0 l
	400	EBW	*Ir*	1♀, 1♂, 5 n, 10 l	0♀, 1♂, 1 n, 0 l	0♀, 0♂, 23 n, 0 l	0♀, 0♂, 0 n, 0 l	0♀, 0♂, 0 n, 0 l	0♀, 0♂, 0 n, 0 l
	600	EBW	*Ir*	0♀, 0♂, 3 n, 0 l	0♀, 0♂, 3 n, 0 l	0♀, 0♂, 13 n, 63 l	0♀, 0♂, 3 n, 6 l	0♀, 0♂, 0 n, 0 l	0♀, 0♂, 1 n, 0 l
			Hc	0♀, 0♂, 0 n, 1 l	0♀, 0♂, 0 n, 0 l	0♀, 0♂, 0 n, 0 l	0♀, 0♂, 0 n, 0 l	0♀, 0♂, 0 n, 0 l	0♀, 0♂, 0 n, 0 l
	800	MEBBLA	*Ir*	0♀, 0♂, 2 n, 6 l	0♀, 0♂, 0 n, 0 l	0♀, 0♂, 27 n, 20 l	0♀, 0♂, 4 n, 3 l	0♀, 0♂, 3 n, 0 l	0♀, 1♂, 5 n, 1 l

Legend: GOK: Gorski Kotar; MED: Medvednica; PAP: Papuk; Sp.: species; S: spring; A: autumn; *Ir*: *Ixodes ricinus*; *If*: *Ixodes frontalis*; *Hc*: *Haemaphysalis concinna*; ♀(female), ♂(male), n (nymph), l (larva); EBDF: European beech forest with Deer fern (*Blechno-fagetum sylvaticae* Ht. 1950); EFDF: European fir forest with Deer fern (*Blechno—Abietetum* Ht. 1950); DBF: Dinaric beech–fir forest (*Omphalodo—Fagetum* Marinčak et al. 1992); MPF: Mixed private forest stand with European beech (*Fagus sylvatica* L.), European silver fir (*Abies alba* Mill.), Sycamore (*Acer pseudoplatanus* L.), European ash (*Fraxinus excelsior* L.) and Common hazel (*Coryilus avellana* L.); POEH: Pedunculate oak forest with European hornbeam (*Carpino betuli—Quercetum roboris typicum* Rauš 1969); EBW: European beech forest with Woodruff (*Galio odorati-Fagetum* Sougnez et Thill 1959); PBF: Pannonian beech–fir forest (*Festuco drymeiae—Abietetum* Vukelić et Baričević 2007); SOFD: Sessile oak forest with *Festuca drymeia* (*Festuco drymeiae—Quercetum petraeae*/Jank. 1968/Hruška 1974); MEBBLA: Mountain European beech forest with balm-leaved archangel (*Lamio orvalae—Fagetum sylvaticae* Ht. 1938).

Seasonal dynamics of ticks at study sites according to altitudes and forest communities are shown in Figures 2–4. During every tick sampling, at all mountain sites, springtime was the season when ticks were collected in the highest numbers (Figures 2–4). The exception from the usual springtime maximum was only recorded during 2021 at 400 m asl. and 600 m asl. sampling sites in Gorski Kotar (Figure 2), as well as at all altitudes (200, 400, 600 and 800 m asl.) on Papuk (Figure 4).

The highest density of ticks per 100 m^2 recorded on Gorski Kotar (207 ticks per 100 m^2) was at 800 m asl. within mixed Dinaric beech–fir forest (*Omphalodo—Fagetum* Marinčak et al. 1992). At the same altitude, the highest tick density was also recorded on Medvednica (109 ticks per 100 m^2) but within European beech forest with Woodruff (*Galio odorati-Fagetum* Sougnez et Thill 1959), while on Papuk, at 200 m asl., in Sessile oak forest with *Festuca drymeia* (*Festuco drymeiae—Quercetum petraeae*/Jank. 1968/Hruška 1974), 42 ticks were collected per 100 m^2 (Table 4).

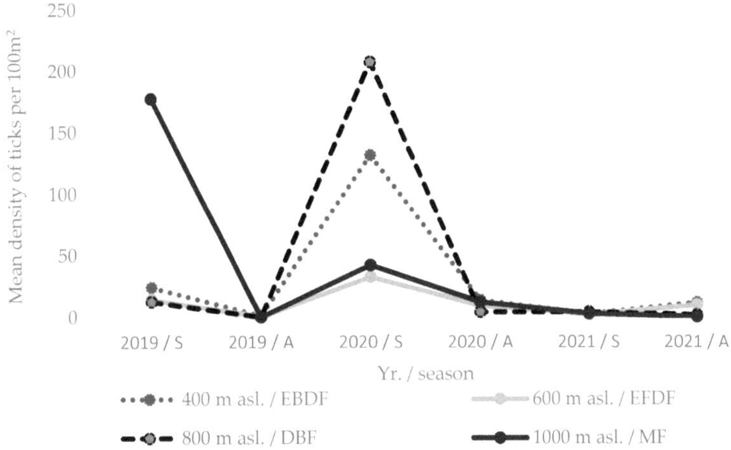

Figure 2. Seasonal dynamics of ticks (*Ixodes ricinus*) sampled at the Gorski Kotar study site (2019–2021) for different altitudes and forest communities (S: spring, A: autumn); EFDF: European fir forest with Deer fern (*Blechno—Abietetum* Ht. 1950); DBF: Dinaric beech–fir forest (*Omphalodo—Fagetum* Marinčak et al. 1992); MPF: Mixed private forest stand with European beech (*Fagus sylvatica* L.), European silver fir (*Abies alba* Mill.), Sycamore (*Acer pseudoplatanus* L.), European ash (*Fraxinus excelsior* L.) and Common hazel (*Corylus avellana* L.).

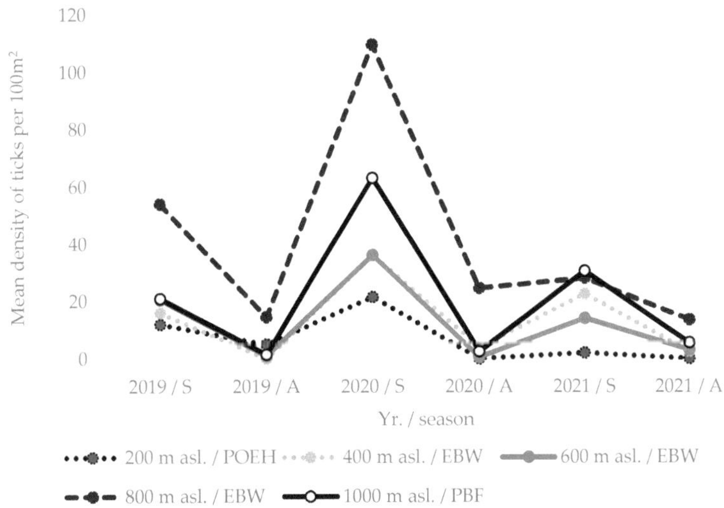

Figure 3. Seasonal dynamics of ticks sampled at the Medvednica study site (2019–2021) for different altitudes and forest communities (S: spring, A: autumn); POEH: Pedunculate oak forest with European hornbeam (*Carpino betuli—Quercetum roboris typicum* Rauš 1969); EBW: European beech forest with Woodruff (*Galio odorati-Fagetum* Sougnez et Thill 1959); PBF: Pannonian beech–fir forest (*Festuco drymeiae—Abietetum* Vukelić et Baričević 2007).

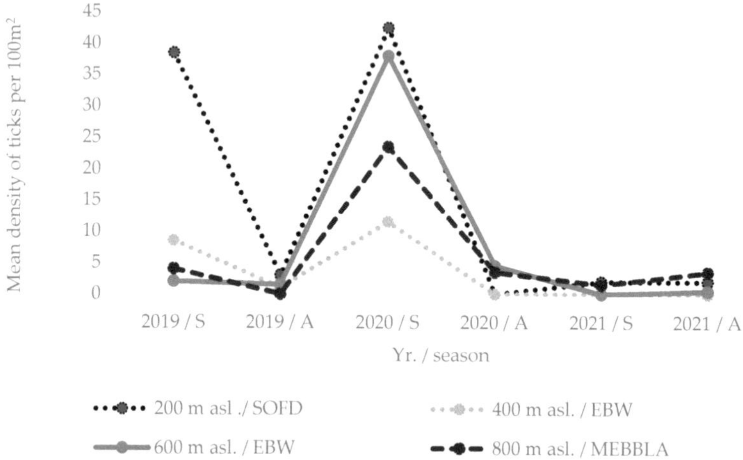

Figure 4. Seasonal dynamics of ticks sampled at the Papuk study site (2019–2021) for different altitudes and forest communities (S: spring, A: autumn); SOFD: Sessile oak forest with *Festuca drymeia* (*Festuco drymeiae—Quercetum petraeae*/Jank. 1968/Hruška 1974); MEBBLA: Mountain European beech forest with balm-leaved archangel (*Lamio orvalae—Fagetum sylvaticae* Ht. 1938).

The overall *B. burgdorferi* s.l. prevalence was 3.7%. In terms of sampling location, the presence of *B. burgdorferi* s.l. was detected in 20% (1/5) of pools containing nymphs and adults collected in the area of Gorski Kotar. *B. burgdorferi* s.l. infection was not detected in samples from the Medvednica and Papuk mountains. Sequencing of the ospA gene and phylogenetic analysis revealed the presence of the *B. burgdorferi* sensu stricto genotype. A phylogenetic tree, constructed using ospA sequences retrieved from GenBank and the sequence identified in this study (14N CRO), is shown in Appendix B (Figure A14).

5. Discussion

As the European Centre for Disease Prevention and Control (ECDC) predicts that the incidence of vector-borne diseases will continue to rise [8,62], it is no surprise that approximately three times as many publications on *Ixodes ricinus* have appeared since 2002 compared with the period between 1980 and 2000 [37]. Despite a large number of faunistic studies on the diversity of tick fauna conducted during the second half of the 20th century along the Adriatic Coast and a few recent studies in northwestern, central and eastern regions of inland Croatia, many areas—including mountains—have not yet been sufficiently studied. In this research, 3 out of 23 previously reported hard tick species (*Ixodes ricinus*, *Ixodes frontalis* and *Haemaphysalis concinna*) were collected in three mountain regions in inland Croatia (Gorski Kotar, Medvednica and Papuk) [48] (Figure 1). At all three mountain study sites, *Ixodes ricinus* was the most prevalent tick species (99.8% of all collected samples) (Table 3). These findings are in line with many studies performed earlier in woodland habitats at lower altitudes in northwestern and eastern parts of Croatia [20,44,46,63] as well as in Europe (e.g., 96% in Southern England, 97.4% in the Serbian capital Belgrade, 90.7% in Germany, etc.) [64–68]. Our results also differ from previous studies performed in pedunculate oak floodplain forests (*Genisto elatae—Quercetum roboris* Ht. 1938) in the Posavina region in Central Croatia [45], where *Ixodes ricinus* was determined in only 3.5% of collected specimens. On the mountain sites of Papuk and Medvednica, besides *Ixodes ricinus*, *Ixodes frontalis* and *Haemaphysalis concinna* were also recorded, but with very small percentages—0.08% and 1.0%, respectively (Table 4). Formerly, *Ixodes frontalis* was sporadically recorded in Croatia (Adriatic Coast and the city of Zagreb) [20], and this record for the Medvednica mountain represents a new locality record, as does the record of *Haemaphysalis concinna* on Papuk Mountain. There were also previous records of *Haemaphysalis concinna* sampled near the Croatian northwestern border, on the island of Krk in the Croatian Littoral [20] and in the eastern parts of the country, where it was present in small numbers, representing only 2.6% of the collected specimens [47]. Even though the genus *Dermacentor* is the second most abundant ixodid taxon in the Western Palaearctic region [69,70], we have collected no representatives in our three-year research, which corresponds to its lowland habitat preferences and its absence from higher mountain regions [45,71]. However, we expected ticks from this genus at low-altitude hill sampling sites or in climatically favorable valleys at higher altitudes [72].

In Central Europe, the altitudinal distribution limit for *Ixodes ricinus* was between 700 and 800 m asl. [73]. However, this limit has been exceeded, as shown by the results of the study on Krkonoše Mountain in the Czech Republic, where *Ixodes ricinus* was recorded even up to a timberline, approximately 1250 m asl. [74]. In Southern Norway, *Ixodes ricinus* was recorded at altitudes of up to 1000 m asl., which represents a considerable increase in altitude relative to earlier findings in this region [75]. In a similar study in Hungary, *Ixodes ricinus* was also found at altitudes of up to 900–1000 m asl. [76]. In this study, *Ixodes ricinus* was found at 1000 m asl. in Gorski Kotar and on the Medvednica mountain, these being the highest finding points of this species in Croatia.

The mean density of hard ticks in the study area was 18.9 per 100 m^2. A significantly lower value was recorded for *Ixodes ricinus* in Northeastern Poland, namely, 9.7 ticks per 100 m^2 [77]. Despite the different types of forest vegetation, the highest abundances and mean densities of ticks collected per 100 m^2 were recorded at high altitudes (800–1000 m asl.) on Gorski Kotar and Medvednica and at lower altitudes (200 m asl.) on Papuk (Table 4), which matches this species' habitat preferences (deciduous coniferous and mixed forests) [78–83]. The highest tick density (207 ticks per 100 m^2) recorded in this study was at 800 m asl. on Gorski Kotar in mixed Dinaric beech–fir forest (*Omphalodo—Fagetum* Marinčak et al. 1992). Such a high tick abundance is partly a result of the overall high prevalence of larvae (76%) collected on Gorski Kotar during this research (Table 4).

The seasonal dynamics of the sampled ticks showed a typical peak during springtime sampling [37], which corresponds to 89.1% of all the tick specimens collected during March, April and May in this study (Figures 2–4). Except for a few exceptions, ticks were mostly sampled at all three developmental stages during every spring and autumn sampling (Table 5). Nymphs were sampled both in spring and in autumn at all mountain sites during the whole research period. Regarding seasonal dynamics, *Ixodes ricinus* larvae are known for their early summer peak [84], which partially corresponds to their dominance recorded during the springtime sampling in our study.

The only tick species found at all life stages (larval, nymphal and adult) at every location in this study was *Ixodes ricinus*, while *Ixodes frontalis* and *Haemaphysalis concinna* were sampled only as subadults (Table 3). Overall, larvae of *Ixodes ricinus* were mainly predominant (53.9%) but closely followed by nymphs (40.1%), while only 6% of the sampled specimens were adults. These results differ from the previous findings of Krčmar et al. (2019), who collected 67% of ticks at an adult stage in eastern woodland parts of the country [47]. At the same time, our data correspond to a high share of subadults (88%) sampled in a forest habitat within Croatia's capital, the city of Zagreb [43]. In floodplain forests of the Posavina region in Central Croatia, only adult *Ixodes ricinus* were sampled from 2011 to 2013 [45]. On the Medvednica mountain, *Ixodes ricinus* nymphs were prevalent at all altitudinal limits (200, 400, 600, 800 and 1000 m asl.), while at the mountain site of Papuk, they were prevalent at 400 m asl. and from 600 to 800 m asl. These data are also partly in agreement with data from Krkonoše Mts., where all ticks collected above 850 m asl. were nymphs [74]. These data are not surprising, since the nymphal stage is the one most frequently recorded when ticks are sampled by the flagging method [74]. In contrast, larval stages were prevalent in collected samples at altitudes above 600 m in Gorski Kotar.

One of the aims of our study was to determine the prevalence of the Lyme disease causative agent among sampled ticks. Of 27 pools, containing 305 ticks, 1 pool (3.7%) was positive for *Borrelia burgdorferi* sensu lato infection. The detected prevalence is in line with a recent study on pathogens of zoonotic importance that revealed 2.2% infection with *Borrelia burgdorferi* s.l. within ticks collected in the Scottish Highlands [85], but it also differs from previous research conducted in urban forests in Croatia's capital which showed high infection rates (55%) within tick populations [26]. If we consider our three research sites, the presence of the pathogen was detected in 20% of the samples from Gorski Kotar, while samples from Medvednica and Papuk showed no infection. Our data partly correspond to prevalences of *Borrelia* spp. within tick populations in urban woodland habitats recorded in Germany (24–34%), Italy (26–36%), Lithuania (25%), Poland (11–27%), Slovakia (6–10%), Switzerland (18%) and Serbia (23.6%) [10,86]. Our findings are also in line with meta-analysis data of *Borrelia* spp. prevalence in ticks in Europe (18.6% in adults and 10.1% in nymphs) [87].

Eight different genospecies of *Borrelia burgdorferi*, specific to different hosts, can be found throughout Europe [85]. Sequencing of the ospA gene and phylogenetic analysis revealed the presence of the *B. burgdorferi* sensu stricto genotype in ticks from the mountain area of Gorski Kotar. These findings cohere with previous reports of *Borrelia afzelii, Borrelia garinii* and also *B. burgdorferi* sensu stricto detected in *Ixodes ricinus* ticks collected in a Lyme borreliosis endemic region of Northern Croatia [27]. Further monitoring of tick fauna, their hosts and the pathogens they carry, along with ecological studies of their habitats, would surely provide vital information on rapidly changing natural and urban environments. Worrying trends of increasing contact between human populations and ticks could make tick surveillance, monitoring and control a high priority in the near future.

6. Conclusions

This is the first report on the altitudinal distributions of hard ticks (Acari: Ixodidae) in mountain regions of Croatia. Three tick species (*Ixodes ricinus, Haemaphysalis concinna* and *Ixodes frontalis*) were recorded in the study areas. In all three areas, *Ixodes ricinus* was the most abundant species.

Ixodes frontalis and *Haemaphysalis concinna* were recorded for the first time both on the Medvednica mountain and on the Papuk mountain.

The highest tick abundances were recorded at higher altitudes (800–1000 m asl.) in Gorski Kotar and on the Medvednica mountain, while on the Papuk mountain the highest abundance was recorded at a lower altitude (200 m asl).

Most ticks were collected during the spring season (March–June)—89.12%—while in autumn (September–December) only 10.88% of ticks were sampled.

The mean tick density of the study area was 18.9 ticks per 100 m^2—Gorski Kotar: 29.8 (range: 0–207.5); Medvednica: 18.8 (range: 0–109.5); Papuk: 8.0 (range: 0–42.5).

The presence of *B. burgdorferi* s.l. was detected in 20% (1/5) of the tick samples from Gorski Kotar, while the samples from Medvednica and Papuk showed no infection.

Findings of *Borrelia burgdorferi* sensu stricto in nymphs and adults of *Ixodes ricinus* from Gorski Kotar indicate the risk of occurrence of Lyme borreliosis at altitudes from 600–800 m asl., where the causative agent of this emerging disease had not been registered before within tick populations in Croatia.

Author Contributions: Conceptualization, M.V. and S.K.; methodology, M.V., S.K., M.B., J.H., V.M.P. and L.B.; software, M.V., S.K., M.B., V.M.P. and L.B.; formal analysis, M.V., S.K., J.H. and V.M.P.; investigation, M.V., M.B. and L.B.; data curation, M.V., M.B. and L.B.; writing—original draft preparation, M.V. and S.K., writing—review and editing, S.K., M.V. and J.M., visualization, M.V., supervision, J.M.; project administration, M.V. and M.B.; funding acquisition, J.M. All authors have read and agreed to the published version of the manuscript.

Funding: This research was funded by the Faculty of Forestry and Wood Technology, University of Zagreb.

Institutional Review Board Statement: The animal study protocol was approved by the Ethic Committee of the Faculty of Forestry and Wood Technology University of Zagreb (Approval Code: EP01-22/23 Approval Date: 19 January 2023.)

Data Availability Statement: The data presented in this study are available on request from the corresponding author.

Acknowledgments: The authors wish to thank Faculty of Forestry and Wood Technology University of Zagreb for all their support and also Croatian Meteorological and Hydrological Service for providing the data from meteorological stations at study sites Gorski Kotar, Medvednica and Papuk.

Conflicts of Interest: The authors declare no conflict of interest.

Appendix A. Tick Sampling Sites

Figure A1. Sampling site 1. Mountain site: Gorski Kotar/altitude: 400 m asl./forest soil: Pseudogley on sloping terrain/forest community: EBDF: European beech forest with Deer fern (*Blechno- fagetum sylvaticae* Ht. 1950).

Figure A2. Sampling site 2. Mountain site: Gorski Kotar/altitude: 600 m asl./forest soil: Dystric cambisol/forest community: EFDF: European fir forest with Deer fern (*Blechno—Abietetum* Ht. 1950).

Figure A3. Sampling site 3. Mountain site: Gorski Kotar/altitude: 800 m asl./forest soil: Dystric cambisol/forest community: DBF: Dinaric beech–fir forest (*Omphalodo—Fagetum* Marinčak et al. 1992).

Figure A4. Sampling site 4. Mountain site: Gorski Kotar/altitude: 1000 m asl./forest soil: Dystric cambisol/forest community: MPF: Mixed private forest stand with European beech (*Fagus sylvatica* L.), European silver fir (*Abies alba* Mill.), Sycamore (*Acer pseudoplatanus* L.), European ash (*Fraxinus excelsior* L.) and Common hazel (*Coryilus avellana* L.).

Figure A5. Sampling site 5. Mountain site: Medvednica/altitude: 200 m asl./forest soil: Dystric cambisol/forest community: POEH: Pedunculate oak forest with European hornbeam (*Carpino betuli—Quercetum roboris typicum* Rauš 1969).

Figure A6. Sampling site 6. Mountain site: Medvednica/altitude: 400 m asl./forest soil: Dystric cambisol/forest community: EBW: European beech forest with Woodruff (*Galio odorati-Fagetum* Sougnez et Thill 1959).

Figure A7. Sampling site 7. Mountain site: Medvednica/altitude: 600 m asl./forest soil: Dystric cambisol/forest community: EBW: European beech forest with Woodruff (*Galio odorati-Fagetum* Sougnez et Thill 1959).

Figure A8. Sampling site 8. Mountain site: Medvednica/altitude: 800 m asl./forest soil: Dystric cambisol/forest community: EBW: European beech forest with Woodruff (*Galio odorati-Fagetum* Sougnez et Thill 1959).

Figure A9. Sampling site 9. Mountain site: Medvednica/altitude: 800 m asl./forest soil: Dolomitic Cambisol/forest community: PBF: Pannonian beech–fir forest (*Festuco drymeiae—Abietetum* Vukelić et Baričević 2007).

Figure A10. Sampling site 10. Mountain site: Papuk/altitude: 200 m asl./forest soil: Luvisol/forest community: SOFD: Sessile oak forest with *Festuca drymeia* (*Festuco drymeiae—Quercetum petraeae*/Jank. 1968/Hruška 1974).

Figure A11. Sampling site 11. Mountain site: Papuk/altitude: 400 m asl./forest soil: Dystric cambisol/forest community: EBW: European beech forest with Woodruff (*Galio odorati-Fagetum* Sougnez et Thill 1959).

Figure A12. Sampling site 12. Mountain site: Papuk/altitude: 600 m asl./forest soil: Dystric cambisol/forest community: EBW: European beech forest with Woodruff (*Galio odorati-Fagetum* Sougnez et Thill 1959).

Figure A13. Sampling site 13. Mountain site: Papuk/altitude: 800 m asl./forest soil: Dystric cambisol/forest community: MEBBLA: Mountain European beech forest with (*Lamio orvalae—Fagetum sylvaticae* Ht. 1938).

Appendix B

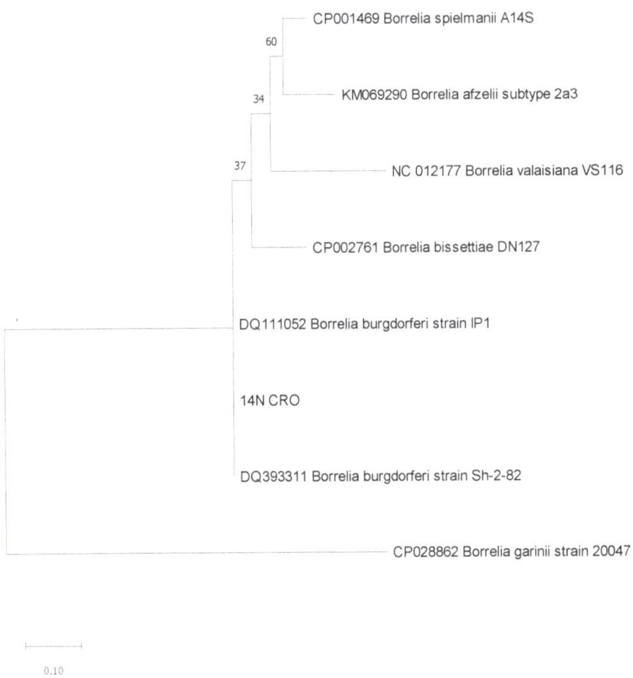

Figure A14. A phylogenetic tree constructed using *ospA* sequences retrieved from GenBank (accession numbers are shown in the tree) and the sequence identified in this study (14N CRO). Evolutionary distances were calculated using the Maximum Likelihood method and Tamura–Nei model in MEGA 11 with 1000 bootstrap replications. The scale bar indicates the nucleotide substitutions per site.

References

1. Smith, T.; Kilbourne, F.L. Investigations into the nature, causation and prevention of Texas or southern cattle fever. *Bull. Bur. Anim. Ind.* **1893**, *1*, 177–304.
2. Dutton, J.E.; Todd, J.L. The nature of tick fever in the eastern part of Congo Free State, with notes on the distribution and bionomics of the tick. *Br. Med. J.* **1905**, *2*, 1259–1260.
3. Dennis, D.T.; Piesman, J.F. Overview of Tick-Borne Infections of Humans. In *Tick-Borne Diseases of Humans*; Goodman, J.L., Dennis, D.T., Sonenshine, D.E., Eds.; ASM Press: Washington, DC, USA, 2005; pp. 3–11.
4. Jongejan, F.; Uilenberg, G. The global importance of ticks. *Parasitology* **2004**, *129*, S3–S14. [CrossRef] [PubMed]
5. Smith, R.; Takkinen, J. Editorial team Collective. Lyme borreliosis: Europe-wide coordinated surveillance and action needed? *Euro Surveill.* **2006**, *11*, 2977. [CrossRef]
6. Bacon, R.M.; Kugeler, K.J.; Mead, P.S. Surveillance for Lyme disease—United States, 1992–2006. *MMWR Surveill. Summ.* **2008**, *57*, SS-10.
7. Mysterud, A.; Jore, S.; Østerås, O.; Viljugrein, H. Emergence of tick-borne diseases at northern latitudes in Europe: A comparative approach. *Sci. Rep.* **2017**, *7*, 16316. [CrossRef]
8. Bajer, A.; Beck, A.; Beck, R.; Behnke, J.M.; Dwuznik-Szarek, D.; Eichenberger, R.M.; Farkas, R.; Fuehrer, H.P.; Heddergott, M.; Jokelainen, P.; et al. Babesiosis in Southern, Central and Northeastern Europe: An Emerging and Re-Emerging Tick-Borne Disease of Humans and Animals. *Microorganisms* **2022**, *10*, 945. [CrossRef]
9. Medlock, J.M.; Hansford, K.M.; Bormane, A.; Derdakova, M.; Estrada-Peña, A.; George, J.C.; Golovljeva, I.; Jaenson, T.G.T.; Jensen, J.K.; Jensen, P.M.; et al. Driving forces for changes in geographical distribution of *Ixodes ricinus* ticks in Europe. *Parasit. Vectors* **2013**, *6*, 1. [CrossRef]
10. Snegiriovaite, J.; Radzijevskaja, J.; Palauskas, A. A brief review: The prevalence of tick borne pathogens in urban and suburban areas. *Biologia* **2020**, *66*, 242–255. [CrossRef]
11. Aguirre, A.A.; Tabor, G.M. Global factors driving emerging infectious diseases: Impact on wildlife populations. *Ann. N. Y. Acad. Sci.* **2008**, *1149*, 1–3. [CrossRef]
12. Spielman, A. The Emergence of Lyme Disease and human Babesiosis in a changing environment. *Ann. N. Y. Acad. Sci.* **1994**, *740*, 146–156. [CrossRef] [PubMed]
13. Gilbert, L.; Maffey, G.L.; Ramsay, S.L.; Hester, A.J. The effect of deer management on the abundance of *Ixodes ricinus* in Scotland. *Ecol. Appl.* **2012**, *22*, 658–667. [CrossRef]
14. Piesman, J.; Gern, L. Lyme borreliosis in Europe and North America. *Parasitology* **2004**, *129*, 191–220. [CrossRef] [PubMed]
15. Hall, J.L.; Alpers, K.; Bown, K.J.; Martin, S.J.; Birtles, R.J. Use of mass-participation outdoor events to assess human exposure to tickborne pathogens. *Emerg. Infect. Dis.* **2017**, *23*, 463–467. [CrossRef] [PubMed]
16. Mrljak, V.; Kuleš, J.; Mihaljević, Ž.; Torti, M.; Gotić, J.; Crnogaj, M.; Živičnjak, T.; Mayer, I.; Šmit, I.; Bhide, M.; et al. Prevalence and geographic distribution of vector-borne pathogens in apparently healthy dogs in Croatia. *Vector Borne Zoonotic Dis.* **2017**, *17*, 398–408. [CrossRef]
17. Estrada-Peña, A. Distribution, abundance, and habitat preferences of *Ixodes ricinus* (Acari: Ixodidae) in northern Spain. *J. Med. Entomol.* **2001**, *38*, 361–370. [CrossRef]
18. Walker, A.R.; Alberdi, M.P.; Urquhart, K.A.; Rose, H. Risk factors in habitats of the tick *Ixodes ricinus* influencing human exposure to *Ehrlichia phagocytophila* bacteria. *Med. Vet. Entomol.* **2001**, *15*, 40–49. [CrossRef]
19. Hansford, K.M.; Fonville, M.; Gillingham, E.L.; Copian, E.C.; Pietzch, M.E.; Krawczyk, A.I.; Vaux, A.G.C.; Cull, B.; Sprong, H.; Medlock, J.M. Ticks and *Borrelia* in urban and peri-urban green space habitats in a city in southern England. *Ticks Tick Borne Dis.* **2017**, *8*, 353–361. [CrossRef]
20. Krčmar, S. Hard ticks (Acari, Ixodidae) of Croatia. *ZooKeys* **2012**, *234*, 19–57. [CrossRef]
21. Gray, J.S. Biology of *Ixodes* species ticks in relation to tick-borne zoonoses. *Wien. Klin. Wochenschr.* **2002**, *114*, 473–478.
22. García-Álvarez, L.; Palomar, A.M.; Oteo, J.A. Prevention and prophylaxis of tick-bites and tick-borne related diseases. *Am. J. Infect. Dis.* **2013**, *9*, 104–116. [CrossRef]
23. Mulić, R.; Petković, B.; Klišmanić, Z.; Jerončić, I. Tick-borne diseases in the Republic of Croatia. *Liječ. Vjesn.* **2011**, *133*, 89–95.
24. Beck, R.; Habrun, B.; Bosnić, S.; Benić, M.; Nemeth-Blažić, T.; Duvnjak, S. Identification of pathogens in *Ixodes ricinus* and *Dermacentor reticulatus* fron public gardens in Zagreb, Croatia. In Proceedings of the 12th International Conference on Lyme Borreliosis and Other Tick-Borne Diseases, Ljubljana, Slovenia, 26–29 September 2010.
25. Golubić, D.; Rijpkema, S.; Tkalec-Makovec, N.; Ruzić, E. Epidemiologic, ecologic and clinical characteristics of Lyme borreliosis in northwest Croatia. *Acta Med. Croat.* **1998**, *52*, 7–13.
26. Barišin, A.; Nemeth Blažić, T.; Jeličić, P.; Gjenero Margan, I.; Capak, K.; Petrović, G. Prikaz istraživanja krpelja na području Grada Zagreba u 2008. godini. *Zbornik radova DDD-ZUPP* **2011**, *1*, 203–211.
27. Rijepkema, S.; Golubić, D.; Molkenboer, M.; Verbeek, D.E.; Kruif, N.; Schellekens, J.F.P. Identification of four genomic groups of *B. burgdorferi* sensu lato in *Ixodes ricinus* ticks collected in a Lyme borreliosis endemic region on northern Croatia. *Exp. Appl. Acarol.* **1996**, *20*, 25–30.
28. Steere, A.C.; Malawista, S.E.; Snydman, D.R.; Shope, R.E.; Andiman, W.A.; Ross, M.R.; Steele, F.M. Lyme arthritis: An epidemic of oligoarticular arthritis in children and adults in three connecticut communities. *Arthritis Rheum.* **1977**, *20*, 7–17. [CrossRef]

29. Parola, F.; Raoult, D. Tick and tickborne bacterial diseases in humans: An emerging infectious threat. *Ticks Tick Borne Dis.* **2001**, *32*, 897–928. [CrossRef]
30. Quine, C.P.; Barnett, J.; Dobson, A.D.M.; Marcu, A.; Marzano, M.; Moseley, D.; O' Brien, L.; Randolph, S.E.; Taylor, J.L.; Uzell, D. Frameworks for risk communication and disease management: The case of Lyme disease and countryside users. *Philos. Trans. R. Soc. B Biol. Sci.* **2011**, *366*, 2010–2022. [CrossRef] [PubMed]
31. Rizzoli, A.; Hauffe, H.C.; Carpi, G.; Vourc'h, G.I.; Neteler, M.; Rosà, R. Lyme borreliosis in Europe. *Euro Surveill.* **2011**, *16*, 19906. [CrossRef] [PubMed]
32. Strnad, M.; Hönig, V.; Růžek, D.; Grubhoffer, L.; Rego, R.O.M. Europe-wide meta-analysis of *Borrelia burgdorferi* sensu lato prevalence in questing *Ixodes ricinus* ticks. *Appl. Environ. Microbiol.* **2017**, *83*, e00609-17. [CrossRef]
33. Bennet, L.; Halling, A.; Berglund, J. Increased incidence of Lyme borreliosis in southern Sweden following mild winters and during warm, humid summers. *Eur. J. Clin. Microbiol. Infect. Dis.* **2006**, *25*, 426–432. [CrossRef]
34. Croatian Institute of Public Health. Statistical Yearbooks (1996–2021). Available online: https://www.hzjz.hr/hrvatski-zdravstveno-statisticki-ljetopis/hrvatski-zdravstveno-statisticki-ljetopis-za-2020-tablicni-podaci/ (accessed on 2 January 2023).
35. Lipozenčić, J.; Šitum, M. Epidemiology of Lyme disease. *Acta Derm. APA* **2001**, *10*, 1–9.
36. Estrada-Peña, A.; Cutler, S.; Potkonjak, A.; Vassier-Tussaut, M.; Van Bortel, W.; Zeller, H.; Fernández-Ruiz, N.; Mihalca, A.D. An updated meta-analysis of the distribution and prevalence of *Borrelia burgdorferi* s.l. in ticks in Europe. *Int. J. Health Geogr.* **2018**, *17*, 41. [CrossRef] [PubMed]
37. Kahl, O.; Gray, J.S. The biology of *Ixodes ricinus* with emphasis on its ecology. *Ticks Tick Borne Dis.* **2023**, *14*, 102114. [CrossRef] [PubMed]
38. Mikačić, D. The ticks of the littoral zone in Yugoslavia. I. The morphology of the species from the genus *Hyalomma*. *Vet. Arhiv.* **1961**, *31*, 305–310.
39. Mikačić, D. The ticks of the littoral zone of Yugoslavia. II. The genus *Haemaphysalis* with reference to the distinction between *H. punctata* and *H. cholodkovskyi*. *Vet. Arhiv.* **1963**, *33*, 133–136.
40. Mikačić, D. Ticks in the littoral belt of Yugoslavia III. Distribution and dynamics of species in the course of the year. *Vet. Arhiv.* **1965**, *35*, 155–170.
41. Tovornik, D.; Brelih, S. Ixodid ticks, the parasites of lizards (Lacertidae) in the karst and other districts of Yugoslavia. *Scopolia* **1980**, *3*, 1–21.
42. Tovornik, D.; Vesenjak-Hirjan, J. A revision of ticks belonging to the *Rhipicephalus sanguineus* complex (Latreille), collected in the Yugoslav Coastal Region. *Biol. Vestn.* **1988**, *36*, 77–84.
43. Vucelja, M.; Bjedov, L.; Boljfetić, M.; Klobučar, A.; Krčmar, S.; Borak, S.; Modrić, M.; Juričić, K.; Peleš, V.; Margaletić, J.; et al. Monitoring of Hard Ticks at Urban Recreational Sites in the City of Zagreb from 2016 to 2018. *Infektol. Glasn.* **2019**, *39*, 33–39. [CrossRef]
44. Mikačić, D. Dinamika pojavljivanja krpelja (Ixodidae) u Sjevernoj Hrvatskoj. *Vet. Arhiv.* **1969**, *39*, 183–186.
45. Vucelja, M. *Rodent Control (Rodentia, Murinae, Arvicolinae) in Pedunculate Oak Forests (Quercus robur L.)—Integrated Approach and Zoonotic Aspect*, 1st ed.; University of Zagreb, Faculty of forestry: Zagreb, Croatia, 2013; pp. 25–210.
46. Borčić, B.; Aleraj, B.; Žutić, M.; Mikačić, D. The role of ticks (Ixodidae) in the maintenance of the tularemia natural focus in central Posavinia. *Vet. Arhiv.* **1978**, *48*, 277–283.
47. Krčmar, S. Diversity, ecology, and seasonality of hard ticks (Acari: Ixodidae) in eastern Croatia. *J. Vector Ecol.* **2019**, *44*, 18–29. [CrossRef]
48. Krčmar, S.; Klobučar, A.; Vucelja, M.; Boljfetić, M.; Kucinić, M.; Madić, J.; Cvek, M.; Bruvo Mađarić, B. DNA barcoding of hard ticks (Ixodidae), notes on distribution of vector species and new faunal record for Croatia. *Ticks Tick Borne Dis.* **2022**, *13*, 101920. [CrossRef]
49. EEA. European Environment Agency Biogeographic Regions Dataset. 2016. Permalink: 2a7b433b388e432d9c3e8cf019af1db0. Available online: https://www.eea.europa.eu/data-and-maps/data/biogeographical-regions-europe-3 (accessed on 8 January 2023).
50. Filipčić, A. Klimatska regionalizacija Hrvatske po Köppenu za standardno razdoblje 1961–1990. u odnosu na razdoblje 1931–1960. *Acta Geograph. Croat.* **1998**, *34*, 1–15.
51. Zaninović, K.; Gajić-Čapka, M.; Perčec Tadić, M.; Vučetić, M.; Milković, J.; Bajić, A.; Cindrić, K.; Cvitan, L.; Katušin, Z.; Kaučić, D.; et al. *Klimatski atlas Hrvatske/Climate atlas of Croatia 1961–1990, 1971–2000*, 1st ed.; Državni Hidrometeorološki Zavod Zagreb: Zagreb, Hrvatska, 2008; pp. 1–200.
52. Šegota, T.; Filipčić, A. Köppenova podjela klima i hrvatsko nazivlje. *Geoadria* **2003**, *8*, 17–31. [CrossRef]
53. Vukelić, J.; Rauš, Đ. *Forest phytocenology and forest communities in Croatia (Šumarska fitocenologija i šumske zajednice u Hrvatskoj)*, 1st ed.; University of Zagreb, Faculty of forestry: Zagreb, Croatia, 1988; pp. 173–177.
54. Vukelić, J. *Croatian Forest Vegetation (Šumska Vegetacija Hrvatske)*, 1st ed.; University of Zagreb, Faculty of forestry: Zagreb, Croatia, 2012; pp. 81–101.
55. Martinović, J. *Management of Forest Soils in Croatia*, 1st ed.; Gračan, J., Ed.; Croatian Forests Limited Liability Company, Ministry of Science and technology: Zagreb, Croatia, 2003; pp. 118–119.

56. Cohnstaedt, L.W.; Rochon, K.; Duehl, A.J.; Anderson, J.F.; Barrera, R.; Yao Su, N.; Gerry, A.C.; Obenauer, P.J.; Campbell, J.F.; Lysyk, T.J.; et al. Arthropod Surveillance Programs: Basic Components, Strategies, and Analysis. *Ann. Entomol. Soc. Am.* **2012**, *105*, 135–149. [CrossRef]
57. Falco, R.; Fish, D. The comparison of methods for sampling the deer tick, *Ixodes dammini*, in a Lyme disease endemic area. *Exp. Appl. Acarol.* **1992**, *14*, 165–173. [CrossRef]
58. Carroll, J.F.; Schmidtmann, E.T. Tick sweep: Modification of the tick drag—Flag method for sampling nymphs of the deer tick (Acari: Ixodidae). *J. Med. Entomol.* **1992**, *29*, 352–355. [CrossRef] [PubMed]
59. Estrada-Peña, A.; Bouattour, A.; Camicas, J.L.; Walker, A.R. *Ticks of Domestic Animals in the Mediterranean Region—A Guide to Identification of Species*, 1st ed.; University of Zaragoza: Zaragoza, Spain, 2004; pp. 1–128.
60. Estrada-Peña, A.; Mihalca, A.D.; Petney, T.N. *Ticks of Europe and North Africa. A Guide to Species Identification*, 1st ed.; Springer International Publishing AG: Cham, Szwitzerland, 2017; pp. 1–404.
61. Guy, E.C.; Stanek, G. Detection of *Borrelia burgdorferi* in patients with Lyme disease by the polymerase chain reaction. *J. Clin. Pathol.* **1991**, *44*, 610–611. [CrossRef] [PubMed]
62. Lindgren, E.; Andersson, Y.; Suk, J.E.; Sudre, B.; Semenza, J.C. Public health. Monitoring EU emerging infectious disease risk due to climate change. *Science* **2012**, *336*, 418–419. [CrossRef]
63. Kovačević, J.; Bučanović, T.; Krčmar, S. Hard tick fauna (Acari: Ixodidae) in different types of habitats in the city of Osijek (Eastern Croatia). *Nat. Croat.* **2020**, *29*, 63–72. [CrossRef]
64. Cull, B.; Vaux, A.G.C.; Ottowell, L.J.; Gillingham, E.L.; Medlock, J.M. Tick infestation of small mammals in an English woodland. *J. Vector Ecol.* **2017**, *42*, 74–83. [CrossRef] [PubMed]
65. Milutinović, M.; Radulović, Ž.; Tomanović, S.; Tomanović, Ž. Seasonal distribution of Borreliae in *Ixodes ricinus* ticks in the Belgrade region, Serbia. *Arch. Biol. Sci. Belgrade* **2006**, *58*, 183–186. [CrossRef]
66. Estrada-Peña, A.; Venzal, J.M.; Sánchez Acedo, C. The tick *Ixodes ricinus*: Distribution and climate preferences in the western Palaearctic. *Med. Vet. Entomol.* **2006**, *20*, 189–197. [CrossRef] [PubMed]
67. Accrosi, A.; Schiavetti, I.; Listorti, V.; Dellepiane, M.; Masotti, C.; Ercolini, C.; Guardone, L.; Razzuoli, E. Hard ticks (Ixodidae) from Wildlife in Liguria, Northwest Italy: Tick Species Diversity and Tick-Host Associations. *Insects* **2022**, *13*, 199. [CrossRef]
68. Bertola, M.; Momtarsi, F.; Obber, F.; Da Rold, G.; Carlin, S.; Toniolo, F.; Porcellato, E.; Falcaro, C.; Mondardini, V.; Ormelli, S.; et al. Occurrence and identification of *Ixodes ricinus* borne pathogens in Northeastern Italy. *Pathogens* **2021**, *10*, 1181. [CrossRef]
69. Rubel, F.; Brugger, K.; Pfeiffer, M.; Chitimia-Dobler, I.; Didyk, Y.M.; Leverenz, S.; Dautel, H.; Kahl, O. Geographical distribution of *Dermacentor marginatus* and *Dermacentor reticulatus* in Europe. *Ticks Tick Borne Dis.* **2016**, *7*, 224–233. [CrossRef]
70. Bilbija, B.; Spitzweg, C.; Papoušek, I.; Fritz, U.; Földvári, G.; Mullett, M.; Ihlow, F.; Sprong, H.; Civáňová Křížová, K.; Anisimov, N.; et al. *Dermacentor reticulatus*—A tick on its way from glacial refugia to a panmictic Eurasian population. *Int. J. Parasitol.* **2023**, *53*, 91–101. [CrossRef]
71. Široký, P.; Kubelová, M.; Bednář, M.; Modrý, D.; Hubálek, Z.; Tkadlec, E. The distribution and spreading pattern of *Dermacentor reticulatus* over its threshold area in the Czech Republic—How much is range of this vector expanding? *Vet. Parasitol.* **2011**, *183*, 130–135. [CrossRef]
72. Földvári, G.; Široký, P.; Szekeres, S.; Majoros, G.; Sprong, H. *Dermacentor reticulatus*: A vector on the rise. *Parasit. Vectors* **2016**, *9*, 314. [CrossRef] [PubMed]
73. Daniel, M. Influence of the microclimate on the vertical distribution of the tick *Ixodes ricinus* (L.) in Central Europe. *Acarologia* **1993**, *34*, 105–113.
74. Materna, J.; Daniel, M.; Danielova, V. Altitudinal distribution limit of the tick *Ixodes ricinus* shifted considerably towards higher altitudes in central Europe: Results of thre years monitoring in the Krkonoše Mts. (Czech Republic). *Cent. Eur. J. Publ. Health* **2005**, *13*, 24–28.
75. Pelsmaeker, N.D.; Korslund, L.; Steifetten, Ø. High-elevational occurrence of two tick species, *Ixodes ricinus* and *I. trianguliceps*, at their northern distribution range. *Parasit. Vectors* **2021**, *14*, 161. [CrossRef]
76. Hornok, S.; Farkas, R. Influence of biotope on the distribution and peak activity of questing ixodid ticks in Hungary. *Med. Vet. Entomol.* **2009**, *23*, 41–46. [CrossRef] [PubMed]
77. Kubiak, K.; Szymańska, H.; Dmitryjuk, M.; Dzika, E. Abundance of *Ixodes ricinus* ticks (Acari: Ixodidae) and the diversity of *Borrelia* species in Northeastern Poland. *Int. J. Environ. Res. Public Health* **2022**, *19*, 7378. [CrossRef] [PubMed]
78. Otranto, D.; Dantas-Torres, F.; Santos-Silva, M.M. Ixodes ricinus (Linnaeus, 1758) (Figs.67–69). In *Ticks of Europe and North Africa: A Guide to Species Identification*, 1st ed.; Estrada-Peña, A., Mihalca, A.D., Petney, T.N., Eds.; Springer International Publishing AG: Cham, Szwitzerland, 2017; pp. 189–195.
79. Paul, R.E.L.; Cote, M.; Le Naour, E.; Bonnet, S.I. Environmental factors influencing tick densities over seven years in a French suburban forest. *Parasit. Vectors* **2016**, *9*, 309. [CrossRef] [PubMed]
80. Estrada-Peña, A.; De La Fuente, J. Species interactions in occurrence data for a community of tick transmitted pathogens. *Sci. Data* **2016**, *3*, 160056. [CrossRef]
81. Schwarz, A.; Maier, W.A.; Kistemann, T.; Kampen, H. Analysis of the distribution of the tick *Ixodes ricinus* L. (Acari: Ixodidae) in a nature reserve of western Germany using geographic information systems. *Int. J. Hyg. Environ. Health* **2009**, *212*, 87–96. [CrossRef]

82. Gray, J.S.; Dautel, H.; Estrada-Peña, A.; Kahl, O.; Lindgren, E. Effects of Climate Change on Ticks and Tick-Borne Diseases in Europe. *Inter. Perspect. Infect. Dis.* **2009**, *212*, 87–96. [CrossRef]
83. Ehrmann, S.; Liira, J.; Gärtner, S.; Hansen, K.; Brunet, J.; Cousins, S.A.O.; Deconchat, M.; Decocq, G.; De Frenne, P.; De Smedt, P.; et al. Environmental drivers of *Ixodes ricinus* abundance in forest fragments of rural European landscapes. *BMC Ecol.* **2017**, *17*, 31. [CrossRef] [PubMed]
84. Petney, T.N.; Pfäffie, M.P.; Skuballa, J.D. An annotated checklist of the ticks (Acari: Ixodidae) of Germany. *Syst. Appl. Acarol.* **2012**, *17*, 115–170. [CrossRef]
85. Olsthoorn, F. The Effect of Woodland Expansion on Tick Populations and Lyme Disease Risk. Ph.D. Thesis, ETH Zürich, Zürich, Switzerland, 2021.
86. Milutinović, M.; Radulović, Ž.; Jovičić, V.; Oreščanin, Z. Population dynamics and *Borrelia burgdorferi* infection rate of *Ixodes ricinus* ticks in the Belgrade area. *Acta Vet.* **2004**, *54*, 219–225.
87. Rauter, C.; Hartung, T. Prevalence of *Borrelia burgdorferi* sensu lato genospecies in *Ixodes ricinus* ticks in Europe: A metaanalysis. *Appl. Environ. Microbiol.* **2005**, *71*, 7203–7216. [CrossRef] [PubMed]

sustainability

MDPI

Article

Participation of Bulgarian Furniture Manufacturing in Global and Local Value Chains as a Factor Supporting Their Innovation Activities

Daniela Ventsislavova Georgieva [1], Nikolay Neykov [2], Andreja Pirc Barčić [3,*], Petar Ćurić [3] and Kristina Klarić [3]

1 International Business School, 1407 Sofia, Bulgaria; danielagr999@gmail.com
2 Faculty of Ecology and Landscape Architecture, University of Forestry, 1797 Sofia, Bulgaria; nkneykov@gmail.com
3 Department of Production Organization, Faculty of Forestry and Wood Technology, University of Zagreb, 10000 Zagreb, Croatia; petar.curic@gmail.com (P.Ć.); kklaric@sumfak.unizg.hr (K.K.)
* Correspondence: apirc@sumfak.unizg.hr

Abstract: Innovations can offer key advantages to companies, but in some EU regions, the design and development of innovation measures are still relatively novel concepts. The aim of this study was to analyze the collaborations of innovative Bulgarian furniture manufacturers with external stakeholders and the used information channels as factors for the development and implementation of innovation and participation in global value chains over their innovation activities. Out of 3890 Bulgarian companies, the number of companies included in the target group was further reduced to 85 firms due to missing information on some variables. The data for the present study were collected using a large-scale questionnaire distributed on the spot during the months of March and April 2022. Logistic regression was used to reveal the real contribution of the collaborations and the information sources to the ability of companies to innovate. The research results indicated that in Bulgaria, the furniture sector is not considered very innovative, and Bulgarian furniture manufacturing companies do not rely on collaboration with the IT and mechatronics sectors. These companies do not want to participate in GVCs, as they refer to them in relation to supply chains. Therefore, they are less dependent on chain shocks. Companies prefer to hide their innovations for further protection, which might be the reason for the lack of cooperation between the furniture manufacturing companies and academia, NGOs, and other relevant institutions. The findings of the study contribute to new insights into the literature on the participation in GVCs as a factor for collaboration with different stakeholders and hence for product and process innovation development within the furniture industry companies.

Keywords: forest-based sector; innovation; global value chains; furniture manufacturers

Citation: Georgieva, D.V.; Neykov, N.; Barčić, A.P.; Ćurić, P.; Klarić, K. Participation of Bulgarian Furniture Manufacturing in Global and Local Value Chains as a Factor Supporting Their Innovation Activities. *Sustainability* **2023**, *15*, 13260. https://doi.org/10.3390/su151713260

Academic Editors: Ivica Papa and Andreja Đuka

Received: 28 July 2023
Revised: 28 August 2023
Accepted: 28 August 2023
Published: 4 September 2023

1. Introduction

Under the pressure of globalization and dynamic technological processes, national and regional economies are putting in increasing efforts in an attempt to improve their competitive positions. The analysis of foreign investment flows and the concentration of production processes at the international level show that the innovation capacity and technological competencies, once concentrated at the head offices of multinational enterprises, are increasingly being outsourced towards their peripheries and are being performed by companies located outside major economic centers. In this context, global value chains (GVCs) are attracting the attention of both the academic community and political organizations as a factor for economic transformation and growth of local economies. According to OECD studies [1], global value chains can generate significant economic benefits for the participating companies, as well as for the national and regional economies where they are located. Due to the COVID-19 pandemic, the concentration of suppliers or customers along the chain may make some firms more vulnerable to shocks from the crisis. Additionally, for industries with high concentrations of exports or imports, fewer possibilities for suppliers or buyers to be replaced exist in the event of a break in the chain [2].

In the literature, the measurement and evaluation of global value chains' innovation potential where global value chains are considered as original innovation systems remain largely unstudied. The most common subject of analysis is the "supply-side" factor led by the indicators of foreign investors (size, corporate governance, technological assets). Other studies focus on the "demand-side" factor, which pays central attention to local businesses and their absorptive capacity, i.e., their capacity to acquire, utilize, transform, and apply technology-driven external knowledge. The analysis of the isolated impact of one factor or another is not sufficient to reveal the complex nature of the impact of global value chains. A comprehensive scientific study that builds conceptually on the symbiosis between geographical location, sectoral affiliation, and innovation potential was not found by the authors of the current research. When applied to global value chains, the analysis of the innovation potential will combine the effects of those factors and will suggest new opportunities to support national and regional competitiveness. Because of that, further study of the interactions between foreign and domestic firms, NGOs, academia, and other stakeholders at different stages of the value chains is needed.

The paper aims to analyze the collaborations of innovative Bulgarian furniture manufacturers with external stakeholders and the use of information channels as factors for (1) the development and implementation of innovation and (2) participation in global value chains over their innovation activities. The subjects of the research are global value chains in the furniture industry in Bulgaria. They will be examined within the context of their impact on the innovation development of furniture enterprises. The adopted research methods are logical, deductive, and comparative methods. Primary data from a survey conducted among innovative furniture enterprises in Bulgaria are presented. To check the authors' hypotheses, logistic regression was used. The logical construction of the study begins with a literature review related to the innovation potential of GVCs, where specific variables are outlined. The articles continue by presenting information on the used methodology, collected data, results, and conclusions.

2. Literature Review and Hypothesis Development

The concept of the "global value chain" refers to the idea that an organization is not considered per se but as part of a common "supply chain", i.e., it is linked to other organizations [3]. Thus, global value chains are seen as a series of stages in the production of goods or services, where each stage adds value. At least two of the stages should be implemented in different countries [4]. An efficient global value chain may have a positive impact on businesses [5] and corporate growth [4]. Competitiveness is linked to the chains in which the companies participate, and, in this respect, the competition is between chains of organizations contributing to the satisfaction of customers [3].

Innovation potential, including in the GVCs, is generally studied at a national level, where sets of indicators have been designed and widely used to capture the specifics of national economies in terms of inputs and outputs of the innovation system performance [6]. With some modifications, these methods have been applied at the sector and regional levels, taking into consideration the technological or local specifics of the countries [7]. However, the indicators used to measure the innovation potential are not appropriate on a country/regional level, or an inappropriate methodology to collect the data is used. Additionally, Bulgarian companies refrain from accounting for their innovation and apply for patents, leading to the phenomenon of "hidden innovation".

The forestry-based sector is considered "low-tech" and less innovative [8–11]. For instance, Pirc and Vlosky [10] showed that applying innovation is becoming more and more important in Croatian furniture companies. Barčić and Motik [2] revealed that, along with small traditional companies, innovative companies in the Croatian furniture industry exist, but due to excessive market opportunities and possibilities, managers, directors, and executive staff still do not recognize which way will take them to one step ahead of the competition.

However, of all the forestry industries in Bulgaria, the most innovative are furniture design, manufacturing, and sales companies. The Bulgarian furniture industry is extremely export-oriented, and because of that, it is very vulnerable to disruptions in external markets. The increase in the prices of basic and auxiliary materials during COVID-19 and the outbreak of the war in Ukraine, as well as problems with the supply of materials, led to a serious delay in orders and hence affected the supply chain. At the same time, during those periods, an increase in house furniture orders occurred as a result of the increase in home purchases in the country. Currently, due to the stagnation of Western markets, turbulences are also expected in the Bulgarian furniture sector, which should significantly affect the GVC. It is thought that to minimize the negative effects on the value chains and their expansion, there is a need to restore "the predictability" of governmental policies. Commercial conflicts and the absence of reforms along the chain also have negative effects on its development and resilience. In addition, the uneven distribution of results along the chain and the risks posed to it by the new technologies and digitalization are among the issues yet to be addressed. For industries with high concentrations of exports or imports, fewer possibilities exist for suppliers or buyers to be replaced in the event of a break in the chain. Furthermore, in some cases, the restructuring of suppliers is not only resource-intensive but also impossible in the context of the specifics of their business [1]. The rational way of choosing products among buyers in the context of COVID-19 reflects on the production of lower-tech and non-innovative products [12]. However, those issues have still not been addressed in the context of the Bulgarian furniture industry.

According to Fagerberg et al. [13], innovation potential is a factor for better participation in the global value chain leading to economic development. Vivek et al. [14] stated that outsourcing some of the production activities benefits the development of innovation, hence helping leading firms optimize their costs and invest more in research and development (R&D). In this way, businesses gain access to new knowledge, ideas, and technology transfer, creating innovation [15,16]. Under the pressure of globalization, national and regional economies focus on improving their competitive positions in the GVC through investments in scientific research and innovation. However, the furniture industry is considered labor-intensive, with fragmented supply chains and a predomination of SMEs [17].

According to Chiu, Hastig, and Sodhi [18,19], the diversity of suppliers in the value chain contributes to businesses when searching for new products. Thus, it will enable the use of new knowledge and technologies. Based on the idea of Humphrey and Schmitz [20], the following economic benefits for companies' participation in GVCs can be outlined: (a) the creation of process innovations, which reduce the cost of the manufactured product or the delivered service by more efficiently converting the input materials and resources into the final product (service); (b) the development of product innovations for higher-quality products and services; (c) the creation of organizational innovations that increase the added value of human labor; and (d) the expansion of the product portfolio by entering into intensive and high-tech industries.

Pisano and Shih [21] considered the idea that the territorial separation of a firm's production and R&D in other countries can limit the creation of micro-level innovations. Outsourcing R&D leads to the creation of dependence on third parties, limitations over the control of the innovation development and management process [22], and worsening communication between stakeholders in non-innovation and innovation activities along the chain. When R&D and production are interdependent, or when production technologies are "immature", outsourcing production is associated with higher costs for the company and a reduction in the added value of innovation. Participating in GVCs can help manufacturers, including furniture ones, to learn from global supply chain leaders [23]. However, global leaders can minimize the transferred knowledge and technology to product information and production techniques, thus limiting the transfer of competencies and R&D along the chain.

Forest product companies can use partnerships and collaborative relationships with other companies to increase their competitiveness. This can be done through developing new value-added products and technologies [24], entering new markets [25], and increasing productivity while decreasing production costs [26]. Hence, essential for the improvement of the value chains is the establishment of new partnerships by identifying the existing and intended partnering practices [27]. Partnerships in the forest product industry can be divided into operational partnerships (partnerships with suppliers and logistics customer service firms), technology partnerships (with technology providers), and financial partnerships [24,27]. Without belittling the importance of all those partnerships for the current study, only financial partnerships with state institutions and technology partnerships with representatives of the ICT (The Information and Communication Technology) sector and mechatronics will be under analysis. The Information and Communication Technology (ICT) sector is indeed one of the most innovative sectors globally. ICTs act as an enabler of innovation, particularly for product and marketing innovation. Additionally, it reduces the barriers to participation in the global economy and hence supports development within GVCs [11]. The selection is based on the stated assumptions that first, the level of digitization of Bulgarian furniture manufacturers shows better performance than that of the national average [28], and second, the sector's innovations are primarily incremental [29,30]. In this respect, the main investments are made for purchasing machinery and equipment [31] and technologies that support marketing and furniture sales. Because of that, clients' demands and financial resources are the main "triggers" for innovations in furniture manufacturers. Additionally, the restricted access of the furniture companies to the market and the lack of cooperation with the main supplier are stated as restrictions to the innovation development in the sector [32]. However, foreign and local customers and suppliers are significant sources of ideas for new projects and technological innovations. Widespread digitalization leads to the establishment of so-called Smart Factories. Hence, companies need to prepare for radical changes attributable to several factors–namely shortened delivery time, flexibility in the volumes produced, the unpredictability of customer demands, the further "branching" and "fragmentation" of the supply chains, and the value added [33,34]. According to Drayse [35], furniture manufacturers use information technology to manage their production process, logistics, and supply. This helps to accelerate the process of globalization and hence participation in GVCs, as it is believed that globalization is being driven by digitization [36]. Jagjit and Lorentz [37] argue that connectivity as a basic form of digitization can be measured by the use of computers and Internet access.

Popova and Georgieva [28] state that a relatively low percentage of Bulgarian furniture companies have web pages even though more than 90% of the companies use computers and the Internet. Predominately Bulgarian furniture companies use the Internet for online interaction with government institutions, suppliers, and customers. ICT and digitization are not seen as factors for achieving competitiveness by furniture manufacturers [38]. However, the stated research does not focus on furniture companies' innovation potential (in terms of different kinds of innovations) and states of digitization as factors for participation in the global value chains, and inversely, processes of optimization, automation, and robotization of production and searches for new raw materials, new energy sources, and energy efficiency improvement have started. Hence, companies will pay more attention to innovation and human capital.

Based on the previously cited literature, three hypotheses were developed as follows:

H1. *The inclusion of furniture companies in Bulgaria in various information channels led to the development of product and process innovations.*

H2. *Furniture companies introduced innovations because of agreement contracts with value chain participants like companies and institutions.*

H3. *The need forprocess and product innovations motivates furniture companies to collaborate with local value chain participants like companies from the mechatronics and IT sectors for the mutual development of processes and products.*

3. Materials and Methods

3.1. Sampling and Data Collection

For the purposes of the study, 3980 Bulgarian companies with NCEA-2008 code 31—furniture manufacturing were identified. Available data from the Bulgarian registry agency were used. Three hundred and thirty of all 3980 furniture manufacturers were selected as a target group. The number of companies included in the target group was further reduced to 85 firms due to missing information on some variables. The target group consisted of representatives of the management of only furniture manufacturers that had implemented innovation during the past 12 months before the survey. A pilot questionnaire was conducted at the beginning of the survey. The data for the present study were collected using a large-scale questionnaire distributed on the spot during the months of March and April 2022.

Based on Boer and During [39], the suggested division of innovation was divided further into product innovation–related to the introduction of new or enhancement of existing products–and process innovation–related to the introduction of new or improving existing activity in the manufacturing process.

Questions were grouped into six sections. Section 1 examined the types of innovations that companies implemented in the last 12 months until April 2022. Sections 2 and 3 explored the collaborations that companies entered with firms from the IT sector. Section 4 explained the common venture with Bulgarian firms from the sector "mechatronics". Section 5 revealed the information sources that questioned companies used to endorse the innovations. Section 6 included questions about the different types of collaborations the companies got into to support the innovation processes. The total number of questions in the questionnaire was 41.

3.2. Data Analysis

For the current study, logistic regression was used to reveal the real contribution of collaborations and the information sources on the ability of companies to innovate. It also analyzed the role of participation in GVC of the target group companies over their innovation activities. Logistic regression is commonly used to analyze innovations in business companies. This methodology was quite appropriate for the current research regarding the categorical data available. Nor et al. [40] used logistic regression to assess the profanities for innovation creation and implementation according to the type of business and various barriers to resource availability. Gerstlberger et al. [41] through logistic regression assessed the role of innovations and efficiency in improving energy efficiency. Świadek and Gorączkowska [42] examined with a logistic regression model the institutional support given to innovation cooperation in the industry. Collaborations that were examined in the current research also included the cooperation of different types, which corresponded to the topical research of these two Polish authors. Collaborations are vital for product innovations according to Odei and Stejskal [43], who again implemented logistic regression to derive empirically based results.

The idea behind the logistic regression model is to calculate the natural logarithm of the odd ratio [44]. The main value of interest could be the probability P of the event Y to have value of Y = 1. The odd ratio, Odd = $P/(1 - P)$, is chance for an event to happen, which means chance for the value of the dependent binomial variable to be $P\{Y = 1\}$ i.e., chance for an innovation to be introduced or collaboration to be built; otherwise, it is the opposite probability $(1 - P)\{Y = 0\}$, and there is no innovation or certain type of collaboration. It measures the effect of explanatory or predictor variables on the outcome or dependent variable [45]. The classical logistic regression model is as follows [44]:

$$\ln\{P/(1-P)\} = b_0 + b_1x \quad (1)$$

where the $\ln\{P/(1-P\}$ is the natural, b_0 is the intercept, and b_1 is the regression coefficient of the variable x. If the model includes numerous variables, each one has its own coefficient denoted by b_i, where i is the number of the variable.

The probability of certain events happening is calculated as follows:

$$P = e^{b0+b1x}/(1 + e^{b0+b1x}) \quad (2)$$

The odds for the appearance of innovation in a certain type and for the collaborative development of processes and new products are calculated as follows:

$$Odd = e^{b_i} \quad (3)$$

where the b_i is the coefficient of the explanatory variable.

The increase of the odds as a result of the explanatory variable positive value of 1:

$$\Delta = |1 - odd| \quad (4)$$

Dependent variables include the companies, which have implemented innovations under hypotheses H1 and H2 and are presented in Table 1. The models developed for the purpose of the current research are the logistic regression–logit models (see [42]) with binary outcomes and binary explanatory variables [14,16] in the context of the survey and the questionnaire sections. The events of interest are variables presented in Tables 2 and 3. If the variable appears as an event, it has a value of one (Y = 1), or otherwise zero (Y = 0).

Table 1. Dependent variables of the logistic regression for H1 and for H2 and independent for H3 (n = 85).

Variable Abbreviation	Description	Questionnaire Part
NPRD	New product development	Introduced Innovation
INSIDE	New processes developed inside the company	
INCOLLAB	New processes developed in collaboration	

Table 2. Explanatory variables for testing the hypotheses H1 and H2 (n = 85).

Variable Abbreviation	Description	Questionnaire Part
CGROUP	Companies from the corporate group	Information Sources and general agreements
CUST	Customers	
VEND	Vendors of raw materials and services	
COMP	Competitors	
CONS	Consultants	
PRESC	Private research institutes	
UNI	Universities	
ASECT	Associations, trade, or sectoral	
SINST	Stated owned institutions	
EUINST	Institutions of the EU	
TRDSHW	Trade shows	
PRNTD	Printed materials	
INT	Internet	
EMEDIA	Electronic media	
OTHER	Other companies from the furniture sector	General agreements
GLOBE	Global companies	
PINST	Private institutions	
FINANC	Financial companies	

Table 3. The dependent variables for testing the H3 hypothesis (n = 85).

Variable Abbreviation	Description	Questionnaire Part
MAUTO	Collaboration with companies from mechatronics sector in production automation	Collaboration with companies from mechatronics sector
MLOG	Collaboration with companies from mechatronics sector in logistics	
MNEWPROD	Collaboration with companies from mechatronics sector in development of new products	
ITSALES	Collaborating with IT companies in the sales	Collaboration with IT companies
ITRECRUIT	Collaborating with IT companies in the recruitment of personnel	
ITACCOUNT	Collaborating with IT companies in the accounting	

The variables in Table 1 are dependent on investigating the relationships between sources of information and the introduced innovations. When the collaborations for mutual innovation development are examined, the variables in the table are explanatory. Independent variables for H1 and H2 are included in Table 2.

For the testing purposes of hypothesis H3, the dependent variables are presented in Table 3.

Table 3's variables reveal the decisions of companies when creating innovations. They interact and develop joint solutions with companies from the IT or mechatronics sectors. In the case of general agreements, collaboration with these companies could be a result of the intentional behavior of the furniture companies to create innovation together with others.

The model for the H1 hypothesis, which investigates the role of the information sources in innovation development, is derived by including the dependent variables from Table 1 and explanatory variables from "Information Sources and general agreements" part of the questionnaire from Table 2. The model is as follows:

$$ln\{P/(1-P)\} = b_0 + b_1CGROUP + b_2CUST + b_3VEND + b_4COMP + b_5CONS + b_6PRESC + b_7UNI + b_8ASECT + b_9SINST + b_{10}EUINST + b_{11}TRDSHW + b_{12}PRNTD + b_{13}INT + b_{14}EMEDIA, \quad (5)$$

where $1/(1-P)$ is the odds, P is the probability of appearance of the dependent variables NPRD, INSIDE, or INCOLLAB, and b_i are the regression coefficients.

Model (5) investigated the probability of the companies introducing innovations (NPRD = 1, INSIDE = 1, or INCOLLAB = 1) as result of the influence of explanatory variables related to the "Information Sources and general agreements" part of the questionnaire.

For the H2 hypothesis, which tested the role of general agreements in innovation development, we developed a second model (6). The model included dependent variables from Table 1 and explanatory variables from "General agreements" part of the questionnaire from Table 2. The model is as follows:

$$ln\{P/(1-P)\} = b_0 + b_1CGROUP + b_2CUST + b_3VEND + b_4COMP + b_5CONS + b_6PRESCUNI + b_7SINST + b_8PINST + b_9OTHER + b_{10}GLOBE + b_{11}FINANC, \quad (6)$$

where $1/(1-P)$ is the odds, P is the probability of appearance of NPRD, INSIDE, or INCOLLAB, and b_i are the regression coefficients.

Model (6) investigated the probability of the companies introducing innovations (NPRD = 1, INSIDE = 1, or INCOLLAB = 1) as result of the influence of explanatory variables related to the "General agreements" part of the questionnaire.

For hypothesis H3, model (1) was transformed with the dependent variables from Table 3, and herein, different from the previous models, the explanatory variables are the dependent ones from Table 1. The model is as follows:

$$ln\{P/(1-P)\} = b_0 + b_1NPRD + b_2INSIDE + b_3INCOLLAB \quad (7)$$

where $1/(1 - P)$ is the odds, P is the probability of the appearance of MAUTO, MLOG, MNEWPROD, ITSALES, ITRECRUIT, or ITACCOUNT, and b_i are the regression coefficients.

Model (7) investigated the probability of the investigated enterprises collaborating with mechatronics or IT companies in different fields of the enterprise's activities. The probability P is for dependent variables (MAUTO = 1, MLOG = 1 MNEWROD = 1, ITSALES = 1, ITRECRUIT = 1, or ITACCOUNT = 1). These variables are considered a result of the influence of explanatory variables (NPRD, INSIDE, and INCOLLAB) related to the "Collaboration with companies from mechatronics sector" part of the questionnaire and the "Collaboration with IT companies" part of the questionnaire.

For all calculations, the product IBM SPSS version 23 was used.

4. Results and Discussion

For this research, the respondents in the innovation survey were 85 furniture-producing companies. Many of them have made at least one type of innovation. The number of companies that have introduced product innovation is 80 (94.1%). As for process innovations, 53 (62.4%) of the companies have developed new and improved production methods and technologies. Out of those, 41 (77.4%) have implemented new process practices with their own resources, and the rest, 12 (22.6%), created process innovations together with IT companies or companies from the mechatronics sector. The percentage of product innovations in the total number of positively answered companies is 94.1%; for process innovations developed inside each company, the share is 62.4%, and the share of the innovations developed with collaboration is 22.6%. The results for model (5), or the information influence model, are presented in Table 4.

Table 4. Results for model (5)–regression coefficients (n = 85).

	NPRD	INSIDE	INCOLLAB
GROUP	−1.35	2.63	−0.61
CUST	−0.02	1.24	−0.016
VEND	−1.49	0.51	1.31
COMP	0.52	−1.17	−0.26
CONS	19.73	1.85	−1.85
PRESC	−1.28	2.21	1.18
UNI	1.82	4.60	0.15
ASECT	18.04	−1.97 *	−2.32
SINST	20.56	27.14	−23.64
EUINST	16.72	3.54	−0.48
TRDSHW	0.55	1.72	−0.18
PRNTD	19.17	0.73	0.45
INT	−0.44	−1.82	0.73
EMEDIA	0.06	−2.66	0.97

* Significant at 0.05 level.

The hypothesis H1 was partly proved. Table 4 reveals that the only significant source of information is sectoral association. The influence of such an association is negative on the development process of innovations inside the companies. The coefficient of −1.97 means that there was an 86% reduction of innovations created inside the furniture companies, without any external help. None of the other information sources have contributed to the creation and the introduction of innovations. The general agreements model (6)'s logistic regression coefficients are presented in Table 5.

Table 5. Regression coefficients of model (6) for general agreements (n = 85).

	NPRD	INSIDE	INCOLLAB
CGROUP	16.92	1.23	2.59
CUST	−2.28	1.67	−0.32
VEND	19.27	−0.25	2.06 *
COMP	17.82	−21.09	24.01
OTHER	−0.51	−1.98	−1.64
CONS	−2.76 *	0.37	2.97 *
GLOBE	−14.23	40.61	−47.12
SINST	−0.81	−61.62	67.48
PINST	2.76	−0.37	−2.97
FINANC	−1.45	21.56	−20.46

* Significant at 0.05 level.

The hypothesis H2 can be considered proven according to certain significant variables. The results for the coefficients show that for new products (NPRD), consultants (CONS) have a significant influence but with a negative sign ($b_5 = -2.76$). Transforming this number into odds means that the usage of consultants leads to a 93% reduction of the possibility of introducing product innovations. Comparing this result with the influence of consultants on the creating process of innovation in collaboration (INCOLLAB), it appears that they increase ($b_5 = 2.97$) the odds, or in other words, the possibility of that by more than 18 times. Vendors (VEND) are also very important. They increase the odds of the introduction of process innovations by more than six times. Regression coefficients for model (3) that tested the collaboration between furniture companies and IT firms for innovation development are presented in Table 6.

Table 6. Regression coefficients of model (7) for the collaboration between furniture-producing companies and IT firms (n = 85).

	ITSALES	ITRECRUIT	ITACCOUNT
NPRD	17.96	17.42	1.03
INSIDE	0.21	−0.62	−0.05
INCOLLAB	0.80	0.09	−0.14

The results in the table reveal that none of the coefficients is significant. The hypothesis H3 was not proven for IT companies i.e., the furniture enterprises do not collaborate with them in innovation development and do not enter into any collaborations related to product or process innovations. The results for the collaborations with mechatronics sector companies are different and are shown in Table 7.

Table 7. Regression coefficients of model (7) for the collaboration between furniture-producing companies and firms from mechatronics sector (n = 85).

	MAUTO	MLOG	MNEWPROD
NPRD	21.00	−0.79	19.92
INSIDE	−0.27	−0.50	1.232
INCOLLAB	2.49 *	0.29	2.99 *

* Significant at 0.05 level.

Regarding these results, the H3 was proven regarding mechatronics companies as collaborators. The results in Table 7 show only two statistically significant coefficients. These are the new automated processes (dependent variable MAUTO) introduced in intentional cooperation with companies (explanatory variable INCOLLAB) from the mechatronics sector and the new commonly developed products (dependent variable MNEWPROD) introduced as a result of process improvement in cooperation with mechatronics companies (explanatory variable INCOLLAB). For the first variable, the results for model (7) presented

$b_3 = 2.49$, which led to Odd = 12.13. That means $\Delta = 11.13$, so it is 11.13 times more likely for furniture companies that want to introduce process innovation to collaborate with mechatronic firms for automation than those that do not plan for process innovation to be developed in collaboration. For the second coefficient from the model (7) $b_3 = 2.99$, the Odd ratio was 20. That means that furniture companies that want to introduce new products are 20 times more likely to collaborate with mechatronics companies regarding the common development of a new product than those that do not plan for process innovation to be developed in collaboration.

In terms of diversifying the supply chain and achieving strategic autonomy, the analyzed companies from the furniture industry do not participate in complex chains that would make them dependent on market shock and changes. The implemented innovations are primarily market-driven and seek ways to optimize the products. None of the companies have outsourced their R&D activities outside the country; hence, the data cannot confirm or reject the idea of Vivek et al. [14] claiming that doing so can benefit the development of innovation and reduce production costs. The research data are indicative that the selected companies have a low level of cooperation with companies from other essential sectors such as IT or mechatronics, or with NGOs and other research organizations. Because of that, we cannot confirm the statements of other scholars [24,26] claiming that partnerships and collaborative relationships with other companies can increase furniture manufacturers' competitiveness. The results in Table 7 revealed that furniture companies rely on firms in the mechatronics sector for automation development. The problem of ensuring that there are enough labor forces pushed the manufacturers to find alternatives. It is interesting to see from the results that the surveyed companies view the innovations as a product of collaboration with mechatronics and associate them with new, exact manufactured products. Manufacturers perceive partnerships with companies from mechatronics as an important source of process and product development. Furniture companies are looking for easily implementable solutions, but in the future, IT companies will be the sources to provide services that keep these solutions up to date. We can accept this process of developing a culture of innovation in the early stages of the Bulgarian furniture sector. The next possible reason why the results in Table 6 are insignificant is the furniture manufacturers' understanding of the role of IT companies. They perceive these companies as supporters of everyday activities. This contributes to the results of Radziwon and Bogers [46], who found that the understanding of innovation and knowledge-based collaboration can differ widely. Still, the current findings show that there is a low share of innovations developed with collaboration, which confirms the statement by Biolcheva [32], which claimed that the lack of cooperation is a restriction to innovation development. Despite the limited scope of the general agreements (Table 5), as sources of innovations, they appear to have some positive relationships with the intensity of innovations in furniture manufacturing. Manufacturers use the agreements to improve their processes. This is an interesting result when compared to the results for agreements with mechatronics and IT companies. General agreement with consulting companies is negatively related to product innovations. This result is in line with the research of Benčiková et al. [47], which proved that t inclusion in a global corporate environment will increase cultural intelligence as well as internal processes like controlling [48,49] investigated enterprises. Manufacturers introduce new processes because of knowledge transfers from consultants and vendors. The results (Table 5) show that manufacturers are cornered on the supply channels. They take into account the consultations along with vendors' recommendations. In this way, the companies improve the operations inside manufacturing on account of developing new products after collaborating with consultants. Comparing the results in Table 5 and those in Table 7, it is obvious that manufacturers develop new products based on process automation. The Internet and electronic media do not contribute to the creation of innovations, which confirms Georgieva's [38] idea that ICT and digitalization [42] are not seen as factors for achieving competitiveness by furniture manufacturers. However, it is not in line with the findings of

Pirc Barčić et al. [50], who stated that that furniture industry companies can benefit from using the Internet in their production process innovation and human resource innovation.

Manufacturers in the Bulgarian furniture industry do not elaborate on innovation solutions themselves. The results in Table 4 contribute to those in other tables. They prove the desire for easiness in implementing innovations. The negative impact of the information coming from sectoral associations can be explained by the role of these associations as mediators between manufacturers and companies for consulting vendors and mechatronics.

However, this study has some limitations that need to be addressed. The main limitation of the study is the fact that no reliable statistical data for the innovation potential of furniture enterprises from the Bulgarian National Statistical Institute could be used for validation or comparison purposes, as the collected and generated data from the institute cover the whole agriculture, forestry, and fishing sectors. A further limitation is the low response rate of the companies, as generally, the Bulgarian forestry sector, including the furniture industry, is considered a closed one and not externally oriented. Still, the current research is a keystone for future analyses of GVCs' impact on the innovation potential of Bulgarian regional competitiveness. Furthermore, the current study does not focus on the vision of the companies regarding the benefits and possibilities of participation in a GVC. This gives the idea that the majority of the surveyed companies refer to the global value chains as supplier chains, not as a method of conducting technology transfer, diffusing production activities, engaging in R&D collaboration, and sharing new knowledge and ideas outside the country of the entity's origin.

5. Conclusions

Innovation goes beyond technology and requires collaboration from many areas to achieve success. Still, innovations can offer key advantages to companies, which can lead to a reduction in prices along the chain. GVCs are a channel for the dissemination of technological knowledge, entrepreneurial culture, and innovation capacity.

- The aim of this study was to analyze the collaborations of innovative Bulgarian furniture manufacturers with external stakeholders and the use of information channels as factors for the development and implementation of innovation and participation in global value chains over their innovation activities.
- The results of the research showed that 94.1% of the researched companies have introduced product innovation, and 62.4% of companies have introduced process innovation, which means that they have developed new and improved production methods and technologies. Out of the companies that have introduced process innovation, 77% have implemented new process practices with their own resources, while 22.6% created process innovations together with IT companies or those from the mechatronics sector.
- An important source of information was sectoral association, which was found to be negative regarding the development the process innovations inside the companies. Regarding the inclusion of furniture companies in Bulgaria, various information channels led to the development of product and process innovations. In terms of respondents introducing innovation because of agreement contracts with value chain participants like companies and institutions, the results showed the influence of consultants on creating process innovation in collaboration. On the other hand, this was not found to be the case when introducing the new products into the company.
- The results showed that furniture manufacturing companies that want to introduce process innovation collaborate with mechatronics firms for automation 11.13 times more than those that do not plan process innovation to be developed in collaboration. However, furniture manufacturing companies that want to introduce new products are 20 times more likely to collaborate with mechatronics companies regarding the common development of a new product than those that do not plan for process innovation to be developed in collaboration. In terms of diversifying the supply chain and achieving strategic autonomy, the analyzed companies from the furniture

industry do not participate in complex chains that would make them dependent on market shock and changes. On the other hand, the results indicated that the companies from the furniture manufacturing sector do not enter into any collaborations with IT companies that are related to product or process innovations.

- The findings of this study contribute to new insights into the literature on participation in GVCs as a factor for collaboration with different stakeholders and hence for product and process innovation development within the furniture industry. Even though the topic cannot be considered new, the Bulgarian case shows that innovative furniture companies do not participate in GVCs and have a low level of cooperation with NGOs, academia, and other essential sectors such as IT and mechatronics. Therefore, they are less dependent on chain shocks. Such data are essential primarily for policymakers, academia, and the EU, as EU policies, regulations, and program frameworks rarely consider national and regional specifics, leading to ineffective and non-sustainable outcomes. According to these results, (the H3 hypothesis failed), national policies are crucial to enforcing collaboration with IT companies for two reasons: the improvement of the information role in innovations and the improvement of the optimality of administrative activities. The policies of the Bulgarian Ministry of Economy and Industry (MEI) are mainly focused on supporting manufacturing processes but not much on the business model creation. In this way, furniture manufacturing enterprises have limited capabilities for cost optimality and competitiveness. The MEI should focus its SMEs Guidance Policy on improving the awareness of the furniture manufacturing enterprises about the benefits they can get from the IT companies in process innovations.
- In Bulgaria, the furniture manufacturing sector is considered not so innovative. Still, because of the phenomenon of hidden innovations, we cannot confirm this statement. Companies prefer to hide their innovations for further protection, which might be the reason for the lack of cooperation between furniture manufacturing companies and academia, NGOs, and other relevant institutions. In this respect, further analysis related to the motives of Bulgarian furniture companies to participate in global value chains is needed, and the lack of cooperation must be further studied. The Bulgarian Ministry of Economy and Industry does not consider the Sectoral associations appropriately. The results from this analysis revealed that only these associations are involved with information sources for innovations. The National Strategy for Small and Medium-sized Enterprises (2014–2020) [51] failed to define the role of sectoral associations in the innovativeness of the enterprises. This is the reason for the enterprises only using them regarding innovations. At the same time, the Bulgarian Branch Chamber of Woodworking and Furniture Industry [52] alone is not capable of orienting entrepreneurs in the external environment. The policies related to furniture manufacturing need to be better implemented by improving the Bulgarian Ministry of Innovation platforms [53].
- The findings of this study indicate several specific policy recommendations to enhance innovation within Bulgaria's furniture sector. Firstly, promoting collaboration among innovative companies, academic institutions, and the IT and mechatronics sectors through incentives and support programs can facilitate knowledge exchange and technological progress. Secondly, encouraging transparency in innovation by developing frameworks that encourage companies to disclose innovations for intellectual property protection can enhance collaboration with academia and stakeholders. Thirdly, creating incentives to participate in global value chains can provide companies access to foreign markets, technological knowledge, and innovation opportunities. Additionally, initiatives promoting knowledge exchange among companies, industry associations, academia, and the IT and mechatronics sectors can drive interaction and cooperation. Adapting existing industrial policies to the furniture sector's needs and sharing experiences through workshops and platforms can further boost collaboration, innovation, and integration into global value chains. Moreover, refreshing research and develop-

ment incentive programs can offer financial support for innovative projects within the furniture sector. Collectively, these policy measures can strengthen collaboration, innovation, and competitiveness in Bulgaria's furniture industry, using research results as guidance for future strategies.

We hope that furniture sectors in other EU member countries may also benefit from aligning overall activities regarding global and local value chains as important segments in supporting the furniture manufacturing companies in their innovation activities and will encourage considerable government support for the furniture sector companies and forest-based sector in Bulgaria and in other EU countries. Additionally, according to the New European Innovation Agenda [54], a key European Union (EU) priority is to generate regional innovation, innovation performance, and innovativeness. However, in some EU regions, the design and development of innovation measures is still a relatively novel concept. It is necessary to strengthen the work between universities and furniture enterprises. Collaboration between science and business for the period of research was left to the initiative of universities and industry associations, which has not led to good results, as the analysis in the present study revealed. Now, the state for the first time supports the enterprises through an integrated policy. The "Research, Innovation, and Digitization for Smart Transformation" national program, which was recently approved by the EU Commission, is a great opportunity for Bulgarian authorities to help furniture companies develop and implement innovations. The program will be able to achieve a significant effect on furniture enterprises if its measures are addressed regarding the issues revealed by the analysis in the current research.

Author Contributions: Conceptualization, D.V.G., N.N., A.P.B., K.K. and P.Ć.; methodology, D.V.G. and N.N.; software, N.N.; validation, D.V.G., K.K. and A.P.B.; formal analysis, D.V.G. and N.N.; investigation, D.V.G., N.N., K.K. and A.P.B.; resources, D.V.G., K.K., A.P.B. and P.Ć.; data curation, D.V.G. and N.N.; writing—original draft preparation, D.V.G., K.K., A.P.B., N.N. and P.Ć.; writing—review and editing, D.V.G., K.K. and A.P.B.; visualization, N.N. and P.Ć.; supervision, D.V.G., K.K. and A.P.B.; project administration, D.V.G. and A.P.B.; funding acquisition, D.V.G., K.K. and A.P.B. All authors have read and agreed to the published version of the manuscript.

Funding: Daniela Ventsislavova Georgieva acknowledges the financial support from the National Science Fund (project title Development and implementation of a model for assessing the innovation potential of global value chains as a basis for increasing regional competitiveness"–contract no. КП-06-H55/8 from 16.11.2021). Kristina Klarić, Andreja Pirc Barčić, and Petar Ćurić wish to thank the University of Zagreb, Faculty of Forestry and Wood Technology Fund for scientific and professional work for the financial support.

Institutional Review Board Statement: The study did not require ethical approval.

Informed Consent Statement: Informed consent was obtained from all subjects involved in the study. All the data was collected face to face and all respondents were informed about their GDPR rights.

Data Availability Statement: The data presented in this article are available on request from the corresponding author.

Conflicts of Interest: The authors declare no conflict of interest. The funders had no role in the design of the study; in the collection, analyses, or interpretation of data; in the writing of the manuscript; or in the decision to publish the results.

References

1. OECD. Global Value Chains: Efficiency and Risks in the Context of COVID-19. 2021. Available online: https://www.oecd.org/coronavirus/policy-responses/global-value-chains-efficiency-and-risks-in-the-context-of-covid-19-67c75fdc/ (accessed on 17 February 2023).
2. Barcic, A.P.; Motik, D. Innovation and innovativeness in medium-low tech/low-tech industries—Wood industry. *Drv. Ind.* **2013**, *64*, 247–256. [CrossRef]
3. Rakovska, M. *Teoretichni Aspekti na Upravlenieto na Verigata na Dostavkite*; Nauchni Trudove na UNSS tom 2; Universitetsko izdatelstvo Stopanstvo: Sofia, Bulgaria, 2009; pp. 247–290.

4. World Bank. *World Development Report: Trading for Development in the Age of Global Value Chains*; World Bank: Washington, DC, USA, 2020.
5. Kim, K.; Park, J.H.; Prescott, J.E. The global integration of business functions: A study of multinational businesses in integrated global industries. *J. Int. Bus. Stud.* **2003**, *34*, 327–344. [CrossRef]
6. European Innovation Scoreboard. 2023. Available online: https://research-and-innovation.ec.europa.eu/statistics/performance-indicators/european-innovation-scoreboard_en (accessed on 11 August 2023).
7. Regional Innovation Scoreboard. 2023. Available online: https://european-union.europa.eu/index_en (accessed on 11 August 2023).
8. Hansen, E.; Kangas, J.; Hujala, T. Synthesis towards future-fittest for mature forest sector multinationals. *Can. J. For. Res.* **2021**, *51*, 871–878. [CrossRef]
9. Weiss, G.; Pettenella, D.; Ollonqvist, P.; Slee, B. *Innovation in Forestry: Territorial and Value Chain Relationships*; CAB International. MPG Books Group: Bodmin, UK, 2011.
10. Pirc, A.; Vlosky, R. Comparison of Innovation in the Furniture Manufacturing Industries of Croatia and the United States. In Proceedings of the International Conference WoodEMA–Present Conditions, Opportunities and New Challenges, Vyhne, Slovakia, 6–8 October 2010; pp. 149–156.
11. Hart, J. Globalization and Digitalization. Academia.edu. 2010. Available online: https://www.academia.edu/1359654/Globalization_and_Digitalization (accessed on 12 August 2023).
12. Ambosa, B.; Brandl, K.; Perri, A.; Scalera, V.; Assche, A.V. The nature of innovation in global value chains. *J. World Bus.* **2021**, *56*, 101221. [CrossRef]
13. Fagerberg, J.; Lundvall, B.Å.; Srholec, M. Global value chains, national innovation systems and economic development. *Eur. J. Dev. Res.* **2018**, *30*, 533–556. [CrossRef]
14. Vivek, S.D.; Richey, R.; Dalela, V. A longitudinal examination of partnership governance in offshoring: A moving target. *J. World Bus.* **2009**, *44*, 16–30. [CrossRef]
15. Levin, D.; Barnard, H. Connections to distant knowledge: Interpersonal ties between more-and less-developed countries. *J. Int. Bus. Stud.* **2013**, *44*, 676–698. [CrossRef]
16. Tzabbar, D.; Vestal, A. Bridging the social chasm in geographically distributed R&D teams: The moderating effects of relational strength and status asymmetry on the novelty of team innovation. *Organ. Sci.* **2015**, *26*, 811–829. [CrossRef]
17. Centre for European Policy Studies (CEPS). The EU Furniture Market Situation and a Possible Furniture Products Initiative. Final Report for European Commission, Brussels, Belgium, 1-306. 2014. Available online: https://www.ceps.eu/system/files/Final%20report_en.pdf (accessed on 13 August 2023).
18. Chiu, Y.C. Balancing exploration and exploitation in supply chain portfolios. Engineering management. *IEEE Trans. Eng. Manag.* **2014**, *61*, 18–27. [CrossRef]
19. Hastig, G.M.; Sodhi, M.S. Blockchain for supply chain traceability: Business requirements and critical success factors. *Prod. Oper. Manag.* **2020**, *29*, 935–954. [CrossRef]
20. Humphrey, J.; Schmitz, H. How does insertion in global value chains affect upgrading in industrial clusters? *Reg. Stud.* **2002**, *36*, 1017–1027. [CrossRef]
21. Pisano, G.; Shih, W. Does America really need manufacturing? *Harv. Bus. Rev.* **2012**, *90*, 94–102.
22. Narula, R. Exploring the paradox of competence-creating subsidiaries: Balancing bandwidth and dispersion in MNEs. *Long Range Planning.* **2014**, *47*, 4–15. [CrossRef]
23. Gereffi, G.; Humphrey, J.; Kaplinsky, R.; Sturgeon, T.J. Introduction: Globalisation, value chains and development. *IDS Bull. Transform. Dev. Knowl.* **2001**, *32*, 1–8. [CrossRef]
24. Chambost, V.; McNutt, J.; Stuart, P. 2009. Partnerships for successful enterprise transformation of forest industry companies implementing the forest biorefinery. *Pulp Pap. Can.* **2009**, *110*, 19–24.
25. Edgington, D. British Columbia's coastal forests, hemlock timber and the Japanese housing market. *Can. J. Reg. Sci.* **2004**, *26*, 415–446.
26. DeLong, D.; Kozak, R.; Cohen, D. Overview of the Canadian value-added wood products sector and the competitive factors that contribute to its success. *Can. J. For. Res.* **2007**, *37*, 2211–2226. [CrossRef]
27. Piltan, M.; Sowlati, P. A review of partnership studies in the forest products value chain: With a focus on developed countries (United States, Canada, and Western Europe). *For. Prod. J.* **2014**, *64*, 4–10. [CrossRef]
28. Popova, R.; Georgieva, V.D. Digitization in forest industry in Bulgaria—State and perspectives, Digitalisation and circular economy: Forestry and forestry-based industry implications. In *Digitalisation and Circular Economy: Forestry and Forestry Based Industry Implications, Proceedings of Scientific Papers*; USB & WoodEMA: Sofia, Bulgaria, 2019; pp. 181–186.
29. Pinkard, E.A.; O'Grady, A.P. Innovation in the forest industry: Ready or not! *Aust. For.* **2016**, *79*, 153–156. [CrossRef]
30. Stefanov, N. The impact of the cap on the sustainability of forestry sector in Bulgaria. *J. Bio-Based Mark.* **2020**, *2*, 29–41.
31. Furniture Global Market Report. COVID-19 Impact and Recovery to 2030. 2021. Available online: https://finance.yahoo.com/news/furniture-global-market-report-2021-114900217.html?guccounter=1&guce_referrer=aHR0cHM6Ly93d3cuZ29vZ2xlLmNvbS8&guce_referrer_sig=AQAAANzVkI4C_ogNIoB8oeEU0hItEc4iMcfIMfHcaQ9dXgLg_njojhsJsGWG97_7VL3VcbhXNXMJO99Yj7gHljcc5kzuESWmT03ETgubGC-a3KoTfEnpdSmfmrZHlmu9sfsY2U4NN2DEnptrgC5K0HzbmfiSP91tSKY_AuwsQuwAJLNp (accessed on 15 June 2023).

32. Biolcheva, P. Barriers before innovations in furniture companies. *Trakia J. Sci.* **2017**, *15*, 125–131. [CrossRef]
33. Andreev, O.; Peneva, G. Problems concerning operations system of the enterprise in the context of Industry 4.0. In Proceedings of the International Scientific Conference E-Governance & ECommunications, Sofia, Bulgaria, 13–14 June 2019; pp. 165–170.
34. Andreev, O.; Peneva, G. A methodological approach to survey the opportunities for implementing Industry 4.0 in Bulgarian Industrial Enterprises. *Manag. Eng.* **2018**, *18*, 739–745.
35. Drayse, M. Globalization and regional change in the U.S. furniture industry. *Growth Chang.* **2008**, *39*, 252–282. [CrossRef]
36. Yadong, L. New OLI advantages in digital globalization. *Int. Bus. Rev.* **2021**, *30*, 101797. [CrossRef]
37. Jagjit, S.; Lorentz, H. Developing design principles for the digitalisation of purchasing and supply management. *J. Purch. Supply Manag.* **2019**, *25*, 78–98. [CrossRef]
38. Georgieva, D.V. Digitization of Bulgarian furniture manufacturers during the COVID-19. *New Knowl. J. Sci.* **2022**, *11*, 22–30.
39. Boer, H.; During, W.E. Innovation, what innovation? A comparison between product, process and organization innovation. *Int. J. Technol. Manag.* **2001**, *22*, 83–107. [CrossRef]
40. Nor, N.G.M.; Bhuiyan, A.B.; Said, J.; Alam, S.S. Innovation barriers and risks for food processing SMEs in Malaysia: A logistic regression analysis. *Geografia* **2016**, *12*, 167–178.
41. Gerstlberger, W.; Knudsen, M.; Stampe, I. Sustainable development strategies for product innovation and energy efficiency. *Bus. Strategy Environ.* **2014**, *23*, 131–144. [CrossRef]
42. Świadek, A.; Gorączkowska, J. The institutional support for an innovation cooperation in industry: The case of Poland. *J. Econ. Econ. Policy* **2020**, *15*, 811–831. [CrossRef]
43. Odei, S.A.; Stejskal, J. Do firms R&D collaborations with the science system and enterprise group partners stimulate Their Product and Process Innovations? *Economies* **2019**, *7*, 43. [CrossRef]
44. Peng, C.Y.; Lee, K.L.; Ingersoll, G.M. An introduction to logistic regression analysis and reporting. *J. Educ. Res.* **2002**, *96*, 3–14. [CrossRef]
45. Nick, T.G.; Campbell, K.M. Logistic Regression. In *Topics in Biostatistics. Methods in Molecular Biology*; Ambrosius, W.T., Ed.; Humana Press: Totowa, NJ, USA, 2007; Volume 404, pp. 273–301. [CrossRef]
46. Radziwon, A.; Bogers, M. Open innovation in SMEs: Exploring inter-organizational relationships in an ecosystem. *Technol. Forecast. Soc. Chang.* **2019**, *146*, 573–587. [CrossRef]
47. Benčiková, D.; Malá, D.; Sedliačiková, M.; Drábek, J.; Kropil, R. Assessment of cultural intelligence as a prerequisite to development of an enterprise within the contemporary global corporate environment. *Ekon. Cas.* **2021**, *69*, 88–109.
48. Sedliačiková, M.; Moresová, M.; Drábek, J.; Kupčák, V. The Significance of Controlling in Enterprises in Emerging Economies. *Cent. Eur. Bus. Rev.* **2021**, *10*, 99. [CrossRef]
49. Sedliačiková, M.; Moresová, M.; Malá, D.; Rowland, Z. Controlling–an empirical study and proposal of a relevant model for sustainable business and development in Slovakia. *J. Bus. Econ. Manag.* **2021**, *22*, 1252–1268. [CrossRef]
50. Barčić, A.P.; Motik, D.; Oblak, L.; Vlosky, R. Management activity linkages to innovation deconstruction: An exploratory study of the furniture industry in Croatia. *BioResources* **2016**, *11*, 3987–4005. [CrossRef]
51. National Strategy for Small and Medium-Sized Enterprises. Available online: https://www.mi.government.bg/en/strategy-policy/national-strategy-for-small-and-medium-sized-enterprises-2014-2020-small-business-act/ (accessed on 27 August 2023).
52. Bulgarian Branch Chamber of Woodworking and Furniture Industry. Available online: https://timberchamber.com/en/ (accessed on 27 August 2023).
53. Bulgarian Ministry of Innovation Platforms. Available online: https://www.mig.government.bg/policies-and-strategies/innovation/?lang=en (accessed on 27 August 2023).
54. European Commission. The New European Innovation Agenda. 2022. Available online: https://research-and-innovation.ec.europa.eu/strategy/support-policy-making/shaping-eu-research-and-innovation-policy/new-european-innovation-agenda_en (accessed on 10 July 2023).

MDPI AG
Grosspeteranlage 5
4052 Basel
Switzerland
Tel.: +41 61 683 77 34

Sustainability Editorial Office
E-mail: sustainability@mdpi.com
www.mdpi.com/journal/sustainability

www.ingramcontent.com/pod-product-compliance
Lightning Source LLC
Chambersburg PA
CBHW041357210926
51850CB00013B/3108
* 9 7 8 3 7 2 5 8 3 6 9 8 7 *